An Integrated Approach to Electrical and Electronics Engineering

An Integrated Approach to Electrical and Electronics Engineering

Editor: Jeremy Giamatti

NYRESEARCH
P R E S S

New York

Published by NY Research Press
118-35 Queens Blvd., Suite 400,
Forest Hills, NY 11375, USA
www.nyresearchpress.com

An Integrated Approach to Electrical and Electronics Engineering
Edited by Jeremy Giamatti

© 2017 NY Research Press

International Standard Book Number: 978-1-63238-540-6 (Hardback)

Cataloging-in-Publication Data

An integrated approach to electrical and electronics engineering / edited by Jeremy Giamatti.
    p. cm.
Includes bibliographical references and index.
ISBN 978-1-63238-540-6
1. Electrical engineering. 2. Electronics. I. Giamatti, Jeremy.
TK145 .I58 2017
621.3--dc23

Printed in the United States of America.

Contents

Preface

The study of electricity and related devices falls under the discipline of electrical engineering. Electronic engineering is a branch of electrical engineering focusing on diverse electrical components for designing advanced devices. This book unfolds the innovative aspects of electrical and electronics engineering which will be crucial for the progress of this field in the future. It strives to provide a fair idea about this discipline and to help develop a better understanding of the latest advances within this area of study. Scientists and students actively engaged in this field will find this book full of unexplored concepts and their applications.

After months of intensive research and writing, this book is the end result of all who devoted their time and efforts in the initiation and progress of this book. It will surely be a source of reference in enhancing the required knowledge of the new developments in the area. During the course of developing this book, certain measures such as accuracy, authenticity and research focused analytical studies were given preference in order to produce a comprehensive book in the area of study.

This book would not have been possible without the efforts of the authors and the publisher. I extend my sincere thanks to them. Secondly, I express my gratitude to my family and well-wishers. And most importantly, I thank my students for constantly expressing their willingness and curiosity in enhancing their knowledge in the field, which encourages me to take up further research projects for the advancement of the area.

Editor

A design solution to reduce DC bus voltage stress in single switch power quality converter

Bindu S. J., C. A. Babu

Department of Electrical and Electronics Engineering, School of Engineering, CUSAT, Kochi, Kerala, India

Email address:

binduspk@gmail.com

Abstract: Power quality is becoming an important issue for electricity consumers at all level of usages. Sensitive equipment and non-linear loads are common in both industrial and domestic environment. Harmonic distortion can result in malfunction of sensitive equipments and generators. Power factor corrected converter is increasingly used in industry to improve input current quality and regulate the output voltage of front end converter. This paper presents a Single Switch Power Quality Converter which achieves both power factor correction and output voltage regulation by using only one switch. This paper deals with the design method in the reduction of DC bus voltage stress during light load, by selecting proper boost inductor using Equal Area Criterion (EAC).

Keywords: Power Factor Correction, BIFRED Converter, Equal Area Criterion, Power Quality, DC Bus Voltage Stress

1. Introduction

Electronic equipments are increasingly being used in everyday life nowadays. A power converter is used as an interface between utility and most of the power electronic equipments. Since these converters draw pulsed current from the supply, which is high in third and fifth harmonic content, line current harmonics are injected to the electrical network. Hence, a power factor correction (PFC) stage is usually inserted to the existing equipment to shape the line current into a sinusoidal waveform and to satisfy necessary standards such as IEEE 519[1] and EN 61000-3-2 [2]. Another reason to limit harmonic currents is to use the full rated current from the available power source. The goal then of a PFC converter is to reduce the harmonic content of the current waveform and keep the phase angle between the current and the voltage as small as possible. In effect the circuit wants to emulate a resistive load.

The new generation of power factor corrected single stage power supplies typically takes the form of a cascaded combination of Discontinuous Conduction Mode (DCM) PFC converter and a DC-DC converter. The two converters share the same controller and switch to regulate the output voltage and to shape the input current. In order to buffer the difference between the instantaneous input power and constant output power an energy storage element is required [3].

Several single stage power factor correction converters have been previously proposed such as Boost Integrated with Fly Back Rectifier / Energy storage / DC- DC (BIFRED) converter which is the integration of a Discontinuous Conduction Mode (DCM) boost converter with Continuous Conduction Mode (CCM) fly back converter. Here DCM boost converter is for input current wave shaping and CCM fly back dc-dc converter is for isolation and load voltage regulation.

Disadvantage of Single Switch power quality converter is that it usually suffers from relatively higher voltage stress at light load [3], [4]. The reason for high DC bus voltage stress during light load is the power unbalance between PFC stage and output stage. Many methods for reducing voltage stress have been reported. One of the proposed solutions for overcoming the dc bus voltage stress is using frequency control [5]. But this approach has short comings which lead to increase in component count and making the control circuit more complex. Another solution proposed; is the concept of series charging and parallel discharging capacitor scheme [6]. The disadvantage of this method is that it increases the number of components in the power circuit. Another approach to this problem is by modulating the predetermined operating frequency of the converter [7] which results in increased complexity in the control circuit. Two elements are common in the design of single switch power quality converter. First, the mode of input inductor must be maintained such that input inductor begins and ends each switch cycle at a ground state. Second, the converter must have an energy storage capacitor which is capable of providing energy when the instantaneous line voltage is near zero.

Fig 1 shows BIFRED converter with negative feedback in the power stage. When Switch S is made ON, the rectified voltage is applied to inductor L_i and the inductor current linearly increases. Therefore, during the ON time interval of the switch, the inductor stores energy independently. When switch S is made OFF, the stored energy of the inductor L_i is transferred to, capacitor C_b and the load. The input power is controlled only by duty cycle and L_i. The PFC stage really does not know whether the load is low or high. So during light load, the energy stored by the PFC stage is same as that of the heavy load causing power unbalance between the input and output [1]. The bulk capacitor stores this unbalanced power leading to increase in the dc bus voltage. If the PFC stage is inherently able to reduce the input power automatically when the load becomes light, then the dc-bus

voltage can be suppressed. One approach to this problem is using proper closed loop controller, which will work in such a way that error between reference voltage and output voltage will be zero, by reducing the ON time interval of the switch, thus maintaining the power balance. Another method is to reduce the voltage across L_i during the switch ON time period, so that the energy absorbed from the line input is also reduced [2].

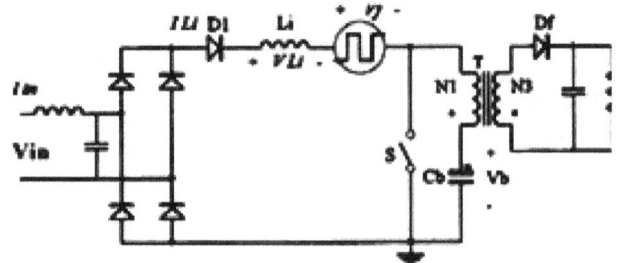

Fig 1. *BIFRED Converter with negative feedback in the power stage*

2. The Proposed Single Switch Power Quality Converter

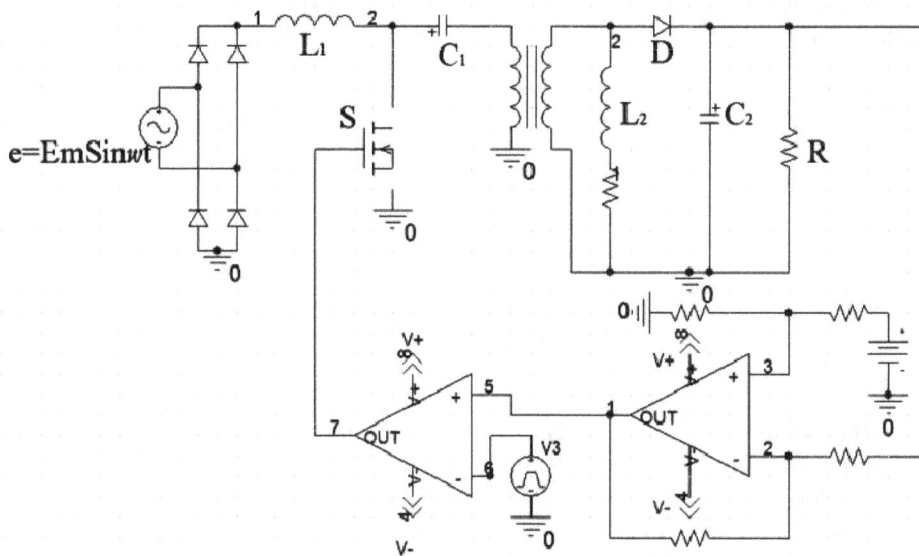

Fig 2. *The Proposed Single Switch Power Quality Converter*

Proposed converter, shown in Fig. 2, is a BIFRED converter, which eliminate diode D_1 and the use of power circuit negative voltage feedback V_f. We should consider the diode D_1 only if the EMI filter is used after the full bridge rectifier. In order to eliminate the problem of DC bus voltage stress at light load, Equal Area Criterion (EAC) is applied for the optimum design of boost inductor and a proportional controller, with duty ratio as manipulated variable and output voltage as controlled variable. To fully explain the circuit, converter operation will be analyzed according to the three operational intervals.

The first interval starts when the switch S turns ON, the input voltage is applied across the inductor. Depending upon the input voltage, energy will be stored in L_1 and Diode D is

reverse biased.

The second interval starts when S turns OFF, which causes the current in inductor L_1 to ramp down and Diode D conducts.

The third interval starts when the current in inductor L_1 remains zero, S remains OFF and Diode D conducts. Load receives energy from the coupled inductor L_2, and C_2 passes energy to the load.

DC bus voltage is eliminated by the selection of inductance using Equal Area Criterion (EAC) along with a proportional controller, by selecting boost inductor in such a way that it should store the energy that matches with the energy required at the output during light load. Here EAC is applied in each switching cycle by equating the area under the inductor current and the area under the reference current. For the design of boost inductor, it is considered that maximum power is delivered at the peak value of input voltage and the duty ratio will be slightly less than 0.5 for maximum power delivery.EAC is applied between theoretical value of input current and the peak inductor current at maximum turn ON time.

3. Design of Power Factor Correction Stage Using EAC

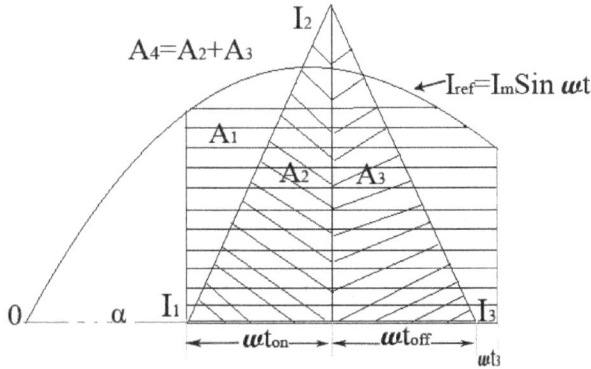

Fig 3. Reference current superimposed on inductor current pulse

Time period T = t_{on}+ t_{off}+ t_3
ωt_{on}- switch on period.
ωt_{off}- switch off period.
ωt_3- dead period
Consider,
α –Angle of switching instant

EAC applied to PFC converter means equalizing the area under the inductor current and the area under sinusoidal reference current over one switching cycle [8]. The reference current I_m Sinωt superimposed on the current through the inductor over a switching interval is shown in Fig. 3. Anticipating sinusoidal input current at unity power factor, an input current is obtained. Peak value of input current during t_{on} is calculated at peak value of sinusoidal reference current.

$$\omega t_{on}+\omega t_{off}+ \omega t_3= \theta \qquad (1)$$

3.1. EAC for the Design

The area A_1 under the reference current I_mSinωt in one switching cycle is given by

$$A_1=\int_\alpha^{\alpha+\theta} ImSin\omega t\ d\omega t \qquad (2)$$

$$A_1= -I_m[-cos\ (\alpha + \theta)-cos\alpha] \qquad (3)$$

When $\alpha = 90^0$,

$$A_1 = I_m Sin\ \theta \approx I_m\ \theta \qquad (4)$$

since θ is very small.
Area A_4 is a triangle

Therefore A4= A2+A3 =½[ωton+ωtoff]I_{2peak} (5)

Current through the boost inductor during ON period is given by

$$EmSin\omega t = L1\ di/dt \qquad (6)$$

$$i= \int \frac{E_m}{L_1} Sin\ \omega t\ dt \qquad (7)$$

Current through the boost inductor during OFF period is given by

$$L1\ di/dt= EmSin\omega t -[Vdc+ nV2] \qquad (8)$$

$$i=\int \frac{E_m}{L_1} Sin\ \omega t\ dt - \frac{V_{dc}+nV_2}{L1}dt \qquad (9)$$

When switch is ON instantaneous current *is given by,*

$$i_r = I_1+ \frac{E_m}{\omega L_1}\ [\ cos\ \alpha-cos\ (\alpha+ \omega t)] \qquad (10)$$

Here , $\alpha<\omega t<\omega t_{on}$
When switch is OFF instantaneous current is given by,

$$i_r =I_2 +\frac{E_m}{\omega L_1}\ (cos\ \alpha+\omega t_{on})-cos(\alpha+\omega t_{on}+\omega t)-\frac{(V_{dc}+nV_2)}{\omega L_1}\omega t \qquad (11)$$

At the beginning $I_1 = 0$
During on time,

$$i_r = \frac{E_m}{\omega L_1}[cos\ \alpha - cos(\alpha + \omega t)] \qquad (12)$$

Where $\alpha < \omega t < \omega t_{on}$

Off mode current becomes zero at $\omega t = \omega t_{off}$

$$I_3 =I_2+\frac{E_m}{\omega L_1}[cos\ (\alpha+\omega t_{on})-cos(\alpha+\omega t_{on}+\omega t)]-\frac{(V_{dc}+nV_2)}{\omega L_1}\omega$$

Where $\alpha < \omega t < \omega t_{off}$ (13)

3.2. Design Consideration of Boost Inductor

The value of this inductor is quite crucial in the performance of the converter, with the small value of this inductor the large switching ripples are injected into supply current, and large value of it doesn't allow shaping the AC mains current in the desired fashion. Therefore the optimum selection of this inductor is essential to achieve satisfactory performance [9].
Design Basis:
1 Assuming zero switching loss.
2 Required power output is obtaining at a low turn ON time or duty ratio (0.26) , by this varying power levels can be achieved under DCM.
3 Find out value of input peak current during turn ON such that area under reference input current in one

switching period made equal to the area under the current pulse as shown in fig. 3

4 Switching instance is considered as $\alpha = 90°$ for the maximum rising and falling slope at the peak of input voltage.

Maximum current (I_2peak) occurs at the end of ON duration.

Using EAC, we have,

$$A_1 = A_4 \tag{14}$$

Hence from (4) & (5)

$$I_m\,\theta = \tfrac{1}{2}[\omega t_{on} + \omega t_{off}]\,I_{2peak} \tag{15}$$

Select reference current in such that-

$$P_{out} = V_{rms} \times I_{rms\,\cdot ref}. \tag{16}$$

From I_{rms} we can find I_m
And using (15) we get the value for I_{2peak}.

Select L_1 such that I_{2peak} occurs maximum duty cycle and $\alpha = 90°$. The OFF duration followed by this I_{2peak} will be minimum and the current at the end of this off duration is zero.

From (12)

$$I_{2peak} = \frac{E_m}{\omega L_1}\sin \omega t_{on} \tag{17}$$

$\sin \omega t_{on} \cong \omega t_{on}$,due to high switching frequency.

$$I_{2peak} = \frac{E_m}{\omega L_1}\omega t_{on} = \frac{E_m}{L_1}t_{on} = \frac{E_m}{L_1}DT \tag{18}$$

$$L_1 = \frac{E_m}{I_{2peak}}DT \tag{19}$$

Here D represents duty cycle

3.3. Expression of DC Bus Voltage, Output Voltage and Duty Ratio

From (13), (17) with $I_3 = 0$ and assuming $\omega t_3 \cong 0$

$$0 = I_{2peak} + \frac{E_m}{\omega L_1}(-\sin \omega t_{on} + \sin \omega t_{on} + \omega t_{off}) - \frac{(V_{dc}+nV_2)}{\omega L_1}\omega t_{off} \tag{20}$$

$$0 = \frac{E_m}{\omega L_1}(\sin \omega t_{on} - \sin \omega t_{on}) + \frac{E_m}{\omega L_1}\sin(\omega t_{on}+\omega t_{off}) - \frac{(V_{dc}+nV_2)}{\omega L_1}\omega t_{off} \tag{21}$$

We have $\sin \omega t_{on} \cong \omega t_{on}$, due to high switching frequency

$$\frac{(V_{dc}+nV_2)}{L_1}t_{off} = \frac{E_m}{L_1}(t_{on}+t_{off}) \tag{22}$$

$$(V_{dc}+nV_2)(t - t_{on}) = E_m T \tag{23}$$

Where $T = t_{on} + t_{off}$,

$$(V_{dc}+nV_2)t - (V_{dc}+nV_2)t_{on} = E_m T \tag{24}$$

$$\frac{t_{on}}{T} = \frac{(V_{dc}+nV_2-E_m)}{(V_{dc}+nV_2)} \tag{25}$$

$$D = 1 - \frac{E_m}{(V_{dc}+nV_2)} \tag{26}$$

Where $D = \frac{t_{on}}{T}$,

$$V_{dc}+nV_2 = \frac{E_m}{1-D} \tag{27}$$

4. Design Example of Single Switch Power Quality Converter

Converter with the following specification is designed:
Input Supply Voltage = 230 V, 50Hz
Output Voltage V_{dc} = 50 V
Output power = 100W
Switching frequency f_s = 20 KHz
Output voltage ripple = 5 %
Duty ratio of the switch = 0.26

4.1. Determining Value of 'L_1' Using EAC

Consider $\alpha = 90°$, as the switching instant

$$D = 0.26$$

Switching frequency f_s =20 KHz
From (19),

$$L_1 = 1.57\text{mH}$$

4.2. Determining Value of Energy Storage Capacitor

The value of DC bus voltage capacitor is quite crucial as it affects the response, cost, stability, size and efficiency. A small value of the capacitor results in large ripple in steady state and big dip and rise in dc link voltage under transient condition. A high value of it reduces the DC bus voltage ripple but increases cost, size, and weight.

From (19) we have,

$$I_{peak} = \frac{DTE_m}{L_1}$$

Energy Stored in inductor = $1/2\ L_1 I^2_{2peak}$

Energy Stored in capacitor = $\frac{1}{2}C_1V^2$

$$C_1V^2 = L_1I^2$$

$$C_1 = 116\mu F$$

5. Control Scheme

Control is necessary for the regulation of output voltage and for improvement of line harmonics. Straight forward duty ratio control is a suitable method of output voltage regulation, due to its simplicity of design using conventional PWM circuitry. Proportional controller is used here, using duty cycle as the manipulated variable, and output voltage as the controlled variable.[10]

6. Simulation Results

To investigate and validate the design the proposed converter was simulated using SABER. The results were found in accordance with the design intends.

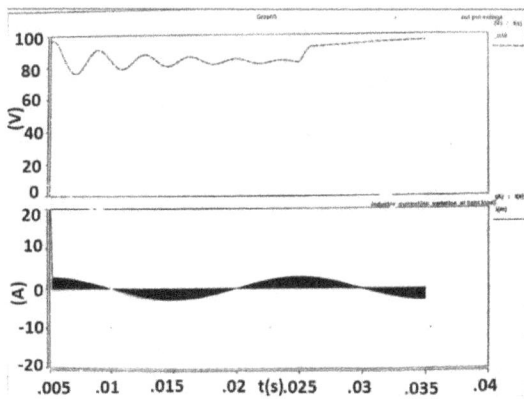

Fig.4. Output voltage and input current under open loop

Open loop simulation was carried out by varying the duty ratio. Output voltage is found linear to ON duty ratio. Input current is sinusoidal and in phase with the input line voltage. Effect of sudden decrease in output load on dc bus voltage was studied. Fig 4 shows that during open loop when load is reduced after 25 ms there is a slight increase in the output voltage, no variation in the input current.Fig.4 also shows that the input converter draws the same current under varying load condition.

Performance under closed loop condition was studied by varying the reference voltage. Output voltage was found varying linear with the reference voltage and input current was found sinusoidal and in phase with the input voltage.

Fig.5 shows the load transient response, the transient overshoot voltage is less than 1%. Fig. 6 shows variation in input line current when the load is reduced after 15 ms. It is observed that in closed loop condition the dc bus voltage stress has been drastically reduced. When output load is decreased suddenly, instantaneous power unbalance will occur, causing the output voltage and dc bus voltage to increase. Proportional controller will immediately detect the increase in output voltage and takes the corrective action by reducing the duty cycle leading to a new energy balance within one to two switching cycles. Fig 7 shows the voltage across switch when load thrown out at 15ms and also we can

observe that the corrective action started within 2 switching cycle, which conform the fast transient response. Fig 8 shows plot between dc bus voltage stress on energy storage capacitor and output power, which confirms the effectiveness in the reduction of dc bus voltage stress. The proposed converter can keep the capacitor voltage between 217-230V for a load change of full load to 10% of load.

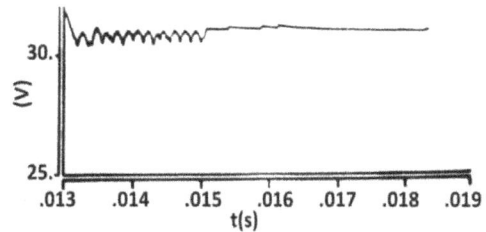

Fig 5. output voltage under closed loop

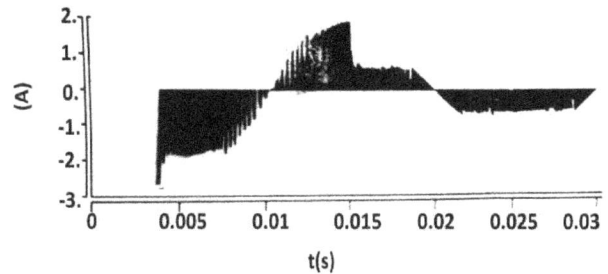

Fig 6. Input current under closed loop

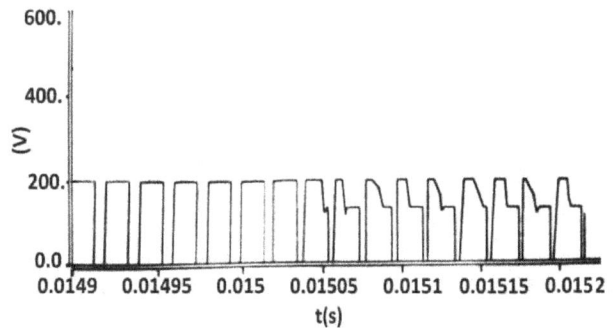

Fig 7. voltage across switch when load thrown out at 15ms

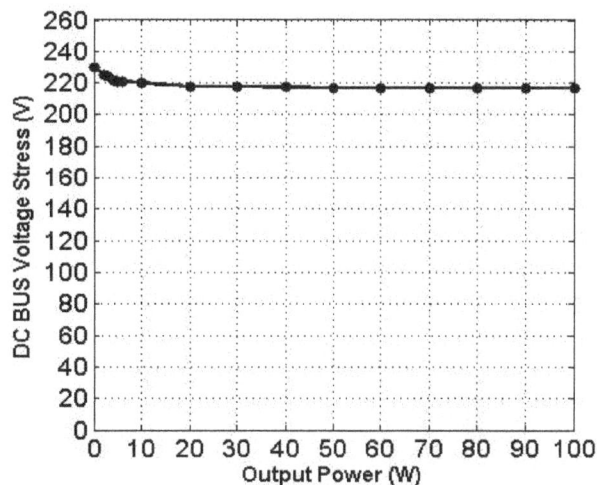

Fig 8. Measured DC Bus voltage Stress against output power

Table I. Measured Harmonic Currents Versus EN Requirements

Harmonic Order n	Measure harmonic Current (mA) @ I_{line}=.627A	Extrapolated harmonic Current(mA) @ I_{line}= 6A	Maximum Permissible current(mA) of EN 61000-3-2
3	60.6	1441	2300
5	18.18	432	1140
7	12.1	288	770
9	8.1	193	400
11	6.5	155	330
13	5.8	138	210
15	5.3	126	150

Table I reveals that proposed converter can provide sufficient margin in harmonic current reduction, even if at line current of 16 A.

7. Experimental Verifications

In order to verify the circuit operation, a 100W, 50V output voltage PFC converter was implemented at constant switching frequency of 20KHz and tested with the following circuit parameters and using MOSFET IRFPF50 as switch. Fig.9 shows input voltage and dc bus voltage. Fig.10 shows input line current is sinusoidal and in phase with the input voltage. The power factor was found to be closed to unity.

Table II. Specifications And Components For The Single Switch Power Quality Converter

Parameter	Value
L_1	1.57mH
C_1	116 μF
n	3.5
C_2	317 μF
T_s	50 μs
D	0.26
R	25Ω
S	IRFPF50
L_1 Core	E 42/21/15
L_1 Winding	116T,13 Wires of SWG 30
Transformer Core	E 65/32/13
Transformer L_2	1.2 mH
Primary Winding	297 T ,SWG 21
Secondary Winding	99 T, SWG 18

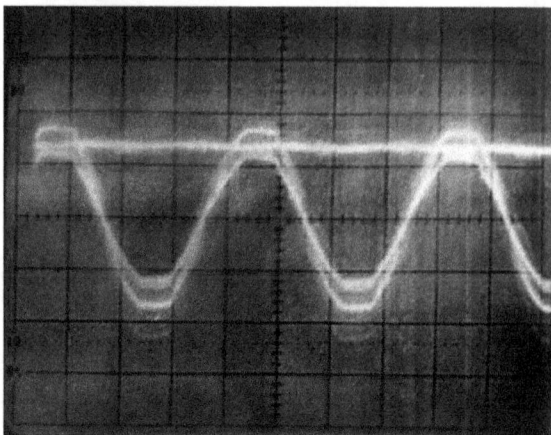

Fig 9. Input voltage and DC bus Voltage

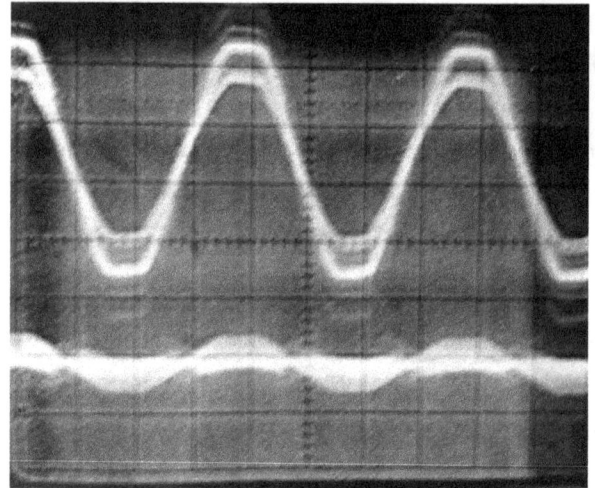

Fig. 10. Input voltage and input current

8. Conclusion

This paper derives a design solution for achieving low voltage stress and unity power factor, in a Single Switch Power Quality converter by optimally selecting the boost inductance using EAC. Fast voltage regulation is achieved by simple PWM control. The performance of the converter with the proportional controller has been verified. The limiting duty ratio for the normal operation of the proposed converter is 0.5. It s observed that by proper selection of the inductance using EAC and with the simple proportional controller, the DC bus voltage at light load is found completely eliminated.

The proposed converter has a simpler power circuit and simpler control circuit, has less component count, and it does not contribute to any additional voltage stress. For cost sensitive application this converter may be preferred.

References

[1]　IEEE Recommended Practices and Requirements for Harmonics Control in Electric Power Systems, IEEEStandard, 519, 1992.

[2]　Limits for Harmonic Current Emissions, International Electrotechnical Commission Standard 61000-3-2,2004.

[3]　M. Madigan, R. Erickson and E. Ismail, "Integrated High Quality Rectifier Regulators" i*n IEEE power ELECTRONICS SPECIALIST CONF......* 1992. p.p. 1043 - 1051.

[4] Jinrong Qian, Fred C. Lee, "Single - Stage Single - Switch p - f-c Ac/Dc converters with DC - Bus voltage feedback for universal line applications *in IEEE transactions on P.E. vol: 13*, No-6 Nov 1998 p.p. 1079 – 1088

[5] Martin H. L. Chow, Yim-Shu Lee, and Chi K. Tse "Single-Stage Single-Switch Isolated PFC Regulator with Unity Power Factor, Fast Transient Response, and Low-Voltage Stress" in *IEEE Transactions on P.E., vol: 15, No-1*, Jan 2000 p.p. 156 - 163

[6] A.K Jha,B.G Fernandes and A.Kishore "A Single Phase Single Stage AC/DC converter with high input power factor and tight output regulation"in Progress in Electromagnetic Research Symposium 2006,Cambridge,USA,March 26-29.pp 322-328

[7] James P. Noon, Alexander Borisovich, "Method to Reduce Bus Voltage Stress in Single-Stage Single Switch Power Factor Correction Circuit", U.S.Patent 6,717,826 B 2, Apr. 6,2004.

[8] Manjusha S. Dawande, Gopal K. Dubey, Programmable Input PFC method for SMR" in *IEEE Transactions on P.E. vol: 11*, No-4 July 1996 p.p. 585 – 591

[9] C Qiao and Keyue M. Smedley, "A Topology Servey of Single - Stage P.F.C. with a Boost Type input Current - shaper." In *IEEE APEC* - 2000 February 6 - 10

[10] Bindu S J, C A Babu, "Analysis and Design of a Single Stage Single Switch Power Factor Converter to reduce Bus Voltage Stress with High Input Power Factor and Fast Output Voltage Regulation", in IEEE 13th workshop on Control and Modelling for Power Electronics (COMPEL-2012),Digital object identifier: 10.1109/COMPEL.2012.6251760

2

Particular Transient Regimes of Asynchronous Motors Supplied by PWM Converters

Monica-Adela Enache, Sorin Enache, Ion Vlad, Gheorghe-Eugen Subtirelu

University of Craiova, Faculty of Electrical Engineering, Craiova, Romania

Email address:

menache@em.ucv.ro (M. A. Enache), senache@em.ucv.ro (S. Enache), ivlad@em.ucv.ro (I. Vlad), esubtirelu@em.ucv.ro (G. E. Subtirelu)

Abstract: This paper presents a few aspects regarding the analysis of some dynamic regimes of asynchronous motors supplied at variable frequency. In this purpose, in the first part of the paper, there are detailed the Simulink blocks of a program for simulating the operation of driving systems with static converters with precomputed commutation moments. Then there are presented the simulations obtained with the help of this program in case when the supply frequency is modified by jump. The accent is laid on the rate of time-dependent variation of the stator current. The results obtained are compared with the experimental results obtained with the help of a high-speed data acquisition board. The paper ends with conclusions and references.

Keywords: Asynchronous Motors, Dynamic Regimes, Simulation, Test

1. Introduction

The problem of optimal control of an asynchronous machine is a very present one; this fact is confirmed by several papers published in outstanding reviews [2], [7], [10] etc. and presented in important international conferences [1], [5], [6], [9], [11], [14] etc.

In comparison with other types of control, PWM control has the advantage that the motor current has a lower content in harmonics, fact materialized in a decrease of the supplementary losses, of the torque pulsations etc.

Along the time there have been formulated a lot of modulation strategies, both analogical and numerical. Among them, the precomputed PWM modulation for harmonics elimination has imposed; in this case, the commutation moments are established analytically, they are memorized and read with variable frequency.

For determining the commutation moments, the line voltage rate is imposed to be symmetrical to $\pi / 2$ and to contain a certain number of pulses characterized by angles α_1, α_2 etc.

Moreover, the amplitude of the fundamental must have the value wanted and as many as possible of the superior harmonics must be null.

It must be emphasized that, owing to the type of the stator winding connection of asynchronous motor, the harmonics which are multiple of three will not occur in the current curve, fact that does not impose supplementary measures for decreasing them.

2. Converter Modelling

The converter with precomputed commutation moments is part of the driving system detailed in figure 1.

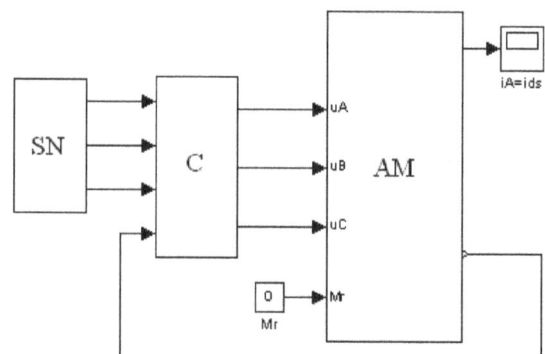

Figure 1. Structure of driving system: SN-supply network, C-converter; AM-asynchronous motor.

A variant of a Simulink model of the analyzed converter will be presented further on.

The Simulink block corresponding to this converter will be composed of other three blocks which simulate the operation of rectifier, filter and inverter from its composition.

2.1. Modelling of Inverter with Precomputed Duration Modulation with Harmonics Elimination

In order to allow the modification of the root-mean-square value of the output voltage of inverter (according to [8]) there is used an overmodulation signal with adjustable filling factor and frequency. Using the notions presented before there has been obtained the Simulink model of the voltage inverter with precomputed commutation moments presented in figure 2.

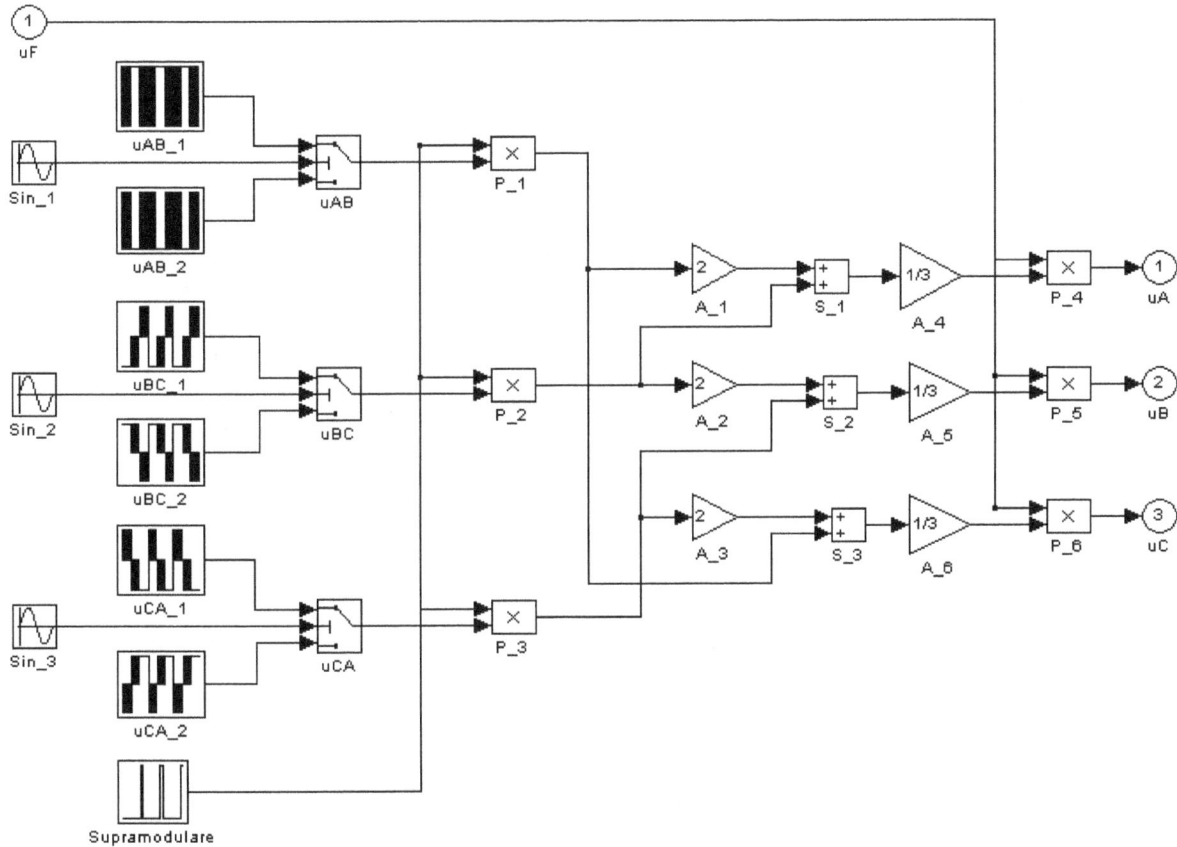

Figure 2. Simulink model of inverter.

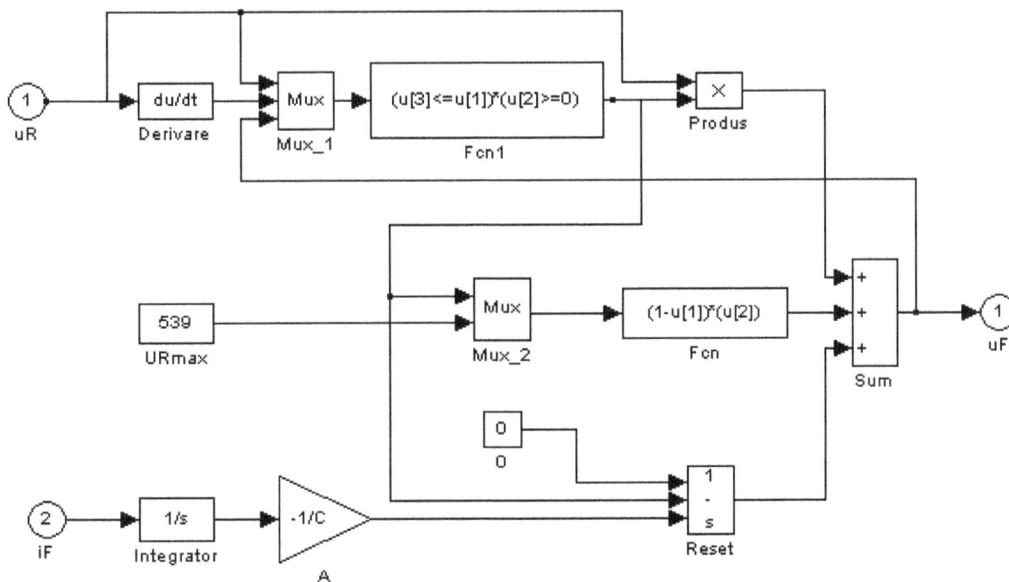

Figure 3. Block scheme of the capacitive filter.

2.2. Simulink Model of the Intermediary Circuit

For carrying out this model, [3] has been used.

In these conditions, the Simulink model of the capacitive filter from the intermediary circuit has the form presented in figure 3.

2.3. Modelling of Rectifier

The rectifier modelling has been carried out starting from one of the observations presented before, that in case of the converter we analyzed, the voltage of the intermediary circuit is practically constant. It results that the rectifier used will always be uncontrolled.

Moreover, this is considered as being a three-phase bridge rectifier, with direct feed to network. Considering that it could be equivalent to two middle-point rectifiers connected in series delivering to the same load, the following block scheme is obtained [4].

3. Motor Modelling

The motor modelling has been carried out starting from the mathematical model written in the two-axes theory.

With the help of this model the block scheme presented in figure 5 has been obtained. The input quantities of this block are the phase voltages (provided by three Simulink blocks from the available programs library) and the resistant torque Mr (considered equal to zero in our simulations).

As output quantities, certain currents, the angular speed and the electromagnetic torque m in dynamic regime are provided.

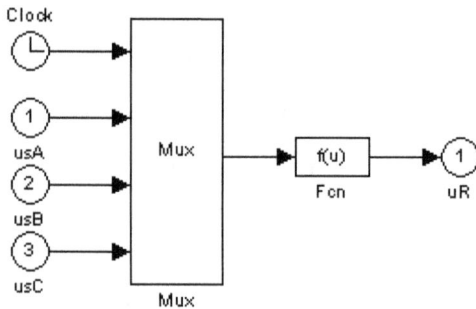

Figure 4. Simulink model of rectifier.

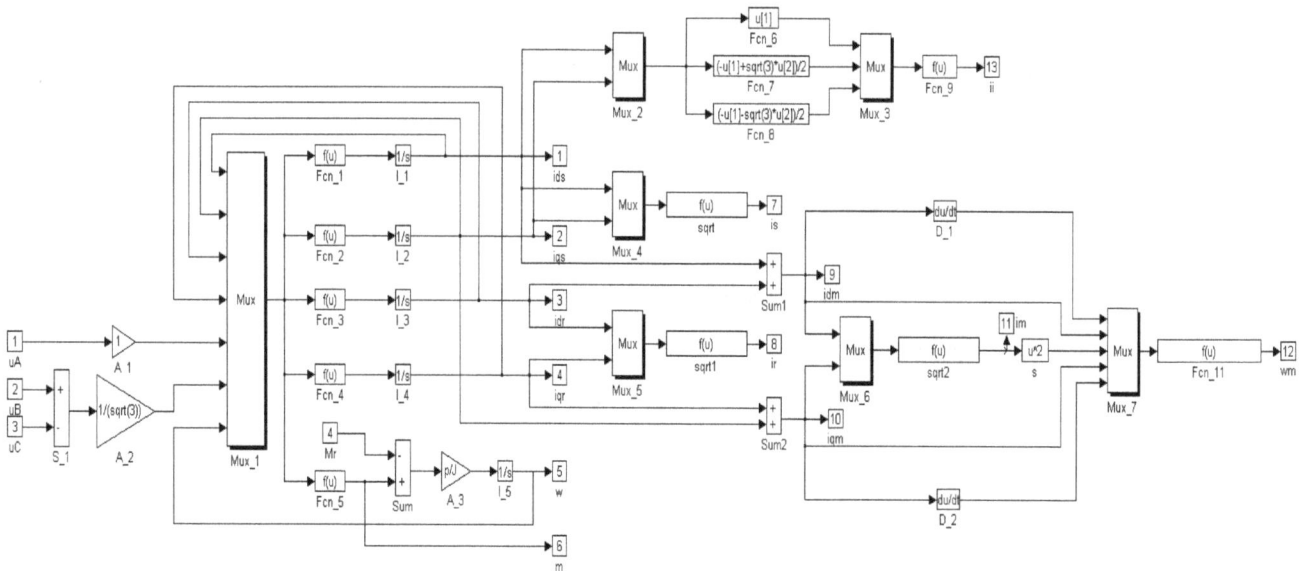

Figure 5. Simulink model of asynchronous motor.

4. Simulations

Running the program detailed before, for several particular values of the parameters of an asynchronous motor rated at 1,2 kW (at 15 Hz), the graphics from figures 6 and 7 have been obtained.

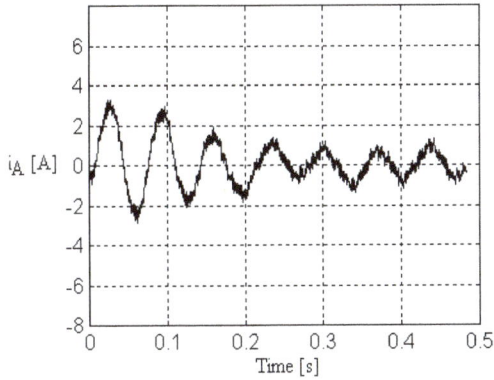

a) Rr=5,5 Ω (cage made of aluminium)

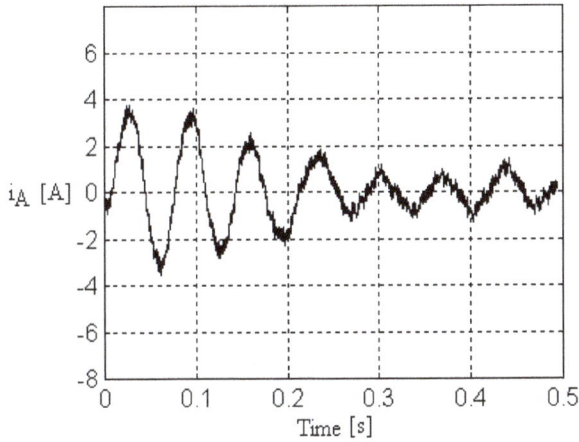

b) Rr=3,2 Ω (cage made of copper)

Figure 6. Time-dependent variations of the current of phase A for cases when the rotor resistance value is modified (J=0,006 Nm).

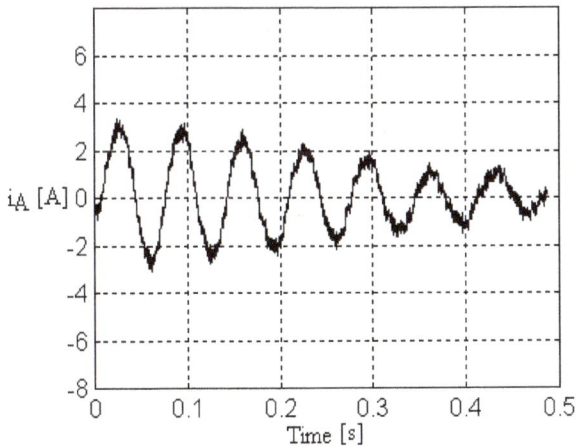

Figure 7. Time-dependent variation of the current of phase A corresponding to the inertia moment modification to the value J=0,012 Nm (Rr=5,5 Ω).

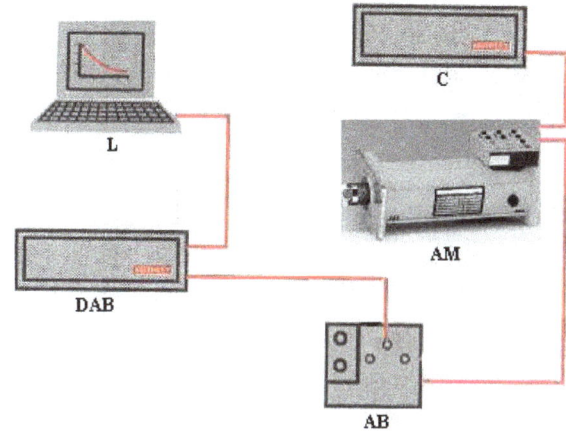

Figure 8. Scheme of assembly.

5. Tests

For carrying out the tests, a squirrel cage asynchronous motor rated at 1,2 kW has been used.

The motor has been supplied by a PWM frequency converter; its main features are detailed in [15].

Using previous experiences [12], for carrying out the measurements there has been used an external data acquisition board connected to the USB terminal of a laptop (Figure 8).

The notations used in the figure 8 are:

C – PWM converter;

AM – asynchronous motor rated at 1,2 kW;

DAB – data acquisition board [16];

L – laptop;

AB – adaptation block.

A photo of this scheme is presented in figure 9, where the elements mentioned in the previous figure can be noticed.

The adaptation of the quantities to be measured (three currents and three voltages) to the values allowed at the data acquisition board input has been achieved by a specialized block with LEM circuits (Figure 10).

This block has been conceived and achieved by the authors of this paper [13].

Figure 9. Photo of the experimental scheme.

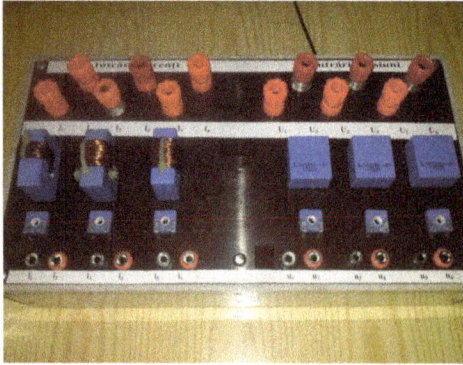

Figure 10. Adaptation block.

With the help of the scheme from figure 8 there has been established the time-dependent variation of current of phase A for case when the reference of speed has been modified by jump (f=15 Hz).

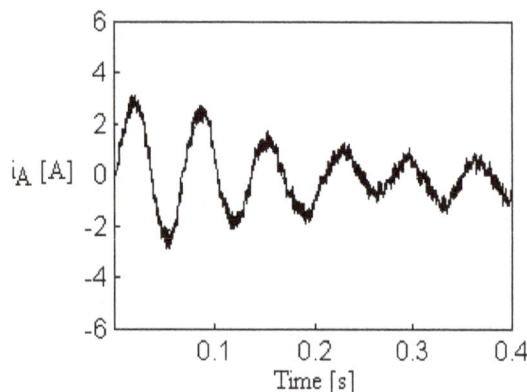

Figure 11. Experimental variation of current.

As it can be noticed, comparing the current graphic obtained experimentally (Figure 11) to the graphics obtained by simulation (Figure 5a), it results that the experiment validates the simulation, from both quantitative and qualitative point of view.

6. Conclusions

The analysis carried out in this paper has aimed at studying the influence of some parameters of asynchronous motor on the dynamic regimes obtained for the case when the supply is ensured by PWM frequency converters.

In this purpose there have been modified in turn the resistance of the rotor cage (by choosing the material it is made of) and the inertia moment (by adding a supplementary inertia weight).

The following conclusions have been obtained from the analysis of the previous graphics:
- the value of the shock current increases in case of cage made of copper;
- the duration of the transient regime of current is a little bit less in case of cage made of aluminium;
- the increase of the inertia moment leads to a strong increase of the duration of the transient regime analyzed

here without affecting the value of the shock current very much;
- the experiment confirmed qualitatively and quantitatively the time-dependent variations of the currents obtained by simulation.

It must be also emphasized that the paper presents a series of blocks built in Simulink which can be used for simulating other dynamic regimes, too.

Moreover, the data acquisition system used here has the advantage that it is mobile and it has a structure which can be easily adapted for other practical situations, too.

Acknowledgements

This work was supported by the strategic grant POSDRU/159/1.5/S/133255, Project ID 133255 (2014), co-financed by the European Social Fund within the Sectorial Operational Program Human Resources Development 2007 – 2013.

References

[1] Benallal, M.N., Ailam, E.H., Moussa, M.A., Mahieddine, A.: Overvoltages caused by the PWM inverter in the stator coils of asynchronous motor, Electric Power Quality and Supply Reliability Conference (PQ), DOI: 10.1109/PQ.2014. 6866820, 2014, pp. 243 - 246.

[2] Contreras-Aguila, L., Garcia, N.: Stability Analyses of a VFT Park Using a Sequential Continuation Scheme and the Limit Cycle Method, IEEE Transactions on Power Delivery, Volume: 26, DOI: 10.1109/TPWRD. 2010.2089703, Publication Year: 2011, pp. 1499 – 1507.

[3] Enache, M.A.: Analysis of some basic processes specific to the operation of reluctance synchronous motor, Thesis for a doctor's degree, Craiova, 2008.

[4] Enache, S., Vlad, I.: Inducction machine – Basic notions – Dynamic processes, Ed. Universitaria, Craiova, ISBN 973-8043-122-2, 2002.

[5] Enache, S., Campeanu, A., Vlad, I., Enache, M.A. : Influence of the filtration circuit parameters of an induction motor drive on stability, 13th International Conference on Optimization of Electrical and Electronic Equipment (OPTIM), ISBN 978-1-4673-1650-7, 2012, pp. 665 – 670.

[6] Han, M., Kawkabani, B., Simond, J.J.: Eigenvalues analysis applied to the stability study of a variable speed pump turbine unit, 2012 XXth International Conference on Electrical Machines (ICEM), DOI: 10.1109/ICElMach.2012.6349984, 2012 , pp. 907 – 913.

[7] Hava, A.M., Çetin, N.O.: A Generalized Scalar PWM Approach With Easy Implementation Features for Three-Phase, Three-Wire Voltage-Source Inverters, IEEE Transactions on Power Electronics, Volume: 26 , Issue: 5, 2011, DOI: 10.1109/TPEL.2010.2081689, pp. 1385 – 1395.

[8] Ivanov, S.: Performant driving systems with asynchronous motor and performant static converters, Thesis for a doctor's degree, Craiova, 1997.

[9] Meiyang, Z., Hui G.: Integrative modeling and simulation analysis of asynchronous motor and control system, 2011 International Conference on Electrical Machines and Systems (ICEMS), Digital Object Identifier: 10.1109/ICEMS.2011.6073845, 2011, pp. 1 – 6.

[10] Niguchi, N., Hirata, K.: Torque-Speed Characteristics Analysis of a Magnetic-Geared Motor Using Finite Element Method Coupled With Vector Control. IEEE Transactions on Magnetics, Vol. 49, Issue 5, DOI: 10.1109/TMAG. 2013.2239271, 2013, pp. 2401 – 2404.

[11] Saman, M., Karaköse, M., Akin, E.: Grey system based on flux estimation for vector control of asynchronous motors, 2011 IEEE 19th Conference on Signal Processing and Communications Applications (SIU), 2011, pp. 666 – 669.

[12] Subtirelu, G.E., Dobriceanu, M., Enache, M.A.: Pulse and Transition Measurements in Dynamic Regimes of Electrical Motors, Annals of the University of Craiova, Year 38, No. 38, ISSN 1842 - 4805, 2014, pp. 221-228.

[13] Vlad, I., Enache, S., Enache, M.A.: Aspects Regarding Operation Characteristics of Brush-less Direct Current Motors, Journal of Electrical and Electronic Engineering, Vol. 2, No. 2, 2014, DOI: 10.11648/j.jeee.20140202.12, pp. 41-46.

[14] Vlad, I., Campeanu, A., Enache, S., Enache, M. A.: Energetic Aspects and Monitoring of Asynchronous Motors Starting, Proceedings of the International Conference on Optimization of Electrical and Electronic Equipment OPTIM 2014, Brasov/Cheile Gradistei, May 22-24, ISSN 1453-7397, 2014, pp. 1-6.

[15] http://www.unisgroup.co.uk/pdf/Telemecanique-Altivar-5-1.1-to-110-kW-User-Manual.pdf.

[16] http://www.distek.ro/ro/Produs/Modul-achizitie-USB-8-canale -50kSa-s-Keitlhey-KUSB-3100-911.

3

Power System Stabilizer Design Using Compressed Rule Base of Fuzzy Logic Controller

Dhanesh Kumar Sambariya

Department of Electrical Engineering, Rajasthan Technical University, Kota, India

Email address:
dsambariya_2003@yahoo.com

Abstract: In this paper, the fuzzy logic controller (FLC) based power system stabilizer (PSS) with compressed / reduced rule is presented. The FLC rule base is generally based on empirical control rules. In this method, the fuzzy system with a large number of fuzzy rules is compressed to a fuzzy system with a reduced number of rules by removing the redundant and inconsistent rules from the rule base which doesn't affect the performance of the fuzzy logic controller. The FLC based PSS has two input signals as speed deviation and derivative of speed deviation with an appropriate number of linguistic variables. The number of compressed rules in the rule base through the proposed dominant rule algorithm is reduced to a number as low in the number of selected linguistic variables to represent input and output signals. The application of the FLC with compressed rules as a power system stabilizer (CR-FPSS) is investigated by simulation studies on a single-machine infinite-bus system (SMIB). The superior performance of this compressed rule based fuzzy PSS (CR-FPSS) as compared to conventional PSS and proves the better efficiency of this new CR-FPSS. The reduced CPU computational time and storage space as compared to the fuzzy power system stabilizer (FPSS), proves its applicability in control.

Keywords: Fuzzy Logic Controller, Rule Base Compression, Compressed Rule FPSS, Single Machine Infinite Bus System, Power System Stabilizer, Dominant Fuzzy Rule Compression

1. Introduction

The electric power demand is increasing with time resulting in extension to the power system network & constraints. These large networks are operated close to their dynamic stability even though it may lead to major system black-outs because of small signal oscillations (0.1-3Hz). These can be damped-out by using conventional power system stabilizers (CPSS) because of their flexibility, easy implementation & of low cost [1, 2]. These CPSSs are the fixed parameter controllers, designed over a nominal operating point to get desired performance at this point as well expect over a wide range of operating conditions and varying system conditions. The CPSSs constitutes a gain amplifier block, lead-lag network and washout block [1]. These CPSS are limited in performance because these may lose effective damping for a wide range of operating conditions within a power system [3].

In 1960 - 70, the classical optimization methods were introduced but not able to converge for non-linear and non-differential engineering problems. Recently, some of the optimization methods such as the particle swarm optimization

(PSO) [4, 5], genetic algorithm (GA) [6, 7], bat algorithm [8] and differential evolution (DE) [9] algorithm, have been applied to complicated and large dimensional power system problems. As a very new optimization method named as Harmony Search Algorithm has been introduced by Geem et al. in 2001 [10], is inspired from the process of the improvisation used by musicians to achieve the harmony. The HS algorithm [11, 12] is a meta-heuristic optimization algorithm which is similar to the PSO [4] and GA [6]. It has been implemented extensively in the fields of engineering optimization [11], in recent years. It became an alternative to other heuristic algorithms like PSO [4], Tabu search [13], simulated annealing [14, 15], evolutionary programming [16], rule based bacteria foraging [17] and type-2 fuzzy power system stabilizer [18, 19].

In case of adaptive PSS, with poor initialization, the performance during learning phase is not satisfactory. Continuity of the objective function is a prerequisite for gradient algorithm used in such applications [20]. As alternative control theories such as the variable structure, the adaptive and linear optimal control theory has been used to design Power System Stabilizers with improved performance in [20, 21]. Fuzzy logic applications to design of the PSS have emerged as a powerful technique which outperforms the CPSS over a wide range of operating conditions of the power system [22]. The special aspect of Fuzzy PSS is its application to ill-defined systems because of its model-free ability as in [22, 23].

In 1970s [24], the applications of fuzzy control in the engineering field were introduced by design of a pilot boiler-steam engine. The application of systematic rule base as a controller is described in [25, 26]. The construction of a fuzzy controller is simpler to its tuning and analysis [27, 28]. The major concern of the researchers is to obtain the high interactivity of the rules, completeness and inconsistencies of the fuzzy rules as reported in [21, 29, 30]. The inconsistency and completeness conviction is being used for the study/analysis of the fuzzy controllers. Therefore, a methodology is carried out to analyze the consistency of rule base for FPSS. The inconsistencies of a rule base are determined by equalities of two fuzzy sets. This method is an alternative to the possibility approach proposed in [31, 32].

The refinement in the rule base is carried out by redundancy approach as proposed in [31]. As the redundant rules do not participate/contribute to the performance analysis of the fuzzy controller, therefore, the rule base can be modified by removing these redundant rules. The main considerable aspects of a fuzzy control are performance criteria and cost factors, which affect the size of rule base [31]. The only problem associated with the fuzzy systems is the large number of rules depending upon the number of inputs. The increased number of rules leads to complexity of quality in terms of poor transparency and unclear/vague interpretation of the fuzzy rules, while the quantitative complexity vests in terms of the high number of rules. The variation of a number of rules with the number of linguistic variables for a particular number of inputs is shown in Figure 1 and may be concluded that the number of rule's increases exponentially with the increase of linguistic variables. If the number of inputs is 'i', if the number of linguistic variables is 'v', then the number of rules 'r' is given by $r=v^i$.

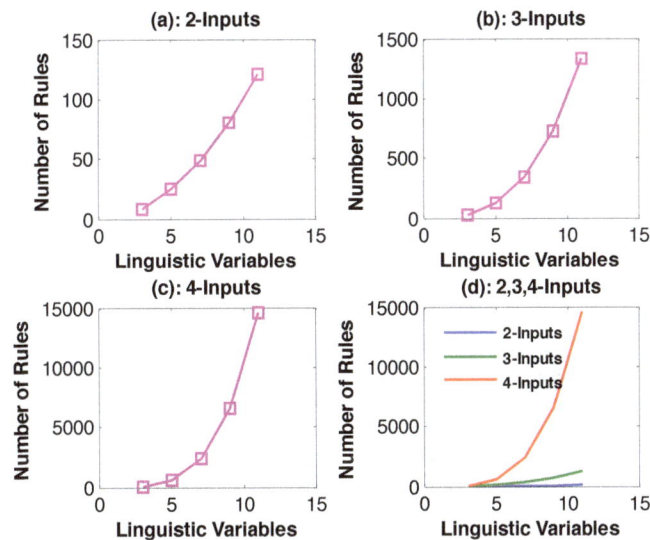

Figure 1. *Variation of number of rules for a fuzzy system with two, three and four inputs.*

It is well established that fuzzy control is applicable to the systems which are not well defined, complex, possesses uncertainties and cannot be described by mathematical modelling [12, 33].The uncertainty is the part of decision making units and is inherent to the systems, where the information is gathered. The reliability of the information-gathering process and the quality of decisions may be seriously affected by these uncertainties. However, it suffers from serious concern to non availability of a standard design procedure, reducibility, optimality and partitioning of the fuzzy rule set. A rule base can be generated by some learning process, identification scheme and more generally from the expert operators can have weakly been contributing rules, redundant rules or inconsistent components of the rules. It is also notable that enlarged rule base suffers from the higher computational time and large storage space. Therefore, an optimal rule base is the requirement of fuzzy systems, thus the

present work is an attempt in this direction by using rule compression strategy as suggested in [12, 29, 32, 34, 35].

This paper presents a new fuzzy PSS with compressed rule base for the system data at a nominal operating condition of the power system and applied to different eight operating conditions with fault based disturbance to examine the robustness of it. The performance with this CR-FPSS is compared to conventional PSS and the original FPSS. The special aspect of this method is to partition the input space in a resilient manner [32, 34].

The rest of this paper is organized as follows. In Section 2, the review of the fuzzy system is incorporated. The rule base compression algorithm is presented in section 3. In section 4, the considered test system for small signal stability improvement using SMIB power system model and dynamics is incorporated. The system is tested with CPSS, with FPSS and with proposed CR-FLPSS. Finally, the observation based conclusion is carried out in section 5.

2. Operation Stages in Fuzzy Systems

The pioneer work on fuzzy logic was introduced in 1960's by Prof. Lofti Zadeh [36]. In fuzzy logic, a programmer deals with the natural, "linguistic sets" of states, such as a large negative medium negative, negative, positive, medium positive, large positive, etc. The main parts to the fuzzy logic process are fuzzification (crisp value to fuzzy input), rule evaluation (fuzzy control), and defuzzification (fuzzy output to crisp value) *[37, 38]*. The FLC has the following main elements as in Figure 2.

1. Rule-Base: It contains fuzzy logic quantification (a set of If-Then rules) of the expert's linguistic description to the system to get good control.
2. Inference Mechanism: It is also called as "inference engine" or "fuzzy inference" module. It emulates the decision making of expert in interpreting and applying to control the plant.
3. Fuzzification Interface: It converts controller inputs into the fuzzy input, which is used by the inference mechanism to activate and apply rules. The crisp value inputs are converted to fuzzy inputs as the basis of

membership functions defined for the fuzzy system.
4. Defuzzification Interface: It converts the output/conclusions of the inference into crisp inputs to the plant.

The purpose of the fuzzy control system is to replace a human expert by a fuzzy rule-based system. The FLC can convert the linguistic control based on expert knowledge into an automatic control strategy. The heart of Fuzzy Systems is a knowledge base, which is formed by the if-then rules of fuzziness.

Fuzzy sets in fuzzy systems are used to map input to output as in Figure 2. The FIS system based on Mamdani as in [24], is considered in this paper. The major stages of mapping involve as fuzzification process, inference process and defuzzification process. The triangular membership functions are used in fuzzification stage [24, 34, 39]. The major parts of an inference system are application, implication and aggregation [34]. The conjunctive process (MIN) is applied in application part, the truncation process on the implication part while a disjunctive process (MAX) is applied in the aggregation part as in Figure 2. There are many defuzzification methods as centroid, bisector, LOM, MOM, SOM but most widely used centroid method is considered in the defuzzification stage [34].

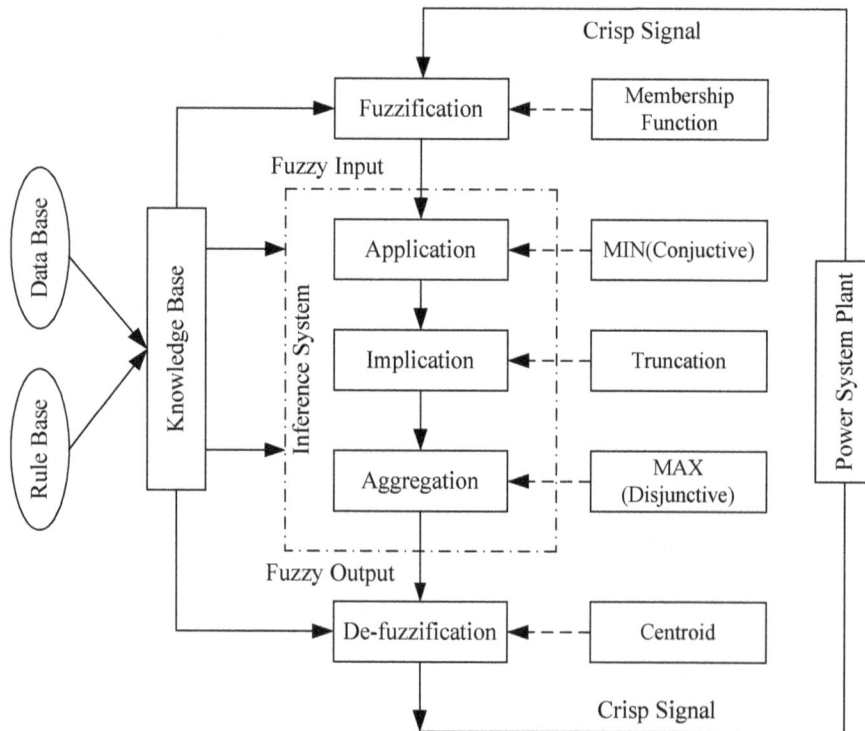

Figure 2. Input to Output mapping in a fuzzy system.

3. Review on Rule Base Compression Algorithm

The efforts are made for the rule base reduction in fuzzy control like interpolation algorithm in [32, 40, 41]. In

interpolation algorithm, only minimal numbers of necessary rules remain at the rule base. Another approach is boolean used for measuring equivalence and inconsistency in the rule base. This method of rule base reduction with application as a robot manipulator control is proposed by Bezine in [42]. Xiong and Litz [43] have proposed an application of genetic algorithm to premise learning methodology for rule base reduction resulting to efficient control of an inverted pendulum.

In this paper, the inconsistency and redundancy as in [31] and dominant rule selection as in [32, 34] are used to eliminate highly redundant rules and design of compressed rule base. It is established that the rule's redundancy or non dominant do not contribute to the performance of the controller as FPSS. Therefore, it is necessary to reduce/compress the rule base for fuzzy inference system (FIS) computations in real time. The term redundancy is the measure of the degree to which the input and output fuzzy rules tend to overlap are introduced in [32, 34]. Therefore, the input and output spaces of any two fuzzy rules closely similar to each other are termed as highly redundant. Thus, the identical rules have redundancy values as 1.0 and totally different rules have a redundancy value as 0 (no value). As in[32], the fuzzy rules in a rule base possess a high degree of redundancy levels may be removed to reduce the size of rule base and maintaining the almost similar performance as with original FIS system.

As the rules having the similar linguistic value of their outputs are called as monotonic rules and are commonly found in fuzzy systems. In the proposed method, monotonic rules are arranged in groups, and the dominant rules are determined for each group. The rules with highest firing strength are called dominant because all other rules from the same group don't have an appreciable impact on the output. The phenomenon is termed as redundancy and would be the basis to rule base compression process. An algorithm for rule base compression can be described as.

Creation of rule base integer table: The rule base integer can be created in two ways: (a) creation by user input, (b) creating from the existing FIS system.

(a) Creation by user input: the user is required to enter information such as the number of inputs and outputs along with the number of linguistic variables for each input and output. The entry for the output values for each possible combination of input linguistic values. According to the entries of input, output and rules, an integer table is created.

(b) Creation from an existing FIS system: in another option, the pre-defined FIS system is loaded and the linguistic variables for inputs and output are assigned by integer numbers. According to the rules within the FIS system, an integer table is created.

1. Input system information: In case of 1(a), the user is required to enter all the information regarding the fuzzy inference system. The information regarding membership function and names for linguistic variables are entered.

2. Creation of the fuzzy rule base with the original rules: In case of 1 (a), according to the entries in step 1and 2, the FIS system is created and saved with innovative details.

3. Rearrange the rules into groups: This step is related to aggregation and is an entirely off-line process. The integer rule table created in step-1 is rearranged in groups and is sorted in ascending order with respect to the chosen output. In a group say output as 2, then in the second group all rules in terms of integers should be accommodated and similarly for others. If a 7×7 fuzzy relation matrix (FRM) is created or loaded, afterwards there would be seven groups with outputs as 1, 2, 3, 4, 5, 6 and 7.

4. Find dominant rule and compressed rules: In this step, dominant rue for each group is determined. Since the dominant rules can be determined after the completion of application and fuzzification stages, therefore, the process is only possible on-line. The determination of dominant rules from each group and arranging in a table is called as reduced/compressed integer rule table.

5. Creation of fuzzy rule base with compressed rules: The reduced rule table found in step 5, for an original rule table either by creating or by loading is saved in this step by different names with *.fis as an extension. Let created/loaded FIS system is pss.fis then the reduced should be saved as pssr.fis.

6. Apply fuzzification, inference and defuzzification: This Step is concerned with the evaluation of the file saved in step six. It evaluates the output of a pssr.fis for given inputs like speed as 0.05, acceleration as 0.005 then the output voltage comes as 0.68.

7. Generate the solution surface: In this step output-surface is generated for original and reduced FIS system by an applied number of points (generally 30-50) for the crisp input values and the defuzzified output values. It validates the compressed system with respect to the original system.

On application of the algorithm as in [32, 34] and following as mentioned above, the monotonic rules remain in the fuzzy rule base. It should be cleared that the number of monotonic rules as the number of linguistic variables or number of groups in the original fuzzy rule base.

It is suggested in [32, 34], that the non-monotonic rules should be completely removed from the new rule base. It should also be taken into consideration that the aggregation process to determine the inconsistent rules should always be off-line, while, that of filtering process for determining non-monotonic rules should be carried out on-line. In above algorithm step 1 is carried out off-line while the steps 2-3 are carried out online. The rule with maximal firing strength out of a group is called as a dominant rule. The other rules don't have adequate effect. If a group has more than one dominant rule than random selection can be done to select powerful rule.

The rule base compression is carried out in the MATLAB Fuzzy Logic Toolbox. The algorithm is applied for 7×7 fuzzy rule matrix with 49 rules. The corresponding rule base is being represented by integer tables in which the linguistic values of inputs and outputs are replaced by integers. The rule-base compression method with the dominant rule method is applied because of its systematic nature and wide applicability. The considered case of 49 rules with speed and acceleration as input and voltage as output being evaluated for rule compression as per algorithm stated above. The procedure is illustrated in the next section.

4. Case Study

4.1. Test System Description

The system under consideration consists of the single machine connected to an infinite bus (SMIB) through a tie-line. The infinite bus can be represented by the Thevenin's equivalent of a large interconnected power system. Synchronous generator control components as schematic are shown in Figure 3.

The combinations of a turbine and governor are being used to drive the synchronous generator. The generator is being excited by an external excitation system. The excitation system is controlled by AVR and PSS to control the low-frequency oscillations arise due to lack of damping of the system's mechanical mode. In case of inadequate damping torque, the system may lose stability. Since the high gain, fast acting AVR leads to generate negative damping, therefore, to provide additional damping supplementary excitation control like PSS is developed. The stability to the system may be improved by adding a control loop with the governor, but is limited because of overheating problem and large time constant. Hence, the PSS is the best choice to improve stability [1, 2, 44].

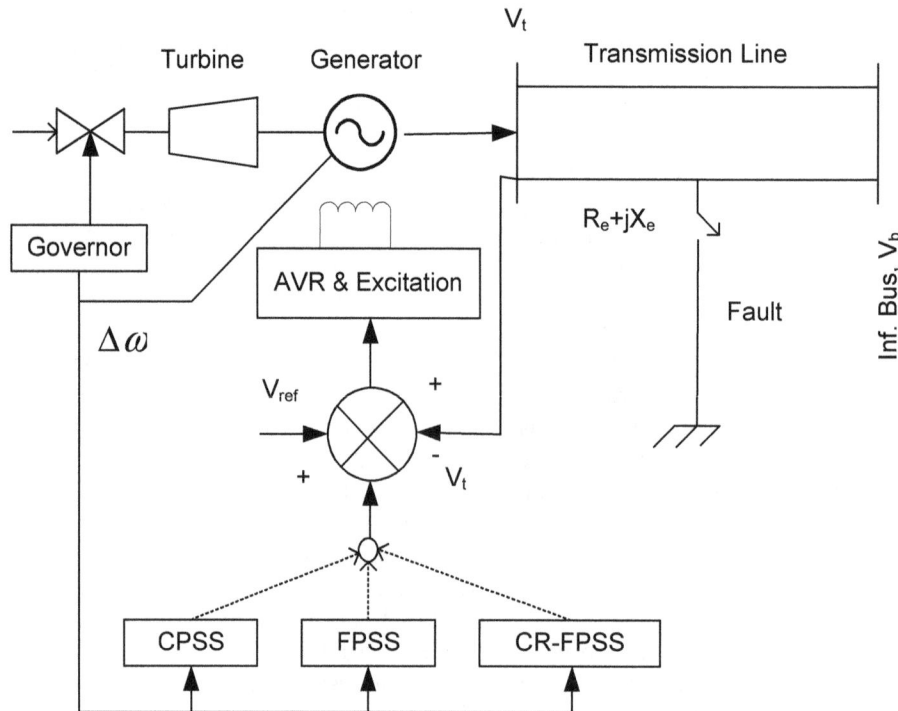

Figure 3. Single Machine Infinite Bus Representation with fault.

Design of conventional PSS is based on a linearized model around a certain operating point. Since the actual power system operates over a wide range of operating conditions and nonlinear characteristics. The tuning of CPSS to cope with most of the operating conditions is very difficult. The change in rotor speed is taken as input to PSS as shown in Figure 4. The structure of PSS is mainly composed by a gain, wash out filter and the phase compensator block [1, 2]. The gain block is used to ensure the required damping factor in the consideration of practical aspects (operating conditions) of the power system as in [2, 45]. The washout filter behaves as a high-pass filter; therefore, the PSS only responds to the speed deviation of generator and not responds to the steady-state operation within the system. The criterion to select the washout time constant T_ω is to pass required PSS signals intact [1, 2, 46]. The main concern of the damping issue is the lag compensation in between excitation input and the air gap electrical torque is provided by third block of CPSS, called as the phase compensator block and showed in Figure 4. The output of the PSS is controlled by the limiting block to prevent the over excitation and concerned block is called as limiting block with bounds generally as ± 0.12 to ± 0.15 pu [1, 2, 47]. The transfer function the conventional PSS (CPSS) is given by.

$$G_{PSS}(s) = K_{pss} \frac{sT_\omega}{1 + sT_\omega} \frac{1 + sT_1}{1 + sT_2} \tag{1}$$

In the SMIB representation, the dynamic interference between the several machines in the power station is neglected, but it is still an appropriate to various types of studies, especially for identical machines, operating at similar load levels. The linearized model of SMIB was the result of a first serious investigation by DeMello and Concordia in 1969 [48]. This model is used in the study of PSS design and is also adopted illustrating the proposed design method. The static exciter of a single time constant automatic voltage regulator (AVR) is connected to the synchronous generator and is represented by.

$$G_{AVR}(s) = \frac{K_A}{1 + sT_A}$$ (2)

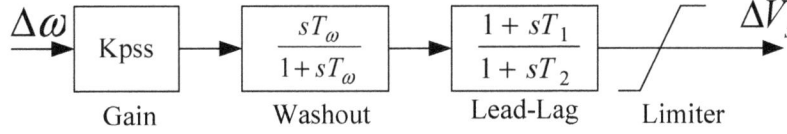

Figure 4. Conventional Power System Stabilizer.

The small perturbation model of the system as in [12], includes the IEEE-type ST1A exciter and the PSS. The linearised differential equations can be written as follows

$$\Delta\dot{\omega} = -\frac{K_1}{2H}\Delta\delta - \frac{K_2}{2H}\Delta E_q'$$ (3)

$$\Delta\dot{\delta} = \omega_0 \, \Delta\omega$$ (4)

$$\Delta\dot{E}_q' = -\frac{1}{K_3 T_{d0}'}\Delta E_q' - \frac{K_4}{T_{d0}'}\Delta\delta + \frac{1}{T_{d0}'}\Delta E_{fd}$$ (5)

$$\Delta E_{fd} = -\frac{K_A K_5}{T_A}\Delta\delta - \frac{K_A K_6}{T_A}E_q' - \frac{1}{T_A}\Delta V_{ref}$$ (6)

By using equation (3)-(6), state space form of the system can be arranged as follows

$$\dot{x}(t) = A\,x(t) + B\,x(t)$$ (7)

$$x(t) = \begin{bmatrix} \Delta\omega & \Delta\delta & \Delta E_q' & \Delta E_{fd} \end{bmatrix}^T$$ (8)

$$A = \begin{bmatrix} -\dfrac{K_D}{2H} & -\dfrac{K_1}{2H} & -\dfrac{K_2}{2H} & 0 \\ 0 & 0 & 0 & 0 \\ 0 & -\dfrac{K_4}{T_{d0}'} & -\dfrac{1}{K_3 T_{d0}'} & \dfrac{1}{T_{d0}'} \\ 0 & -\dfrac{K_A K_5}{T_A} & -\dfrac{K_A K_6}{T_A} & -\dfrac{1}{T_A} \end{bmatrix}$$ (9)

$$B = \begin{bmatrix} 0 & 0 & 0 & \dfrac{K_A}{T_A} \end{bmatrix}$$ (10)

The FLPSS rules are determined by using the proposed rule compression method. Eight operating points of SMIB power system (i.e power system pants as 01–8 and shown in Table 1). The plants of SMIB power system with FLPSS are simulated in an SIMULINK working environment of MATLAB.

Table 1. SMIB Power System Plants.

Power System Plant	Active Power, Pg0	Reactive Power, Qg0	Reactance, Xe
1	0.50	0.0251	0.2
2	0.75	0.0566	0.2
3	1.00	0.1010	0.2
4	1.25	0.1588	0.2
5	0.50	0.0505	0.4
6	0.75	0.1152	0.4
7	1.00	0.2087	0.4
8	1.10	0.2550	0.4

4.2. Development of Compressed Rule Table

The number of Linguistic Variables is taken as 7 to each input. The input signals are taken as Speed Deviation (Error) and rate of change of Speed Deviation (derivative of Error) while the output is taken as Change in Voltage. For simplicity, these three signals are being represented as Speed, Acceleration and Voltage. The seven linguistic variables are defined by LN, MN, N, Z, P, MP, and LP as in Table 2.

Table 2. Representation of Linguistic Variables (7×7 FRM).

S. No.	Symbol	Linguistic Variables	Integer
1	LN	Large Negative	1
2	MN	Medium Negative	2
3	N	Negative	3
4	Z	Zero	4
5	P	Positive	5
6	MP	Medium Positive	6
7	LP	Large Positive	7

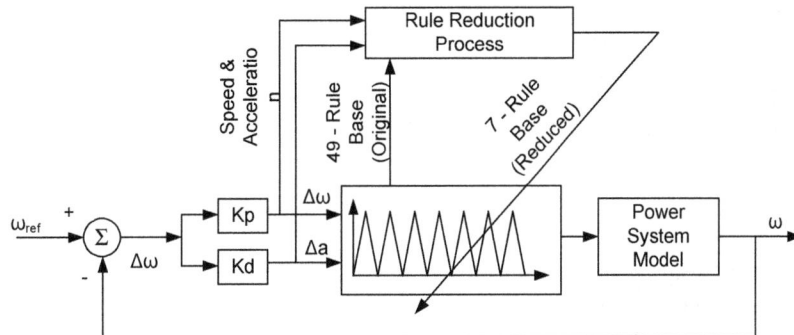

Figure 5. Rule reduction process.

Table 3. *Input and Output Linguistic Variables.*

Acceleration		LN	MN	N	Z	P	MP	LP
Speed Deviation	LN	LN	LN	LN	MN	MN	N	Z
	MN	LN	LN	MN	MN	N	Z	P
	N	LN	MN	LN	N	Z	P	MP
	Z	MN	MN	N	Z	P	MP	MP
	P	MN	N	Z	P	LP	MP	LP
	MP	N	Z	P	MP	MP	LP	LP
	LP	Z	P	MP	MP	LP	LP	LP

As per the Table 3, the number of output rules is 49 and the table is called as fuzzy relation matrix (FRM) or simply rules table. As per step-5, the dominant rules (only in online process) for selected inputs, i.e. speed as 0.008929 pu and accelerations as 0.08929 pu are determined from the groups as in Table 4. The procedure of speed and acceleration is determined as in Figure 5. The rules are written in the form of IF-THEN rules. The rule table for the original FIS system is arranged by an integer as in step-3 and is shown Table 5(a). The same integer table is being re-arranged into a group rule table and is represented in Table 5(b). The number of groups is equal in the number of linguistic variables. Generally, one dominant rule is selected from each group and resulting in the compressed rule number as 7 in Table 6. A case may exist that a particular group is having more than one powerful rule then a single rule should be selected on a random basis. As stated above, the reduced rule table for 7×7 FRM in Table 6 can be written as the linguistic variables as in Table 7. The dominant linguistic variables enlisted as a result of the online process and resulting rules as in Table 7 is shown by bold and italic in Table 3.

Table 4. *Speed and Acceleration for creation of compressed rule base for plant 1 to 8.*

Plant	Speed	Acceleration
Plant-1	0.004464	0.04464
Plant-2	0.006696	0.06696
Plant-3	0.008929	0.08929
Plant-4	0.01116	0.1116
Plant-5	0.004465	0.04465
Plant-6	0.006697	0.06697
Plant-7	0.008929	0.08929
Plant-8	0.009821	0.09821

The representation of the bold-italic in Table 3 is same in terms of output linguistic variable as in Table 6 (in integer form) and Table 7 (in LV form). It is the rule base saved with *.fis extension and used for the performance analysis of FPSS using reduced rule FIS. To look into the validation of the primary and decreased rule table, the output surface is generated and the similarity of the output surfaces proves the mimics operation of the reduced rule system to its first counterpart.

4.3. Test System Response Analysis

The system considered in the PSS design using a reduced number of rule base of 7×7 FRM is applied to plants 1-8 as in Table 1. In this case, the plants being tested with Fuzzy PSS (reduced rule base) and compared to the system response with CPSS and with FPSS; presented in Figure 6 to 9. In all responses, the system under different operating conditions is stable with CR-FPSS. The computation time taken by CPU, with CPSS, FPSS and proposed CR-FPSS is recorded in Table 8; resulting least with CR-FPSS as compared to FPSS for all plant conditions (lightly loaded to heavy loaded). Here, it should be noted that the computational time in case of CPSS is least in comparison to FPSS and CR-FPSS, but is not able to stabilize on heavy loading conditions like plant-8, shown in Figure 9(b). For further clarification of the system response, the generator speed, control voltage, terminal voltage, air-gap electric torque, angle and excitation voltages are shown in Figure 10(a), 10(b), 11(a), 11(b), 12(a) and 12(b), respectively for nominal operating conditions of the SMIB power system as in plant 3.

Table 5. *Integer Table for Original PSS rule base.*

(a) Complete integer Rule table				(b) Grouped integer Rule table			
Rule	Speed	Acc.	Vol.	Rule	Speed	Acc.	Vol.
1	1	1	1	1	1	1	1
2	1	2	1	2	1	2	1
3	1	3	1	3	1	3	1
4	1	4	2	4	2	1	1
5	1	5	2	5	2	2	1
6	1	6	3	6	3	1	1
7	1	7	4	7	3	3	1
8	2	1	1	8	1	4	2
9	2	2	1	9	1	5	2
10	2	3	2	10	2	3	2
11	2	4	2	11	2	4	2
12	2	5	3	12	3	2	2
13	2	6	4	13	4	1	2
14	2	7	5	14	4	2	2
15	3	1	1	15	5	1	2
16	3	2	2	16	1	6	3
17	3	3	1	17	2	5	3
18	3	4	3	18	3	4	3
19	3	5	4	19	4	3	3
20	3	6	5	20	5	2	3
21	3	7	6	21	6	1	3
22	4	1	2	22	1	7	4
23	4	2	2	23	2	6	4
24	4	3	3	24	3	5	4
25	4	4	4	25	4	4	4
26	4	5	5	26	5	3	4
27	4	6	6	27	6	2	4
28	4	7	6	28	7	1	4
29	5	1	2	29	2	7	5
30	5	2	3	30	3	6	5
31	5	3	4	31	4	5	5
32	5	4	5	32	5	4	5
33	5	5	7	33	6	3	5

(a) Complete integer Rule table				(b) Grouped integer Rule table			
Rule	Speed	Acc.	Vol.	Rule	Speed	Acc.	Vol.
34	5	6	6	34	7	2	5
35	5	7	7	35	3	7	6
36	6	1	3	36	4	6	6
37	6	2	4	37	4	7	6
38	6	3	5	38	5	6	6
39	6	4	6	39	6	4	6
40	6	5	6	40	6	5	6
41	6	6	7	41	7	3	6
42	6	7	7	42	7	4	6
43	7	1	4	43	5	5	7
44	7	2	5	44	5	7	7
45	7	3	6	45	6	6	7
46	7	4	6	46	6	7	7
47	7	5	7	47	7	5	7
48	7	6	7	48	7	6	7
49	7	7	7	49	7	7	7

Table 6. *Reduced Rule Integer Table for PSS.*

Initial Parameters of FPSS for Compressed Rules		
Speed	Acceleration	Voltage
1	1	1
1	4	2
1	6	3
4	4	4
5	4	5
3	7	6
5	5	7

Table 7. *Reduced Rule Table in LV form for PSS.*

Rule No.	Linguistic Rules
1	If (speed is LN) and (acceleration is LN) then (voltage is LN)
2	If (speed is LN) and (acceleration is Z) then (voltage is MN)
3	If (speed is LN) and (acceleration is MP) then (voltage is N)
4	If (speed is Z) and (acceleration is Z) then (voltage is Z)
5	If (speed is P) and (acceleration is Z) then (voltage is P))
6	If (speed is N) and (acceleration is LP) then (voltage is MP)
7	If (speed is P) and (acceleration is P) then (voltage is LP)

Table 8. *CPU Computational Time.*

Power System Model	CPU time for CPSS	CPU time for Fuzzy PSS	CPU time for CFPSS (Prop.)
Plant-1	0.7958	130.1118	23.0398
Plant-2	0.7256	127.3183	22.1662
Plant-3	0.7159	128.7899	22.3365
Plant-4	0.7486	155.4953	24.8997
Plant-5	0.5609	137.367	21.6231
Plant-6	0.6083	140.6716	22.2323
Plant-7	0.73	138.4199	25.0446
Plant-8	1.3983	141.1053	26.4898

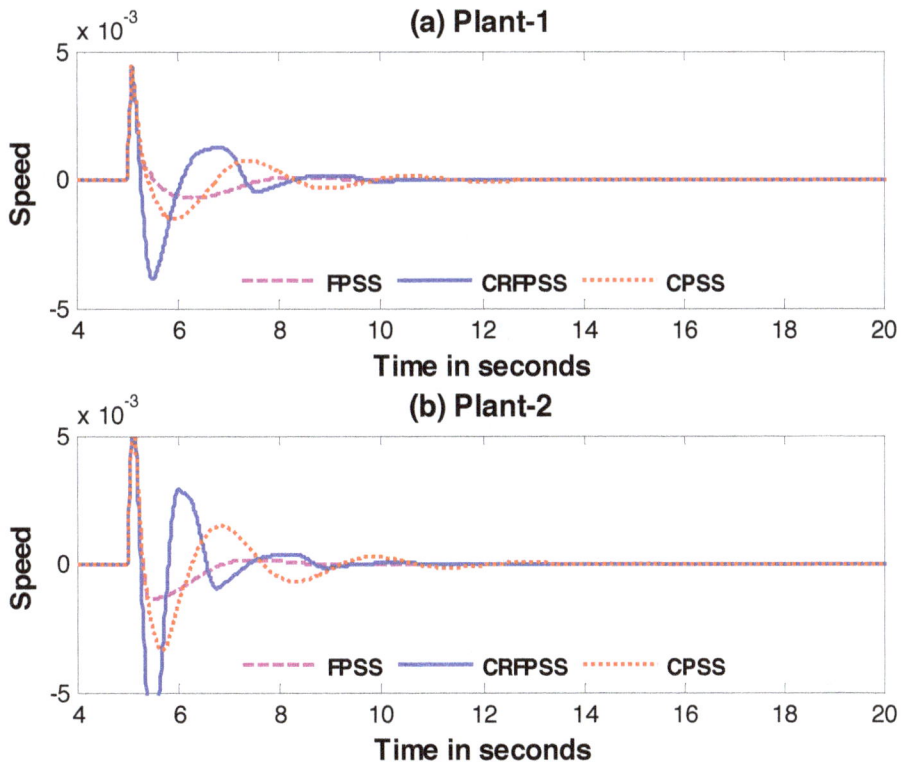

Figure 6. *Speed response with CRFPSS, FPSS and CPSS for (a) Plant-1, and (b) Plant-2.*

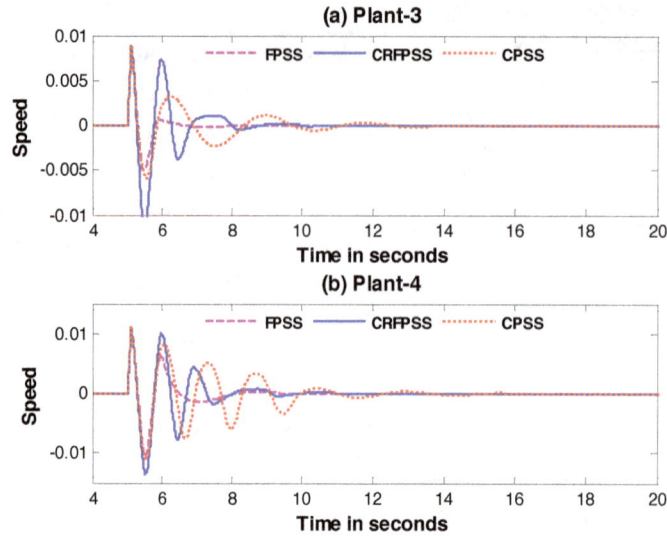

Figure 7. Speed response with CRFPSS, FPSS and CPSS for (a) Plant-3, and (b) Plant-4.

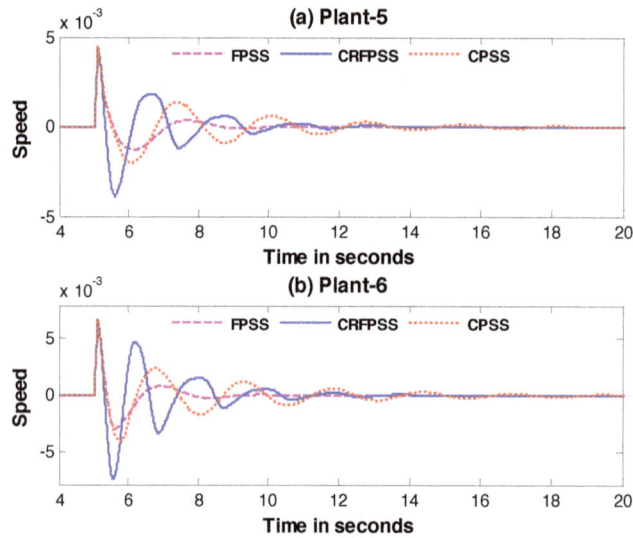

Figure 8. Speed response with CRFPSS, FPSS and CPSS for (a) Plant-5, and (b) Plant-6

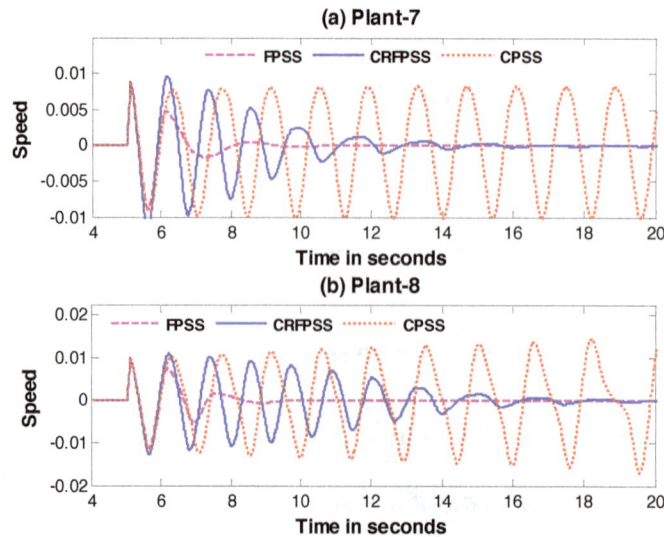

Figure 9. Speed response with CRFPSS, FPSS and CPSS for (a) Plant-7, and (b) Plant-8.

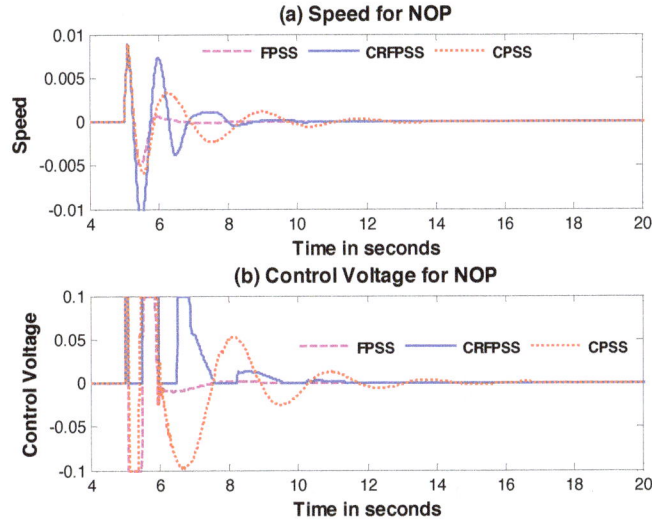

Figure 10. *Response with CRFPSS, FPSS and CPSS at nominal operating point for (a) Speed, and (b) Control Voltage.*

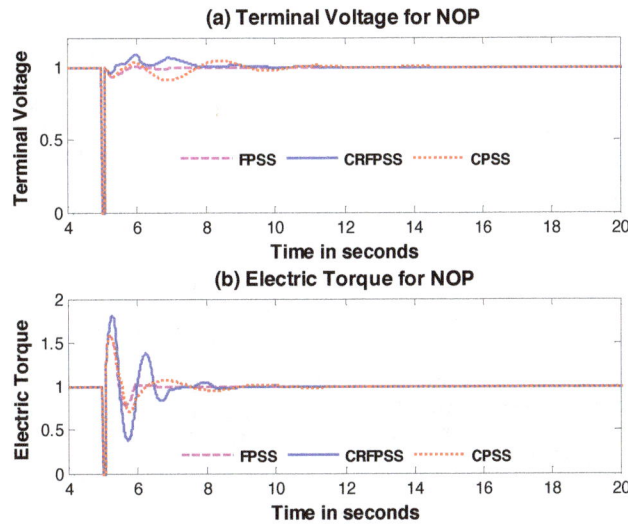

Figure 11. *Response with CRFPSS, FPSS and CPSS at nominal operating point for (a) Terminal Voltage, and (b) Electric air-gap torque.*

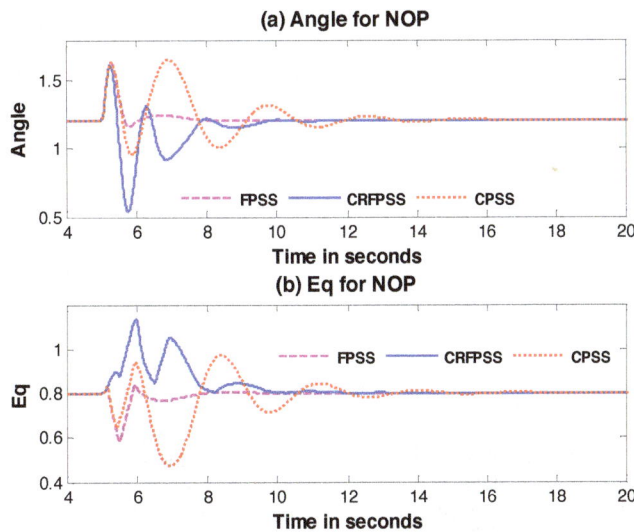

Figure 12. *Response with CRFPSS, FPSS and CPSS at nominal operating point for (a) Angle, and (b) Excitation Input (Eq').*

5. Conclusion

A novel rule reduction algorithm is applied to design a fuzzy logic based power system stabilizer. The reduced rule based FLC is applied to a single-machine infinite-bus system as the power system stabilizer for small signal stability improvement.

The proposed method (dominant rule reduction) based designed CR-FPSS is applied by considering 7×7 FRM with 49 rules, which is compressed to 7-rules. The application of the reduced tables of FIS system can stabilize the SMIB system for wide operating conditions (as plant 1-8) with decreased settling time as compared to CPSS, while as reduced computation time and memory storage as compared to FPSS. The superior performance of this compressed rule based fuzzy PSS (*CR-FPSS*) as compared to conventional PSS and proves the better efficiency of this new *CR-FPSS*.

This method is applicable to the Mamdani type of fuzzy systems and pertinent irrespective of the number of inputs, outputs, linguistic values, membership functions and concerned rules. The dominant rule compression algorithm manages the complexity in the fuzzy systems. In this paper, the algorithm is applied only for a single cycle from original FIS system creation/loading of the reduced FIS system. It can be applied to another number of cycles, but the complexity would be increased. The complexity associated to an increase in cycles is a further scope of research.

References

[1]   P. Kundur, M. Klein, G. J. Rogers, and M. S. Zywno, "Application of power system stabilizers for enhancement of overall system stability," IEEE Transactions on Power Systems, vol. 4, pp. 614-626, 1989. doi:10.1109/59.193836

[2]   E. V. Larsen and D. A. Swann, "Applying Power System Stabilizers Part I: General Concepts," IEEE Transactions on Power Apparatus and Systems, vol. PAS-100, pp. 3017-3024, 1981. doi:10.1109/tpas.1981.316355

[3]   M. Nambu and Y. Ohsawa, "Development of an advanced power system stabilizer using a strict linearization approach," IEEE Transactions on Power Systems, vol. 11, pp. 813-818, 1996. doi:10.1109/59.496159

[4]   S. M. Abd-Elazim and E. S. Ali, "A hybrid Particle Swarm Optimization and Bacterial Foraging for optimal Power System Stabilizers design," International Journal of Electrical Power & Energy Systems, vol. 46, pp. 334-341, 2013. doi:http://dx.doi.org/10.1016/j.ijepes.2012.10.047

[5]   H. M. Soliman, E. H. E. Bayoumi, and M. F. Hassan, "PSO–based Power System Stabilizer for Minimal Overshoot and Control Constraints," Journal of Electrical Engineering, vol. 59, pp. 153–159, 2008. doi:http://iris.elf.stuba.sk/JEEEC/data/pdf/3_108-6

[6]   H. Alkhatib and J. Duveau, "Dynamic genetic algorithms for robust design of multimachine power system stabilizers," International Journal of Electrical Power & Energy Systems, vol. 45, pp. 242-251, 2013. doi:10.1016/j.ijepes.2012.08.080

[7]   S. Panda, "Multi–Objective Non–Dominated shorting Genetic Algorithm–II for Excitation and TCSC–based Controller Design," Journal of Electrical Enginnering, vol. 60, pp. 86-93, 2009. doi:http://iris.elf.stuba.sk/JEEEC/data/pdf/2_109-6

[8]   D. K. Sambariya and R. Prasad, "Robust tuning of power system stabilizer for small signal stability enhancement using metaheuristic bat algorithm," International Journal of Electrical Power & Energy Systems, vol. 61, pp. 229-238, 2014. doi: http://dx.doi.org/10.1016/j.ijepes.2014.03.050

[9]   S. Panda, "Robust coordinated design of multiple and multi-type damping controller using differential evolution algorithm," International Journal of Electrical Power & Energy Systems, vol. 33, pp. 1018-1030, 2011. doi:10.1016/j.ijepes.2011.01.019

[10]   Zong Woo Geem, Joong Hoon Kim, and G. V. Loganathan, "A New Heuristic Optimization Algorithm: Harmony Search," Simulation, vol. 76, pp. 60-68, February 1, 2001 2001. doi:10.1177/003754970107600201

[11]   Z. W. Geem, "Harmony Search Applications in Industry," in Studies in Fuzziness and Soft Computing, Volume 226, Soft Computing Applications in Industry. vol. 226, B. Prasad, Ed., ed: Springer-Verlag Berlin Heidelberg, 2008, pp. 117-134. doi:10.1007/978-3-540-77465-5_6

[12]   D. K. Sambariya and R. Prasad, "Design of Harmony Search Algorithm based tuned Fuzzy logic Power System Stabilizer," International Review of Electrical Engineering (IREE), vol. 8, pp. 1594-1607, October 2013 2013. doi: http://dx.doi.org/10.15866/iree.v8i5.2117

[13]   Y. L. Abdel-Magid, M. A. Abido, and A. H. Mantawy, "Robust tuning of power system stabilizers in multimachine power systems," in IEEE Power Engineering Society Winter Meeting, 2000, p. 1425 vol.2. doi:10.1109/pesw.2000.850180

[14]   M. A. Abido, "Robust design of multimachine power system stabilizers using simulated annealing," IEEE Transactions on Energy Conversion, vol. 15, pp. 297-304, 2000. doi:10.1109/60.875496

[15]   M. A. Abido, "Simulated annealing based approach to PSS and FACTS based stabilizer tuning," International Journal of Electrical Power & Energy Systems, vol. 22, pp. 247-258, 2000. doi:10.1016/s0142-0615(99)00055-1

[16]   M. A. Abido and Y. L. Abdel-Magid, "Optimal Design of Power System Stabilizers Using Evolutionary Programming," IEEE Transactions on Energy Conversion, vol. 17, pp. 429-436, 2002. doi:10.1109/mper.2002.4312476

[17]   S. Mishra, M. Tripathy, and J. Nanda, "Multi-machine power system stabilizer design by rule based bacteria foraging," Electric Power Systems Research, vol. 77, pp. 1595-1607, 2007. doi:10.1016/j.epsr.2006.11.006

[18]   D. K. Sambariya and R. Prasad, "Evaluation of interval type-2 fuzzy membership function & robust design of power system stabilizer for SMIB power system," Sylwan Journal, vol. 158, pp. 289-307, April 2014 2014. doi:http://sylwan.ibles.org/archive.php?v=158&i=5

[19]   D. K. Sambariya and R. Prasad, "Power System Stabilizer design for Multimachine Power System using Interval Type-2 Fuzzy Logic Controller," International Review of Electrical Engineering (IREE), vol. 8, pp. 1556-1565, October 2013 2013. doi: http://dx.doi.org/10.15866/iree.v8i5.2113

[20] P. S. Bhati and R. Gupta, "Robust fuzzy logic power system stabilizer based on evolution and learning," International Journal of Electrical Power & Energy Systems, vol. 53, pp. 357-366, 2013. doi:http://dx.doi.org/10.1016/j.ijepes.2013.05.014

[21] T. Hussein, M. S. Saad, A. L. Elshafei, and A. Bahgat, "Robust adaptive fuzzy logic power system stabilizer," Expert Systems with Applications, vol. 36, pp. 12104-12112, 2009. doi:10.1016/j.eswa.2009.04.013

[22] K. A. El-Metwally and O. P. Malik, "Fuzzy logic power system stabiliser," IEE Proceedings-Generation, Transmission and Distribution, vol. 142, pp. 277-281, 1995. doi:10.1049/ip-gtd:19951748

[23] K. A. El-Metwally and O. P. Malik, "Application of fuzzy logic stabilisers in a multimachine power system environment," IEE Proceedings-Generation, Transmission and Distribution, vol. 143, pp. 263-268, 1996. doi:10.1049/ip-gtd:19960193

[24] E. H. Mamdani and S. Assilian, "An experiment in linguistic synthesis with a fuzzy logic controller," International Journal of Man-Machine Studies, vol. 7, pp. 1-13, 1975. doi:http://dx.doi.org/10.1016/S0020-7373(75)80002-2

[25] N. Hosseinzadeh and A. Kalam, "A rule-based fuzzy power system stabilizer tuned by a neural network," IEEE Transactions on Energy Conversion, vol. 14, pp. 773-779, 1999. doi:10.1109/60.790950

[26] D. K. Sambariya, R. Gupta, and A. K. Sharma, "Fuzzy Applications to Single Machine Power System Stabilizers," Journal of Theoretical and Applied Information Technology, vol. 5, pp. 317-324, 2009. doi:http://www.jatit.org/volumes/research-papers/Vol5No3/9Vol5No3.pdf

[27] T. Hiyama, "Application of neural network to real time tuning of fuzzy logic PSS," in Neural Networks to Power Systems, 1993. ANNPS '93., Proceedings of the Second International Forum on Applications of, 1993, pp. 421-426. doi:10.1109/ann.1993.264311

[28] R. Gupta, D. K. Sambariya, and R. Gunjan, "Fuzzy Logic based Robust Power System Stabilizer for Multi-Machine Power System," in IEEE International Conference on Industrial Technology, ICIT 2006., 2006, pp. 1037-1042. doi:10.1109/icit.2006.372299

[29] T. Hussein, M. S. Saad, A. L. Elshafei, and A. Bahgat, "Damping inter-area modes of oscillation using an adaptive fuzzy power system stabilizer," Electric Power Systems Research, vol. 80, pp. 1428-1436, 2010. doi:10.1016/j.epsr.2010.06.004

[30] T. Hussein, A. L. Elshafei, and A. Bahgat, "Design of a hierarchical fuzzy logic PSS for a multi-machine power system," in Mediterranean Conference on Control & Automation (MED '07') 2007, pp. 1-6. doi:10.1109/med.2007.4433681

[31] M. K. Ciliz, "An advanced tuning methodology for fuzzy control: with application to a vacuum cleaner," in Control Applications, 2003. CCA 2003. Proceedings of 2003 IEEE Conference on, 2003, pp. 257-262 vol.1. doi:10.1109/cca.2003.1223320

[32] A. Gegov, "Complexity Management in Fuzzy Systems," Springer, Berlin, 2007.

[33] L. Wang, R. Yang, P. M. Pardalos, L. Qian, and M. Fei, "An adaptive fuzzy controller based on harmony search and its application to power plant control," International Journal of Electrical Power & Energy Systems, vol. 53, pp. 272-278, 2013. doi:http://dx.doi.org/10.1016/j.ijepes.2013.05.015

[34] A. Gegov and N. Gobalakrishnan, "Advanced Inference in Fuzzy Systems by Rule Base Compression," Mathware & Soft Computing, vol. 14, pp. 201-216, 2007.

[35] D. K. Sambariya and R. Gupta, "Fuzzy Applications in a Multi-Machine Power System Stabilizer," Journal of Electrical Engineering & Technology, vol. 5, pp. 503-510, 2010. doi:http://www.jeet.or.kr/ltkpsweb/pub/pubfpfile.aspx?ppseq=100

[36] L. A. Zadeh, "Fuzzy Sets," Information Control, vol. 8, pp. 338-353, 1965.

[37] D. Flynn, B. W. Hogg, E. Swidenbank, and K. J. Zachariah, "Expert control of a self-tuning automatic voltage regulator," Control Engineering Practice, vol. 3, pp. 1571-1579, 1995. doi:10.1016/0967-0661(95)00167-s

[38] K. J. Åström and B. Wittenmark, "On self tuning regulators," Automatica, vol. 9, pp. 185-199, 1973. doi:10.1016/0005-1098(73)90073-3

[39] J. Dombi, "Membership function as an evaluation," Fuzzy Sets and Systems, vol. 35, pp. 1-21, 1990. doi:10.1016/0165-0114(90)90014-w

[40] Z. Q. Wu, M. Masaharu, and Y. Shi, "An improvement to Kóczy and Hirota's interpolative reasoning in sparse fuzzy rule bases," International Journal of Approximate Reasoning, vol. 15, pp. 185-201, 1996. doi:http://dx.doi.org/10.1016/S0888-613X(96)00054-0

[41] L. Kóczy and K. Hirota, "Interpolative reasoning with insufficient evidence in sparse fuzzy rule bases," Information Sciences, vol. 71, pp. 169-201, 1993. doi:http://dx.doi.org/10.1016/0020-0255(93)90070-3

[42] H. Bezine, N. Derbel, and A. M. Alimi, "Fuzzy control of robot manipulators: - some issues on design and rule base size reduction," Engineering Applications of Artificial Intelligence, vol. 15, pp. 401-416, 2002. doi:10.1016/s0952-1976(02)00075-1

[43] N. Xiong and L. Litz, "Reduction of fuzzy control rules by means of premise learning – method and case study," Fuzzy Sets and Systems, vol. 132, pp. 217-231, 2002. doi:http://dx.doi.org/10.1016/S0165-0114(02)00112-4

[44] S. Kamalasadan and G. Swann, "A novel power system stabilizer based on fuzzy model reference adaptive controller," in IEEE Power & Energy Society General Meeting (PES '09'), 2009, pp. 1-8. doi:10.1109/pes.2009.5275897

[45] P. Shamsollahi and O. P. Malik, "Design of a neural adaptive power system stabilizer using dynamic back-propagation method," International Journal of Electrical Power & Energy Systems, vol. 22, pp. 29-34, 2000. doi:10.1016/s0142-0615(99)00032-0

[46] K. R. Padiyar, Power System Dynamics Stability and Control. Hyderabad, India: BS Publications, 2008.

[47] M. Ramirez-Gonzalez, R. Castellanos B, and O. P. Malik, "Application of simple fuzzy PSSs for power system stability enhancement of the Mexican Interconnected System," in IEEE Power and Energy Society General Meeting-2010, 2010, pp. 1-8. doi:10.1109/pes.2010.5589526

[48] F. P. Demello and C. Concordia, "Concepts of Synchronous Machine Stability as Affected by Excitation Control," IEEE Transactions on Power Apparatus and Systems, vol. PAS-88, pp. 316-329, 1969. doi:10.1109/tpas.1969.292452

4

Improved fuzzy c-means algorithm for image segmentation

Xuegang Hu[1, 2], Lei Li[1]

[1]College of Computer Science and Technology, Chongqing University of Posts and Telecommunications, Chongqing, China
[2]Research Center of System Science, Chongqing University of Posts and Telecommunications, Chongqing, China

Email address:
marblelilei@163.com (Lei Li), huxg62461211@126.com (Xuegang Hu)

Abstract: In order to preserve more image details and enhance its robustness to noise for image segmentation, an improved fuzzy c-means algorithm (FCM) for image segmentation is presented by incorporating the local spatial information and gray level information in this paper. The modified membership function and clustering center function are more mathematically reasonable than those of the FLICM, so the iterative sequence can converge to a local minimum value of the improved objective function. The new fuzzy factor grants the algorithm a novel balance between robustness to noise and effectiveness of preserving the details. The revised algorithm flow has significantly accelerated the processing procedure. Through these improvements, the experiments on the artificial and real images show that the proposed algorithm is very effective.

Keywords: Clustering, Image Segmentation, Fuzzy C-Means, Local Minimum Value, Gray Level Information

1. Introduction

Image segmentation plays an important role in a variety of applications such as machine vision, image analysis and image understanding, so it is a hot topic in image processing in recent years [1]. Currently, the methods of image segmentation are broadly divided into four categories: threshold, clustering, edge detection and region extraction. Among the clustering-based methods, the fuzzy c-means algorithm (FCM) is one of the most popular methods of image segmentation, which was firstly proposed by Dunn [2] and improved later by many other scholars [3]. But the classic FCM algorithm and its improved ones are not suitable for images corrupted by noise, outliers and other imaging artifacts [4].

In recent years, incorporating local spatial information and gray level information to compensate the drawback above mentioned is becoming more and more popular. By introducing local spatial information in the objective function of the FCM, Ahmed *et al.* proposed an improved FCM algorithm (FCM_S) [5]. The main advantage of this method is good performance of noise-immunity, but the disadvantage is summarized as follows: First, the method lacks sufficient robustness to noise. Second, it can't present a non-Euclidean structure of the image data. Finally, it increases running time. Tolias *et al.* imposed space constraints on clustering results to modify the objective function of FCM and obtained some positive effects [6]. Pham introduced space term into the

objective function of the FCM and significantly improved its noise-immunity capability [7]. In order to improve the anti-noise performance, robustness and reduce the processing time, Chen and Zhang proposed FCM_S1 and FCM_S2 based on FCM_S [8]. But the disadvantage is summarized as follows: it firstly lacks some robustness to noise and image speckle, it then needs some parameters chosen empirically, so it limits its application. Finally, the processing time depends on the size of the segmented image. To solve the above problems, Cai *et al.* put forward a generalized fast FCM (FGFCM) [9], the algorithm overcomes the above mentioned drawbacks of FCM_S to a certain extent and obtains better clustering performance, but the algorithm can't directly segment the color image, and some parameters need to be selected manually. In 2010, Krinidis and Chatzis proposed an improved FCM segmentation algorithm (FLICM) by integrating local spatial information and gray level information in the energy function [10]. It not only has better segmentation performance, but also doesn't need manual preselected parameters. However, Celik pointed out that the iterative sequence in the energy function doesn't converge to the minimum value because of the defects of FLICM, therefore, the FLICM segmentation results are not optimal [11]. Gong *et al.* presented a variant of FLICM algorithm (RFLICM) in 2012, whose spatial distance was replaced by local variable coefficient in the energy function [12]. Although the RFLICM algorithm exploits more local texture information, it ignores the relationship between the central pixel and its neighbor pixels. In 2013, Gong *et al.* further introduced weigh factors and nuclear distance parameters into its objective function and proposed the KWFLICM algorithm, this method have improved the segmentation results [13], but its processing time is significantly higher than that of the RFLICM.

To further improve the accuracy of image segmentation and reduce time consumption, using the new constraint factor instead of fuzzy constraint factor of the FLICM, we presented an improved fuzzy c-means algorithm for image segmentation. This algorithm enhances its robustness to noise and preserves more image details for image segmentation. Experimental results have showed that this method can quickly and accurately segment images such as synthetic and natural images, which has good anti-noise performance at the same time.

The rest of this paper is organized as follows. In Section 2, the classical fuzzy c-means clustering algorithm is briefly described. The proposed algorithm and our motivation are introduced in Section 3. Experimental results have showed in Section 4. Conclusions will be drawn finally.

2. Fuzzy C-Means Algorithm

The FCM algorithm for image segmentation is a clustering algorithm based on the most optimal function, its objective function is as follows

$$J_m = \sum_{i=1}^{N}\sum_{j=1}^{c} u_{ji}^m d^2(x_i, v_j), \tag{1}$$

where $X = \{x_1, x_2, \cdots, x_N\} \subseteq R^{N \times D}$ is the dataset in the images to be segmented. v_j denotes the center value of cluster j. $d^2(x_i, v_j)$ stands for a distance measure between dataset x_i and cluster center v_j. N is the number of gray levels of the image. c is the number of clusters with $2 \leq c < N$. The parameter m determines the amount of fuzziness of the result classification between 1.5 and 2.5 [14], u_{ji} is the degree of membership of x_i in the jth cluster satisfying

$$\sum_{j=1}^{c} u_{ji} = 1, \forall i \in \{1, 2, \cdots, N\}. \tag{2}$$

To calculate the possible extremum according to (2), an Euler-Lagrange function is used as follows

$$L(u_{ji}, v_j) = \sum_{i=1}^{N}\left(\sum_{j=1}^{c} u_{ji}^m \left\| x_i - v_j \right\|^2 + \lambda_i \left(1 - \sum_{j=1}^{c} u_{ji} \right) \right),$$

where λ_i is a parameter. And we can obtain the one order partial derivative of u_{ji} and v_j. According to (2), the values of membership and cluster centers can be calculated as follows

$$u_{ji}^* = 1 \left/ \sum_{k=1}^{C}\left(\frac{\left\| x_i - v_j \right\|}{\left\| x_i - v_k \right\|} \right)^{2/(m-1)} \right. \tag{3}$$

and

$$v_j^* = \sum_{i=1}^{N} u_{ji}^m x_i \left/ \sum_{i=1}^{N} u_{ji}^m \right. . \tag{4}$$

The classical FCM algorithm can acquire good segmentation effect, but it is very sensitive to noise. Therefore, many scholars are developing its anti-noise capability.

3. Fuzzy Local Information C- Means Cluster Algorithm

Because the classical FCM algorithm only considers the gray value of pixels and doesn't take into account the relationship between the center pixel and its neighbors. So it is not suitable for images corrupted by noise. The FLICM algorithm takes advantage of neighborhood information, which has some anti-noise performance. It not only reduces the accuracy of the FCM algorithm, but also exists some mathematically unreasonable functions. Motived by these considerations, we propose a modified method by reducing fuzzy constraint factor values of the FLICM to enhance the accuracy of image segmentation, so we can get a new balance between robustness to noise and effectiveness of preserving the details. The new algorithm's objective function is defined as follows

$$J_m = \sum_{i=1}^{N}\sum_{j=1}^{c}\left\{ u_{ji}^m \left[\sum_{l=1}^{d}\left(x_{il} - v_{jl} \right)^2 \right] + \sum_{\substack{p \in N_i \\ p \neq i}} \frac{1}{d_{ip}+1} u_{jp}^m e^{\frac{-\sum_{l=1}^{d}(x_{pl}-v_{jl})^2}{\delta}} \right\}, \tag{5}$$

where d denotes the image dimension. d_{ip} is the spatial Euclidean distance between pixels i and j. x_{pl} is the neighborhood information of data item x_{il}. u_{jp} represents the fuzzy membership of the jth pixel lying within a window around x_{il} with respect to cluster j, N_i is the dataset of neighbors falling into a window around x_{il}. δ is a distance variance that denotes the degree of aggregation around the cluster. Its definition shows as follows

$$\delta = \left(\frac{1}{N-1}\sum_{i=1}^{N}\left(d_i - \bar{d} \right)^2 \right)^{\frac{1}{2}}.$$

Where $d_i = \left\| x_{il} - \bar{x} \right\|$, it is the distance from data item xil to the mean value of all pixels falling into the window. The mean distance of di is $\bar{d} = \sum_{i=1}^{N} d_i / N$.

To calculate the possible extremum according to (2), an Euler-Lagrange function is used as follows

$$L(u_{ji}, v_j) = \sum_{i=1}^{N}\sum_{j=1}^{C}\left(u_{ji}^m \left[\sum_{l=1}^{d}\left(x_{il} - v_{jl} \right)^2 \right] + \right.$$

$$\sum_{p \in N_i, p \neq i} \frac{1}{d_{ip}+1} u_{jp}^m \exp\left(-\frac{1}{\delta}\Sigma_{l=1}^{d}\left(x_{pl}-v_{jl}\right)^2\right) + \lambda_i\left(1-\sum_{j=1}^{C}u_{ji}\right)\right),$$

we can obtain the one order partial derivative of u_{ji} and v_j. Let $\frac{\partial L}{\partial u_{ji}}=0$ and, $\frac{\partial L}{\partial v_{jl}}=0$ we have

$$\frac{\partial L}{\partial u_{ji}} = mu_{ji}^{m-1}\left\|x_i-v_j\right\|^2 - \lambda_i = 0$$

and

$$\frac{\partial L}{\partial v_{jl}} = \sum_{i=1}^{N}\left\{u_{ji}^m\left(v_j-x_i\right)+\frac{1}{\delta}\sum_{\substack{p \in N_i \\ p \neq i}}\frac{1}{d_{ip}+1}u_{jp}^m\left(x_p-v_j\right)e^{-\frac{\|x_p-v_j\|^2}{\delta}}\right\}=0.$$

According to (2), the values of membership and cluster centers can be calculated as follows

$$u_{ji}^* = 1 \bigg/ \sum_{k=1}^{C} \frac{\left\|x_i-v_j\right\|^{\frac{2}{m-1}}}{\left\|x_i-v_k\right\|^{\frac{2}{m-1}}} \qquad (6)$$

and

$$v_j^* = \frac{\sum_{i=1}^{N}\left(u_{ji}^m x_i - \frac{1}{\delta}\sum_{\substack{p \in N_i \\ p \neq i}}\frac{1}{d_{ip}+1}x_p u_{jp}^m e^{\frac{\|x_p-v_l\|^2}{\delta}}\right)}{\sum_{i=1}^{N}\left(u_{ji}^m - \frac{1}{\delta}\sum_{\substack{p \in N_i \\ p \neq i}}\frac{1}{d_{ip}+1}u_{jp}^m e^{\frac{\|x_p-v_l\|^2}{\delta}}\right)}. \qquad (7)$$

Thus, the above algorithm is summarized as follows

Step 1: Fix the number m, ε, and c.

Step 2: Initialize the clustering center and calculate the initial membership matrix as described in Eq. (3).

Step 3: Set the loop counter b=0.

Step 4: Calculate the membership matrix as described in Eq. (6).

Step 5: Update the clustering center as described in Eq. (7).

Step 6: If max$\{V_{new}-V_{old}\}<\varepsilon$ then stop, otherwise, set b=b+1 and go to step 4.

4. Experimental Results and Analysis

In this section, we compare the accuracy, the efficiency and the robustness to noise of our algorithm with the FLICM algorithm. In the experiments, the code of the FLICM was provided by the authors.

Fig. 1 shows a synthetic image Synthetic's segmentation effect after mixed Speckle noise, the noise level is 15%. The size of Synthetic is 256*256 pixels. The gray level value of the left is 20, and the right is 120. We generally set the parameters to be $c=2$, $\varepsilon=10^{-5}$, $m=1.7$.

Although the two algorithms worked well in the segmentation experiment, we still found that there are more "glitches" in Fig. 1 (c) than that in Fig. 1 (d). According to Fig. 1, we know that our method achieves better performance

under Speckle noises than FLICM.

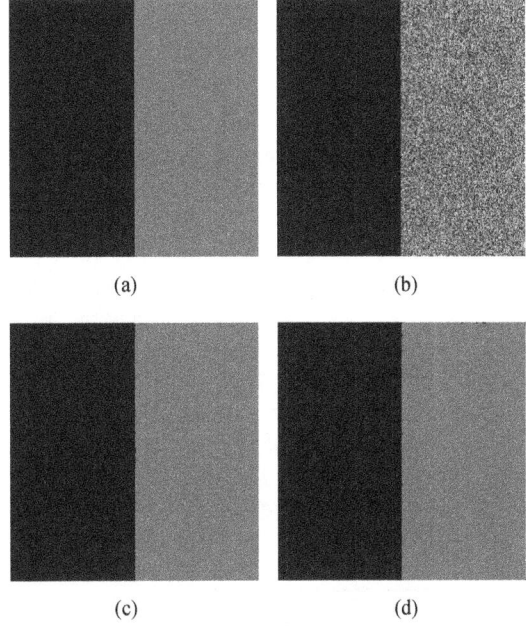

Fig. 1. Segmentation results on Synthetic image. (a) Original image, (b) the same image corrupted by the Speckle noise (0.15), (c) FLICM result, (d) our result.

Fig. 2. Segmentation results on Cameraman image. (a) Original image, (b) FLICM result on (a), (c) our result on (a), (d) the same image corrupted by Speckle noise (0.05), (e) FLICM result on (d), (f) our result on (d).

Fig. 2 presents a comparison of segmentation results on a real image. In this experiment, the Cameraman image was divided into four categories-coats, trousers, skin and others. So the parameters was set to be $c=4$, $\varepsilon=10^{-5}$ and $m=1.7$.

In Fig. 2 (b), an obviously false segmentation area can be found on the right side, which should be a part of the sky information misclassified to another clustering, and there is no such a mistake in Fig. 2 (c). On the other hand, Fig. 2 (c) preserves more details than Fig. 2 (b) such as the hand and the edge of face. Due to the noise, the effect on Fig. 2 (e) is poor. But Fig. 2 (f) shows that the result from our method has much clearer edge and smoother regions while clearing almost added noises. This experiment suggests that our proposed algorithm can accurately segment images. Furthermore, it has good anti-noise performance at the same time.

Fig. 3 presents a comparison of segmentation results on a nature image. In this experiment, the Flower image was divided into two categories-flower and background. So the parameters was set to be $c=2$, $\varepsilon=10^{-5}$ and $m=1.7$.

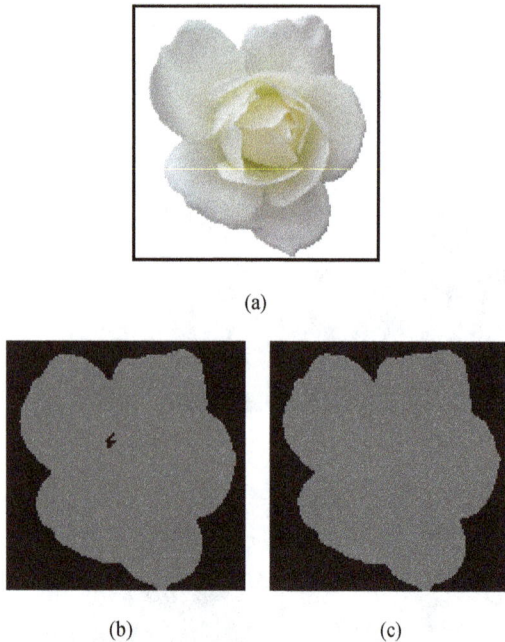

(a)

(b)   (c)

Fig. 3. *Segmentation results on Flower image. (a) Original image, (b) FLICM result, (c) our result.*

In Fig. 3 (b), an obvious area is divided to the wrong area belonging to the flower area. But Fig. 3 (c) avoids this error generated. So our algorithm achieves better segmentation effect than the FLICM.

Furthermore, Fig. 4 shows some segmentation results on real images. The left column shows the original images, while the right column shows the segmentation results which obtained by the proposed algorithm.

Celik *et al.* found that the iterative sequence of the FLICM can't mathematically achieve to converge to a local minim because of the improper membership function and clustering center function [11], although it achieve good clustering result. In this paper, we have modified those functions by mathematical means. This is the main reason why the

proposed algorithm achieves better segmentation effect than the FLICM.

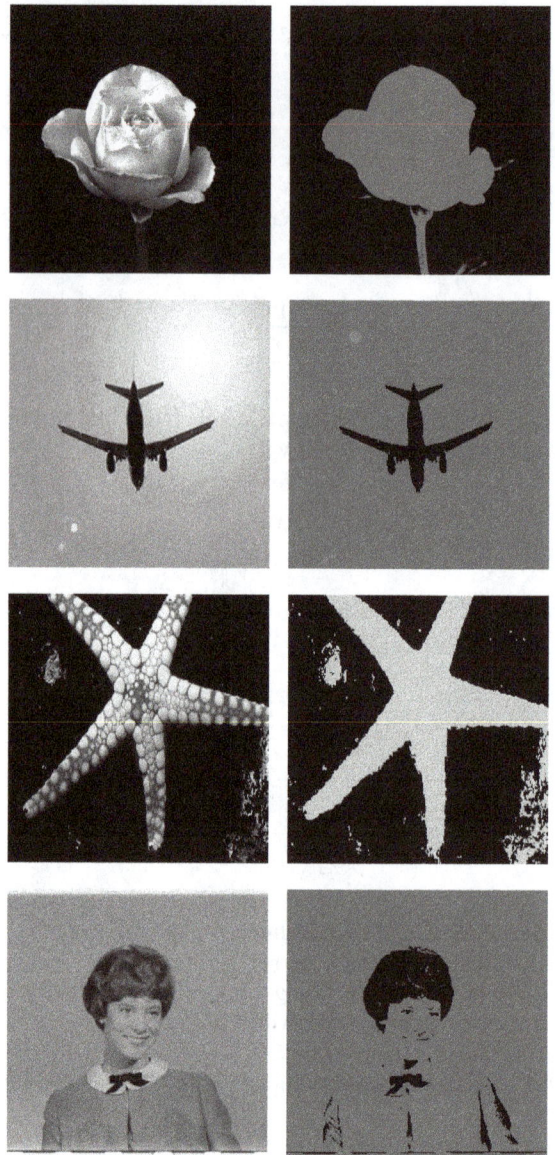

Fig. 4. *Segmentation results on real images by the proposed algorithm*

Finally, Table 1 shows the comparison of the running time on different size images using the FLICM and our proposed algorithm. In this experiments, we generally set the parameters to be $c=3$, $\varepsilon=10^{-3}$, $m=2$ and $\max Iter = 300$.

Table 1. *Running Time (in seconds) of the Two Algorithms*

	200*200	300*300	600*600	1024*1024
FLICM	23.5	27.6	155.8	592.1
Our method	12.6	13.4	84.1	164.9

All experiments performed on an AMD Athlon(tm) 64 X2 Dual Core Processor 5400+ (2.8 GHz) workstation under Windows 7 Ultimate using MATLAB R2009a. As the result of running time is shown above, our algorithm is much faster than FLICM, especially, when the size of image is 1024*1024 pixels, the consumption time is only 27% of that of FLICM computed. This must owe to the novel algorithm flow. We initialize the membership matrix using Eq. (3) instead of initializing randomly it, so the value of initial membership matrix is more approximate to the final value. This modification has greatly decreased the iterative numbers.

5. Conclusions

In this paper, an improved algorithm based on FCM is proposed by modifying the fuzzy factor FLICM used. Through this way, we get a novel balance between robustness to noise and effectiveness of preserving the details. The results reported in this paper show that our method is effective to synthetic images, real images and nature images. The experiment results suggest that the proposed algorithm obviously improves the performance of image segmentation, which has good anti-noise performance at the same time.

So the major advantages of the proposed algorithm over the FLICM are summarized as follows:
- ✓ Its computational time is less;
- ✓ It preserves more image details;
- ✓ Its iterative functions are more mathematically reasonable.

Acknowledgements

The author would like to thank Dr. Krinidis for providing FLICM source codes for comparison, and the anonymous reviewers for their constructive comments.

References

[1] X. Muñoz, J. Freixenet, X. Cufi, *et al*, "Strategies for image segmentation combining region and boundary information," Pattern recognition letters vol. 24, no. 1, pp. 375-392.

[2] J. Dunn, "A fuzzy relative of the ISO-DATA process and its use in detecting compact well separated clusters," J. Cybern., vol. 3, no. 3, pp. 32-57, 1974.

[3] J. Bezdek, "Pattern recognition with fuzzy objective function algorithms," New York: Plenum, 1981.

[4] Y. Liu, X. Wang, H. Yu, W. Zhang, "Brain tumor segmentation based on morphological multiscale modification and fuzzy c-means clustering," Journal of Computer Applications, vol. 34, no. 9, pp. 2711-2715, 2014.

[5] M. Ahmed, S. Yamany, N. Mohamed, et al, "A modified fuzzy c-means algorithm for bias field estimation and segmentation of MRI data," IEEE Trans. Med. Imag., vol. 21, no. 3, pp. 193-199, 2002.

[6] Y. Tolias and S. Panas, "Image segmentation by a fuzzy clustering algorithm using adaptive spatially constrained membership functions," IEEE Trans. Syst., Man, Cybern., vol. 28, no. 3, pp. 359-369, Mar. 1998.

[7] D. Pham, "Fuzzy clustering with spatial constraints," in Proc. Int. Conf. Image Processing. New Work, 2002, vol. II, pp. 65-68.

[8] S. Chen, D. Zhang, "Robust image segmentation using FCM with spatial constraints based on new kernel-induced distance measure," IEEE Trans. Syst., Man, Cybern., vol. 34, pp. 1907-1916, 2004.

[9] W. Cai, S. Chen, D. Zhang, "Fast and robust fuzzy c-means clustering algorithms incorporating local information for image segmentation," Pattern Recognition, vol. 40, no. 3, pp. 825-838, Mar. 2007.

[10] S. Krinidis and V. Chatzis, "A robust fuzzy local information C-means clustering algorithm," IEEE Trans. Image Process., vol. 19, no. 5, pp. 1328-1337, May 2010.

[11] T. Celik and H. Lee, "Comments on "A Robust Fuzzy Local Information C-Means Clustering Algorithm"," IEEE Trans. Image Process, vol. 22, no. 3, pp. 1258-1261, 2013.

[12] M. Gong, Z. Zhou, J. Ma, "Change detection in synthetic aperture radar images based on image fusion and fuzzy clustering," IEEE Trans. Image Process, vol. 21, no. 4, pp. 2141-2151, 2012.

[13] M. Gong, Y. Liang, J. Shi, *et al*, "Fuzzy c-means clustering with local information and kernel metric for image segmentation," IEEE Trans. Image Process, vol. 22, no. 2, pp. 573-584, 2013.

[14] Pal, R. Nikhil and C. James, "On cluster validity for the fuzzy c-means model," IEEE Trans. Fuzzy Syst. vol. 3, no. 3, pp. 370-379, 1995.

5

A Study and Design of a Rat-Race Coupler Based Microwave Mixer

Asif Ahmed

Department of EEE, American International University Bangladesh, Dhaka, Bangladesh

Email address:
aahmed@aiub.edu

Abstract: A microwave mixer circuits has been designed and fabricated, having frequency range of 2 GHz- 4 GHz, with optimum operating point at 3 GHz. The underlying theory of rat-race coupler was used to achieve frequency up-conversion and down-conversion process with the help of a nonlinear element - Schottky diode. The simulation of both the rat-race coupler and the mixer circuit has been done with Agilent Advanced Design System (ADS) software, and the fabricated mixer circuit has been tested using Vector Spectrum Analyzer (VSA). The measured results of the mixer has shown perfect consistency with the simulated outputs.

Keywords: Rat-Race Coupler, Microwave Mixer, Non-Linear Mixing

1. Introduction

A microwave mixer [1-8] is a device that up-concerts low frequency signal to microwave signal, and down-converts microwave signal to a corresponding base-band signal. The most usual application of mixers is in transceiver where two frequency beat together to generate two different frequencies, i.e., sum and difference of the two frequencies.

The history of mixers dates back to 1900s when it was used for radio reception circuits. During the time of Second World War microwave mixers were very commonly used for military radar. Now-a-days, microwave mixers gained its interest for both civil and military purposes, for example, radars, transceivers for WLAN, satellite, guided weapon and radio astronomy, transportation, etc. The application of the microwave mixers ranges from one to several hundred gigahertz. Research and Development (R&D) labs has been improvised microwave mixers by shifting from simple lumped components deployment to incorporation of nonlinear components, e.g., diodes. The continuous endeavor has been progressed through the traditional lumped component approach to the hybrid microwave integrated circuits (MIC) and the monolithic microwave integrated circuit (MMIC). The challenges of suppressing intermodulation products (IP) has been an important characteristic attributes of more complex mixer circuits.

This paper will show a design and fabrication of a mixer with Schottky diode as nonlinear component, and the corresponding measurement results to compare with simulated expectations.

2. Theory Behind Frequency Mixers

The effort of development of mixers has been given us many variations [9] in the kind of mixers circuits.

Single-Ended Mixer (SEM): is the most preliminary form of mixer that has only one port to mix where the signal are combined externally to feed to it.

Single-Balanced Mixer (SBM): combines two single ended mixers, and local oscillator (LO) noise side-band is balanced out by via a 4-port 3 dB coupler.

Double-Balanced Mixer (DBM): incorporates two baluns and mixing components as many as four in quantity in the form of a ring, bridge, star or quad. Inherently, it cancels LO amplitude modulated (AM) noise, and has high bandwidth.

Double-Double Balanced Mixer (DDBM): has eight mixing components and it separates RF, LO and intermediate frequency (IF) rendering improved (higher) dynamic range.

Image Rejection Mixer (IRM): employs two identical mixers either SEM, SBM or DBM in a phase alignment so as to reject the images.

Image Recovery (Enhancement) Mixer: is a kind of IRM where image power from one mixer is converted to IF in other mixer, and vice versa.

Researches have been conducted to improve linearity [10]

of a mixer. By the term linearity, it means that the input and output changes should have same value. The linearity is also an important attribute of mixers as without linearity there would be unwanted AM noise at the output.

Another common attribute of a mixer is its conversion efficiency, and many efforts have been expended in making the conversion efficiency (CE) [11] of a mixer optimum. The conversion efficiency is the difference between the RF input power and IF output power.

$$CE = RF\ Power\ (dBm) - IF\ Power\ (dBm).$$

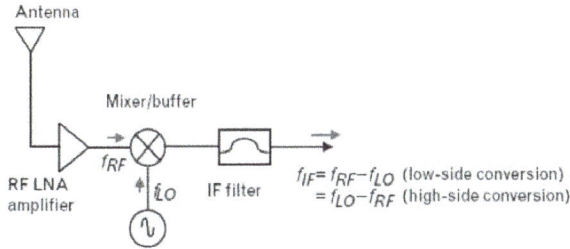

Figure 1. Basic function of a mixer.

In general, it can be described briefly about mixer that it is a non-linear transducer having 3-ports, namely, RF (Radio Frequency) port, LO (Local Oscillator) port and IF (Intermediate Frequency) port. The RF is the transmission frequency that is converted to IF for the purpose of improved selectivity. Figure 1 shows a basic functional mixer employed in a receiver.

Another attribute of a frequency mixer is its image frequency (IM), and in [12], it shown how to characterize a mixer by 3-port conversion matrix for RF, IF and IM signals.

At the IF port of mixer all frequencies: $\omega = \pm(N.\omega_{LO} \pm M.\omega_{RF})$ can exist. The general expression of IF for down conversion is following:

$$\omega_{IF} = \pm(\omega_{LO} - \omega_{RF}) \qquad (1)$$

Conversely, the expression of up-conversion is

$$\omega_{IF} = \pm(\omega_{LO} + \omega_{RF}) \qquad (2)$$

The IF and IM signals at the IF port for the down-conversion are defined as follows:

$$\omega_{IF} = \omega_{RF} - \omega_{Lo},$$

$$\omega_{IM} = \omega_{RF} + \omega_{Lo}.$$

Recent literature [13-15] shows other attributes of mixers with different working principles.

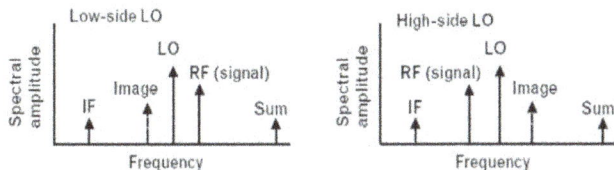

Figure 2. Converted frequency spectrum.

Figure 2 shows sketches of converted frequencies in frequency domain.

Figure 3 illustrates the block diagram of a harmonic mixer. In this paper a rat-race coupler has been used for adding and subtracting the signals and diode for implementing the nonlinear block.

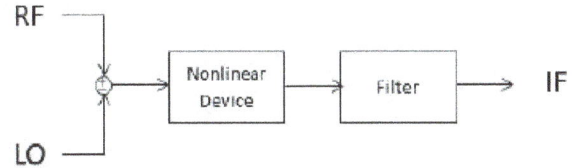

Figure 3. Harmonic Mixer.

Harmonic mixers produce more than just the sum and difference frequencies. The intermodulation products are given by $= N \times RF \pm M \times LO$, and their levels, relative to the desired output of $RF \pm LO$, for a common mixer.

2.1. Rat-Race Coupler

Rat-race coupler is a 3 dB coupler consists of four transmission lines with impedance of $\sqrt{2}Z_0$, three with length of $\lambda/4$ and one with length of $3\lambda/4$. There are four terminals with impedance of Z_0.

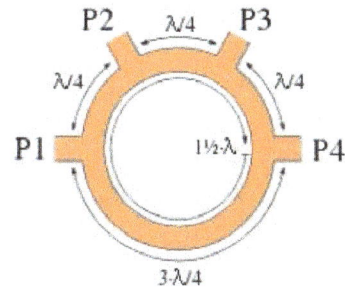

Figure 4. Rat-race coupler.

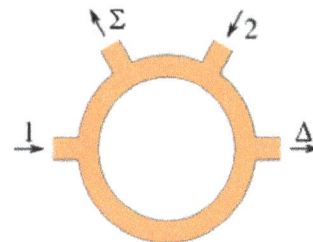

Figure 5. Function of the rat-race ports.

Figure 4 depicts a basic form of a rat-race coupler. The S-parameter matrix of the coupler is following:

$$S = -\frac{j}{\sqrt{2}}\begin{bmatrix} 0 & 1 & 0 & -1 \\ 1 & 0 & 1 & 0 \\ 0 & 1 & 0 & 1 \\ -1 & 0 & 1 & 0 \end{bmatrix}$$

This coupler can divide the input power in halves but the output waves are in 0 or 180 degree phase difference. If we apply two inputs in the first and third ports we would have the sum of the waves from the second, and differences in the fourth port as shown in figure 5.

2.2. Nonlinear Elements

If we apply the sum and differences of the rat-race coupler to a nonlinear device, we would have different harmonics of their multiplications which is the base of the operation of harmonic mixers. By filtering out the undesired harmonics, we can have the demodulated signal from the carrier.

The diode used for mixing can be modeled at the RF frequency as a resistor and capacitor in parallel. The resistor is usually in a range of 50 to 150Ω, and the capacitor between 1 and 1.5 times the junction capacitance. The IF output impedance is usually between 75 ohms and 150 ohms. At low IF frequencies the output impedance is almost purely resistive. Therefore, a matching network between coupler and the diode is needed for the particular frequency in order to deliver the maximum power in that frequency. Well-matched diodes can reduce the overall noise figure (NF) too.

For selecting the mixing diode, one has to look for the cut-off frequency of the diode, for series resistance R_s and junction capacitance C_j.

In order to ground the RF and LO signal in the output of nonlinear device these frequencies should be grounded by a

small capacitor, and also it can damp the IF noise at the output.

3. Design and Fabrication with ADS

Schematics of Rat-race Coupler

The schematic of the rat-race coupler has been drawn in Agilent's ADS software. As we already know it includes four pieces of $\sqrt{2}Z_0$ transmission lines (three with $\lambda/4$ and one with $3\lambda/4$) for the circle, 4 T-junctions and 4 pieces of Z_0 transmission line for the terminals. Multiple $\lambda/4$ has been used for the case of $3\lambda/4$ and some of them are curved in order to have a shape close to the circle.

Figure 6 schematic part of the coupler part of the mixer. It includes the 4 ports and transmission lines that produce the coupler. Substrate parameters for micro-strip are also shown. The width and length of the transmission line has been calculated with LineCalc tool included in the ADS software.

Using S-parameter simulation, the plot of S_{11}, S_{12}, S_{21} and S_{22} has been obtained between the frequency 1 GHz and 6 GHz as shown in figure 7. As depicts in the figure 7, for S_{11} and S_{12} there is more than 40 dB attenuation, and for the other ports the attenuation is about 4 dB. Comparing the S-parameter matrix for rat-race, -40 dB represents the zero and -4 dB represents 1. The layout of the rat-race is shown in figure 8.

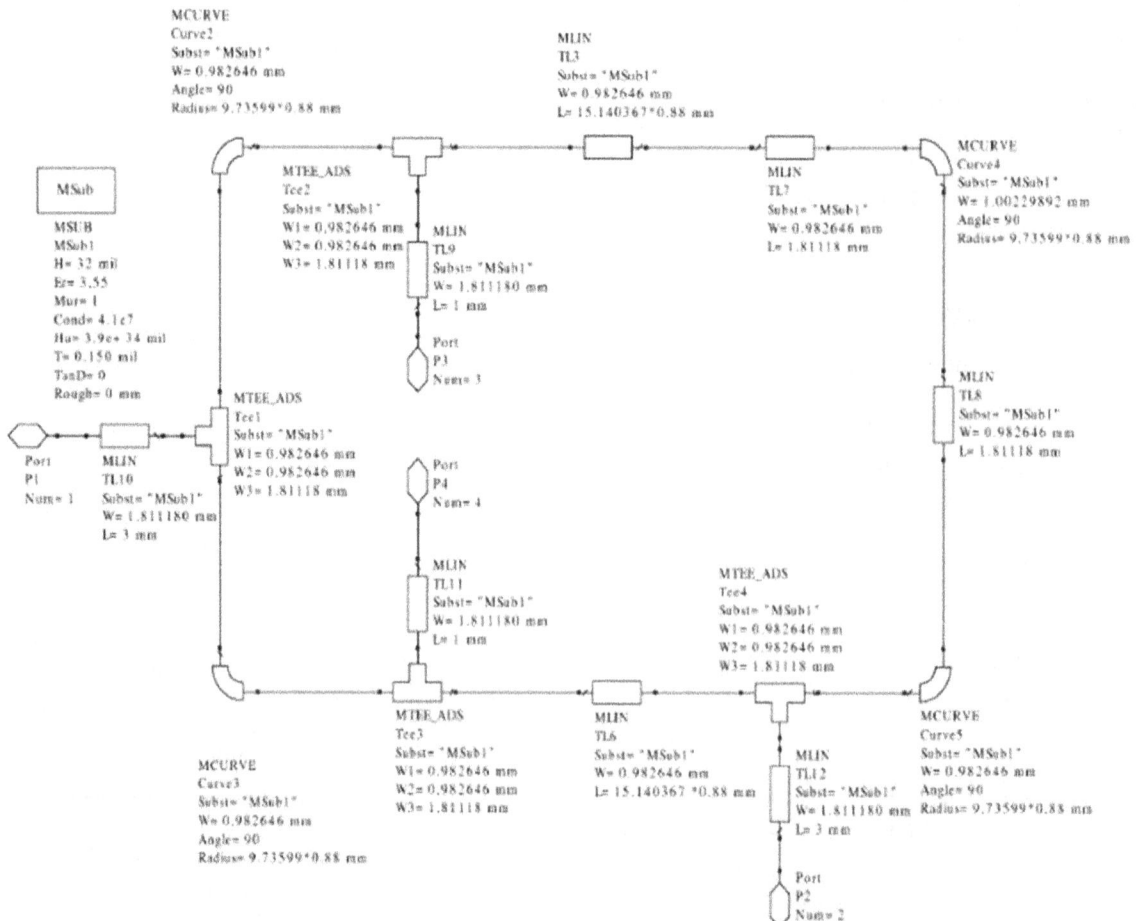

Figure 6. Rat-race Schematic.

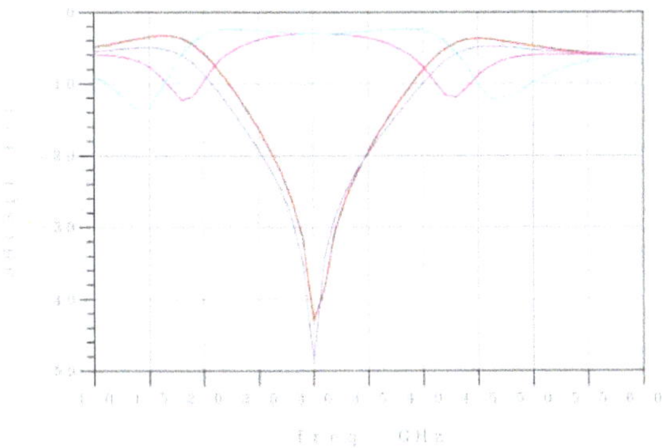

Figure 7. *Rat-race S-parameter plot.*

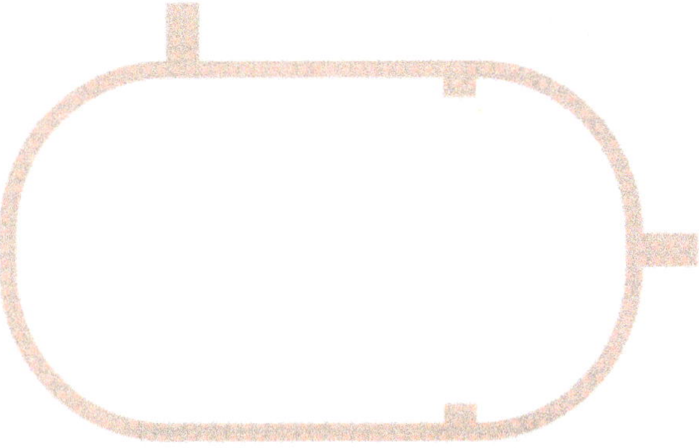

Figure 8. *Rat-race layout.*

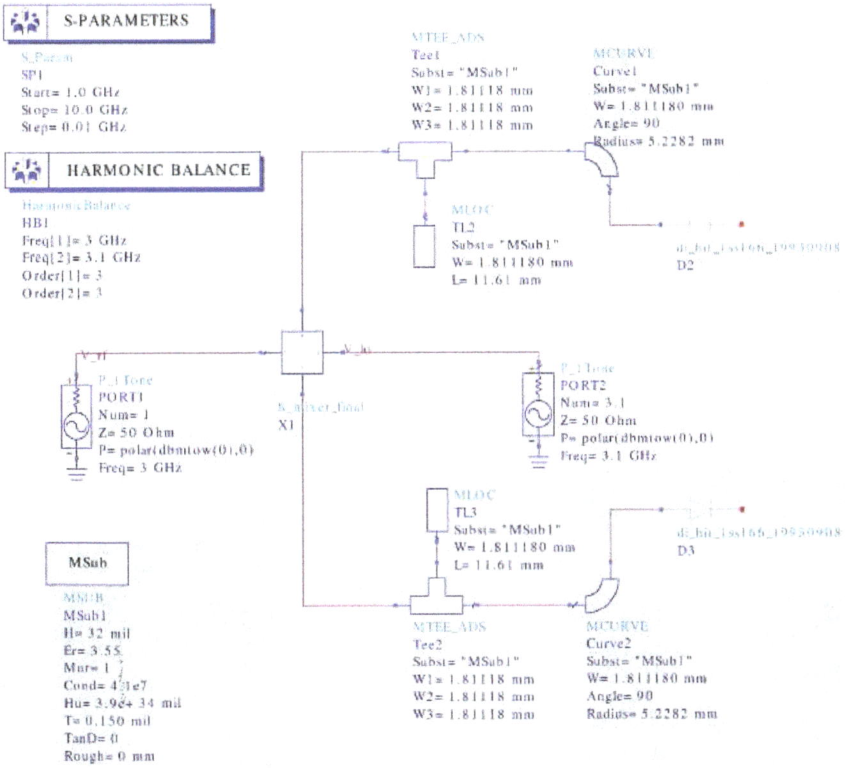

Figure 9. *Complete Mixer Schematic.*

After this the nonlinear components, i.e., diodes were added for achieving the frequency mixing, but since the input impedance of diode is not 50 Ω, so we need to design matching network. Complete schematic including the rat-race and the matching network along with the nonlinear devices is presented in Figure 9.

Figure 10 shows the final layout of the complete mixer which later were fabricated with chemical etching process, and were tested for conformity to the simulated output.

The simulation of the S-parameters has been set between 1 GHz to 10 GHz with 0.01 GHz steps.

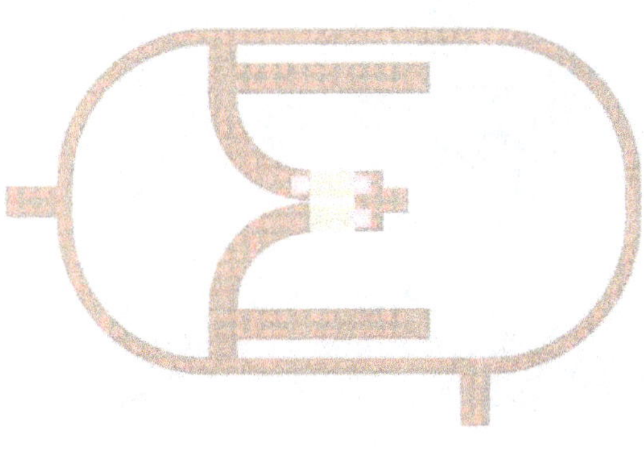

Figure 10. *Complete Mixer layout.*

4. Results: Fabrication and Measurement

The layout has been obtained from the ADS, and was saved in GDSII file format for it to be opened in LayoutEditor software which gives us the actual size of the layout graphics by printing it in pdf file. Using the printed graphics of mixer layout, a mask was created to develop the circuit on the copper clad board which later on was etched to develop the complete mixer layout. Figure 11 shows the fabricated layout of the complete mixer.

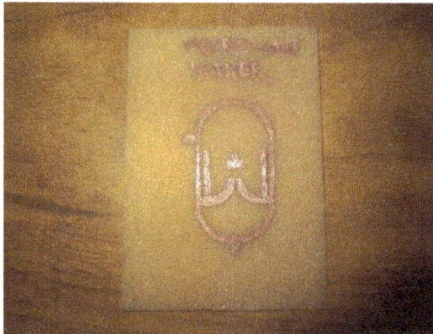

Figure 11. Fabricated Mixer layout on copper board.

The development of the complete circuit has been conducted by connecting and soldering the nonlinear components. Two SMD (Surface Mount Device) diodes were connected and soldered as shown in figure 12. Three PCB mount SMA-RF (jack) connector were connected to the RF, LO and IF ports.

(a)

(b)

Figure 12. Completed Microwave Mixer.

For testing the developed mixer circuit, two signal has been applied in RF and LO port of frequency 3.1 GHz and 3 GHz respectively. A 40 pF capacitor was used as LPF (Low Pass Filter) at the output. The output magnitude for both high and low frequency is presented in figure 13 and figure 14 respectively.

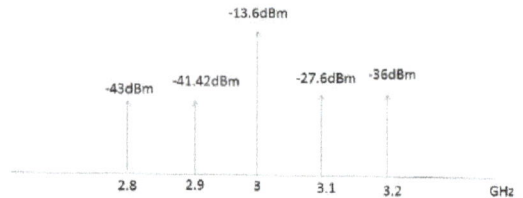

Figure 13. High Frequency output.

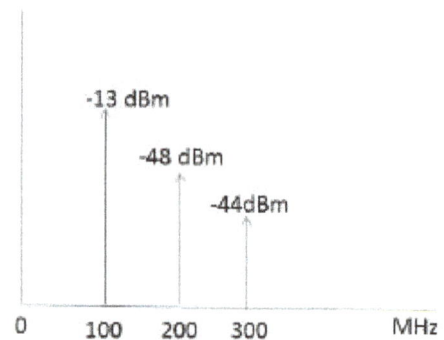

Figure 14. Low Frequency output.

The measured outputs from the spectrum analyzer is shown in the figure 15 and figure 16 for high and low frequency respectively.

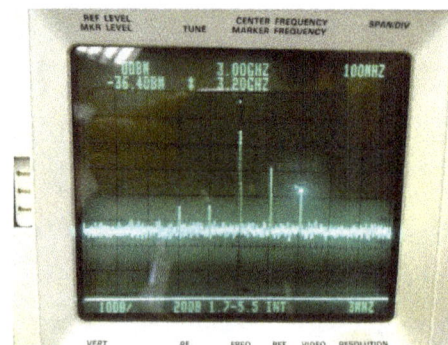

Figure 15. High Frequency Spectrum.

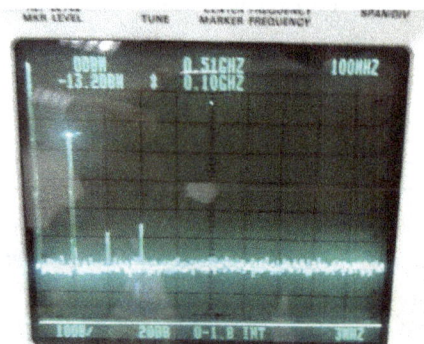

Figure 16. Low Frequency Spectrum.

Furthermore, the RF port has been fed with an AM modulated signal with central frequency of 3.1 GHz and modulating signal frequency of 200 kHz. The output was exactly as expected as shown in figure 17.

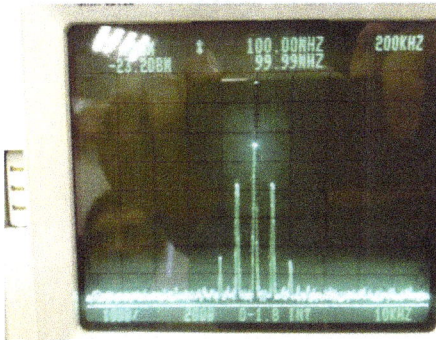

Figure 17. LF spectrum of down-converted modulated signal.

5. Discussion and Conclusion

The design of the microwave mixer involved designing and calculation of line length, width for the conformance to required line impedances, and LineCalc application integrated in ADS software was used carefully for that purpose.

The whole circuit needed to match to the nonlinear diodes that was applied at the end part of the design process. The output (IF) has been taken from the inside part of the coupler circle so as to make the whole device more compact and easy to handle.

For the design requirements T-connectors were used, and those were needed to be included carefully and calculatedly as they could have been changed the response of the coupler.

The resonant frequency of the whole circuit would change if the T-connectors, line width and lengths is changed or modified incorrectly.

References

[1] H. E. M. Barlow, K. V. G. Krishna, "A Hall-effect microwave mixer" in *Proc. of IEE- Part B: Electronic and Communication Engineering*, 1962, vol. 109, no. 44, pp. 131-136.

[2] H. J. O'Neill, "Image-frequency effects in a microwave crystal mixer," in *Proc. of the Institution of Electrical Engineers*, 1965, vol. 112, no. 11, pp. 2019-2024.

[3] T. Oxley, F. Hilsdent, "The performance of backward diodes as mixers and detectors at microwave frequencies" *Radio and Electronic Engineer*, 1966, vol. 31, no. 3, pp. 181-191.

[4] W. J. Moroney, Y. Anand, "Reliability of Microwave mixer diodes" in *10th Annual Reliability Physics Symposium*, 1972, pp. 57-63.

[5] O. S. A. Tang, C. S. Aitchison, "Practical performance of a microwave distributed MESFET mixer" *Electronic Letters*, 1985, vol. 21, no. 5, pp. 172-173.

[6] G. Tomassetti, "An Unsual microwave mixer" in *16th European Microwave Conference*, 1986, pp. 754-759.

[7] M. I. Sobhy, F. Bassirato, "Non-ideal modelling and design of microwave mixers" in *IEEE MTT-S International Microwave Symposium Digest*, 1988, pp. 1111-1114.

[8] H. Zaghloul, T. H. T. van Kalleveen, et. al., "A simple method for the evaluation of microwave mixer diodes" *IEEE Transactions on Instrumentation and Measurement*, 1990, vol. 39, no. 6, pp. 928-932.

[9] T. H. Oxley, "50 years development of the microwave mixer for heterodyne reception" *IEEE Transactions on Microwave Theory and Techniques*, 2002, vol. 50, no. 3, pp. 867-876.

[10] K. Holland, J. Howes, "Improvements to the microwave mixer and power sensor linearity measurement capability at the National Physical Laboratory" in *IEE Proc.- Science, Measurement and Technology*, 2002, vol. 149, no. 6, pp. 329-332.

[11] Gan Chee Tat, Lim Koon Tin, "A study and design of microwave mixer for Conversion Efficiency (CE) performance" in *Asia-Pacific Conference on Applied Electromagnetics*, 2007, pp. 1-5.

[12] Cao Haibo, R. Weber, "Three-port conversion scattering parameters characterization for microwave mixers" in *51st Midwest Symposium on Circuits and Systems*, 2008, pp. 414-417.

[13] C. Bohemond, T. Rampone, A. Sharaiha, "Performance of a Photonic Microwave Mixer based on cross-gain modulation in a semiconductor optical amplifier" *Journal of Lightwave Technology*, 2011, vol. 29, no. 16, pp. 2402-2409.

[14] Yuding Wang, Lv Zhiqing, et. al., "Back door effects of High Power microwave on microwave mixer" in *International Conference on Microwave and Millimeter Wave Technology (ICMMT)*, 2012, vol. 4, pp. 1-4.

[15] A. Kazemipour, M. Salhi, et. al., "Novel method to measure the conversion-losses (C.L.) of microwave and mm-wave mixers" in *Asia-Pacific Microwave Conference Proc. (APMC)*, 2013, pp. 731-733.

6

Performance Analysis of DWDM System with Optical Amplifiers in Cascade Considering the Effect of Crosstalk

Abu Jahid[*], Sanwar Hossain, Raziqul Islam

Dept. of Electrical and Electronic Engineering, Bangladesh University of Business and Technology (BUBT), Dhaka, Bangladesh

Email address:

setujahid@gmail.com (A. Jahid), sanwar_05eee@yahoo.com (S. Hossain), raziqul@yahoo.com (R. Islam)

Abstract: In this paper, an analytical approach is presented to evaluate the performance of DWDM system with intensity modulation direct detection (IM/DD) due to the effects of amplified spontaneous emission (ASE) noise of optical amplifier, optical receiver noises and crosstalk. And a system has been proposed with an optimum number of amplifiers with higher gain and an improved optical receiver has been proposed. We have investigated the effects of optical amplifiers and optical receiver in the presence of crosstalk on the overall performance; in particular, Signal-to-Noise Ratio (SNR) and Bit-Error-Rate (BER) performance of a DWDM system. The system performances are evaluated for varying different amplifier gains, number of optical amplifiers, hop length, number of hops, receiver bandwidth and receiver gain considering the relationship between crosstalk and the number of wavelengths channel spacing. It is found that, the performance is highly degraded due to crosstalk and noises. As a result the system suffers significant power penalty at a given BER.

Keywords: ASE, IM/DD, BER, Crosstalk, Power Penalty

1. Introduction

Dense Wavelength Division Multiplexing (DWDM) is a rapidly maturing transmission technique to satisfy substantial increase of telecommunication and optical network capacity. In recent year, tremendous rate of traffic growth due to the demands for multimedia services urges the development of wavelength routing technologies in place of the point-to-point multi-wavelength transmission. In a WDM system, the channel capacity can be increased by assigning each input signals to a particular wavelength of a light. An appropriate channel filtering and proper switching scheme reduce delay time of WDM signals at the receiver output and the resultant capacity is the aggregate of carrying capacity of each signals. By using WDM technology in optical networks, 50THz link capacity can be achieved. [1] In order to access the huge bandwidth capacity and to utilize it effectively, DWDM is one of the promising techniques over Coarse WDM (CWDM) technology. In DWDM, the wavelength spacing (0.8 nm) is much closer than CWDM (20 nm) and therefore DWDM has greater capacity.

Now a days DWDM system supports 160 wavelengths transmitting at a speed of 10Gbps. So the total system capacity is 1.6Tbps whereas CWDM provides only 50Gbps. [2] One of

the significant key features of DWDM is the ability to amplify all the incoming signals with different wavelengths at once without any optical to electrical conversion. It implies that DWDM combines multiple optical signals so that they can be amplified as a group and transmitted over a single fiber to increase the capacity. Another important feature is that DWDM link can carry each signal in a different rate and in a different format. [3]

In order to increase the data transmission efficiency of DWDM systems optical amplifiers plays a vital role for optical communication system. In order to compensate fiber losses and to achieve successful data transmission multiple optical amplifiers are installed along the fiber link. An optical amplifier has the ability to amplify, regenerate and to synchronize data of optical signals up to destination. The key advantages of optical amplifiers is to provide large bandwidth and to exhibit low noise, low insertion loss at third transmission window (1550 nm range). [4] Erbium-Doped Fiber Amplifier (EDFA) operates in C-band region in the range of 1540 nm to 1565nm wavelength. EDFA enables to amplify all wavelengths to overcome loss over long spans of fiber, high passive losses and to minimize dispersion effects. [5, 6] EDFA can be employed in where the application requires high bit rates, low noise and high output power. Due to wide amplification spectrum band EDFA is widely used. [7, 8]

Amplified Spontaneous Emission (ASE) noise accumulates to the signal during its amplifications, which deteriorates the signal-to-noise ratio (SNR) and hence increase the bit error rate (BER). ASE beat noise is the dominant over the thermal noise and other receiver noises too. [9] Unequal gain spectrum of optical amplifiers is the major limitation during amplification. This mean that all wavelengths are not equally amplified which limits the performance. [10]

For a successful and efficient data reception the transmitted power should be large enough. The receiver sensitivity is the minimum amount of power requirement from the transmitter and is used to separate each single bit of 1's and 0's from the original input signal at receiving end. [11, 12] The acceptable BER is 10^{-9} in optical communications. Despite the attenuation along the optical fiber the transmitted power should be high enough to maintain the target BER. The large amount of transmitted power generates channel impairments and nonlinearities which are not expected. As a result, a high input power does not ensure that the reception would be successful. Thus the input power value should meet the minimum power requirement at the receiver. [13, 14] The receiver must be operated in the dynamic range for the incoming optical signals. However, the transmission length increases with the increase in the bit rate and the parameters have the capability of absenting in the network. [15]

Crosstalk is the power leakage from other channels at the same or different nominal wavelength on the signal channel. It occurs when signals from one channel to arrive in another channel and can be considered as interference. Crosstalk occurs due to the optical filters, wavelength multiplexers and demultiplexers, switches, optical amplifiers, and the fiber itself in a WDM link. Crosstalk has serious effect on the system performance because it reduces SNR and hence BER is increased. Crosstalk is one of the major impairments in optical networks. Theoretical result shows that for a given value of input power crosstalk increases with the increase of number of wavelength. Thus it limits the allowable number of channels per fiber. Crosstalk can be classified into two categories as intra-band or homodyne and inter-band or heterodyne crosstalk.

[16-19]

In this paper, BER performance of DWDM system with optical amplifiers in cascade incorporated with optical crosstalk has been analyzed. Simulations are performed to evaluate the impact of crosstalk and noises on the SNR and BER. The paper has been organized as following, section I describes the introduction, section covers the system model, mathematical analysis is presented in section III, section IV discusses the results and discussion and finally conclusion has been presented in section V.

2. System Model

A typical block diagram of Dense Wavelength Division Multiplexing (DWDM) system with various components is shown in Fig. 1. It consists of M number of transmitters, an optical waveguide which would be the medium of transmission with cascading a numbers of optical amplifiers and N numbers of receiver to reproduce that transmitted signal. Most practical optical communication links use laser diodes as transmitters and P-i-N or Avalanche Photodiodes (APD) as receivers. A transmitter converts an electrical signal into an optical signal whereas an optical receiver does the reverse operation. These devices modulate and detect solely the intensity of carriers not their phase which implies that all transmitted signal intensities are non-negative. The optical signals from different WDM channels are multiplexed by a WDM multiplexer. Before sending the resultant WDM signals to the transmission fiber link, a power amplifier is used to increase its output power. A number of line amplifiers are installed in cascade along the fiber link to compensate fiber losses. The definite interval can be defined as the distance between two successive line amplifiers and this length is typically 60-120 km. The individual signal can be extracted from the multiplexed channel by using appropriate channel tuning. To provide high receiver sensitivity and high gain, optical amplifiers are used in front of demultiplexer. Therefore, a faithful and successful reception from the demultiplexed signals can be obtained.

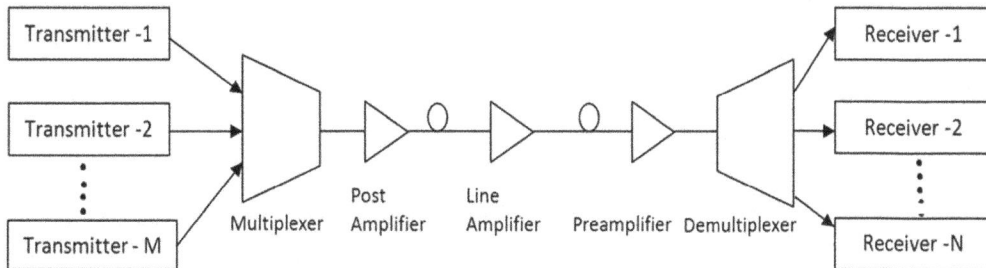

Figure 1. DWDM system with optical amplifiers in cascade.

3. Mathematical Model

Crosstalk is a major limiting factor to the implementation of cascaded optical amplifier in DWDM systems. It is further noticed that other receiver noises like shot noise, dark current noise, thermal noise and Amplified Spontaneous Emission

(ASE) noise are also accumulated during the optical transmission through the fiber link.

The discrete nature of electrons in photodiode generates shot noise. It can be expressed as: [20]

$$i_s = \sqrt{2e(Ip)B} \qquad (1)$$

Where i_s is the shot noise current, Ip is the photocurrent and B is the photodiode bandwidth.

Spontaneous fluctuations due to thermal interaction between free electrons and vibrating ions inside an optical waveguide create thermal noise and it can be defined as-

$$i_t = \sqrt{4kTB/R_L} \qquad (2)$$

Where i_t is thermal noise current, k is the Boltzmann constant $(1.38*10^{-23}$ J/K), R_L is the load resistance and T (K) is the absolute temperature. The effect of thermal noise can be minimized by increasing the value of load resistance and the value of receiver bandwidth should be limited.

A very small amount of reverse leakage current flows in the receiver when there is no optical power is incident on the receiver is named as dark current noise. It is given by-

$$i_d = \sqrt{2e(I_D)B} \qquad (3)$$

The spontaneous recombination of electrons and holes in the amplifier medium is mainly responsible for ASE noise. This recombination generates a broad spectral background of photons that gets amplified along with the optical signal. The Power Spectral Density (PSD) of the ASE noise is- [14]

$$N_{sp}(f)= n_{sp}(G-1)hf= Khf \qquad (4)$$

Where,

n_{sp}=Spontaneous emission factor, G=Amplifier gain, h=Plank's constant, f=Frequency of radiation.

In order to compensate fiber losses, optical amplifiers in cascaded form are needed to be used along the fiber link despite the ASE noise is generated. The gain of optical amplifier should be such that it can balance the losses along the optical fibers. The span length solely depends on number of optical amplifiers. ASE noise limits the hop length as well as the number of optical amplifiers. In order to receive a strong signal at the output, optimum numbers of amplifiers are used with larger gain. Thus it makes the receiver sensitivity higher and minimizes the cost.

If the amplifiers gain (G) is adjusted to compensate for the total losses, then

$$G \text{ (dB)} = P_L \text{ (dB)} = (\alpha fc + \alpha j_) L \text{ (dB)}$$

Where, α_{fc}=Fiber cable loss(dB/km), α_j=Joint loss(dB/km), L=Hop length (in km)

Then total amplified spontaneous emission noise at the input of the receiver is,

$$P_{ASE} = NKhfB \qquad (5)$$

Where N is the total number of amplifiers and N = L_t/L, [L_t=Total transmission distance in km]

With the addition of crosstalk, the total noise current is now

$$In = (i_s^2 +i_d^2+ i_t^2 + P_c)1/2 \qquad (6)$$

Where P_c is the crosstalk power.

The signal to noise ratio is defined as the ratio of signal power to the noise power. It is given by the following equation-

$$SNR = \frac{I_p^2}{I_n^2} \qquad (7)$$

The SNR for the p–i–n photodiode receiver may be obtained by summing the noise contributions from Equation 1, Equation 2, Equation 3 and Equation 5. It is given by from Equation 7:

$$\frac{S}{N} = \frac{I_P^2}{2eB (I_P + I_D)+\frac{4kTBF_n}{R_L}+P_{ASE}+P_C} \qquad (8)$$

Where F_n represents the noise figure.

It is observed from equation (8), without internal avalanche gain thermal noise is dominates over the shot noise and dark current noise.

In the Avalanche Photodiode (APD) the signal current is increased by a factor M and hence the overall SNR is increased as thermal noise remains unaffected. But the multiplication factor (M) increases the dark current noise and quantum noise which may be a limiting factor. The overall signal-to-noise ratio for APD receiver can be expressed as- [20]

$$\frac{S}{N} = \frac{I_P^2M^2}{2eB (I_P+I_D)M^{2+x}+\frac{4kTBF_n}{R_L}+ P_{ASE}+P_C} \qquad (9)$$

Where the factor x ranges between 0 and 1.0 and depending on photodiode material.

It is apparent from Eq. (9) that the first term in the denominator increases with increasing. For lower M the combined thermal and amplifier noise term dominates and the total noise power is virtually unaffected when the signal level is increased, giving an improved SNR. However, when M is large, the thermal and amplifier noise terms become insignificant and the SNR decreases with increasing M at the rate of M^x. Therefore, an optimum value of the multiplication factor M should be used.

Therefore Bit Error Rate (BER) in terms of SNR can be written as:

$$BER= 0.5 \text{ erfc } \left[\frac{\sqrt{\frac{S}{N}}}{2\sqrt{2}} \right] \qquad (10)$$

The BER depends on noise as well as other impairments on the system. It should be noted that a high SNR means a higher power signal compared to the noise power and a lower the bit error rate means less error probability.

4. Results and Discussions

From the mathematical model discussed in section III, the performance of DWDM system in the presence of crosstalk has been analyzed using graphical representation. The system performance has been demonstrated varying different parameters like number of amplifiers, amplifier gains, number of hops and hop length. We also quantified the results in terms of optical receiver type and system bandwidth. We used MATLAB tool for this simulation. Theoretical analysis has been carried out for the system for a given set of parameters and finally the system performance has been plotted for different parameters. The set of parameters are as follows for which we have used some fixed values in our experiment:

Wavelength (λ) = 1550 nm, Spontaneous Emission factor (n_{SP}) = 1.85, Load resistance (R_L) = 4 kΩ, Dark current (i_D) = 3 nA, Crosstalk power density = 10^{-14}, Noise figure (F_n) = 3 dB.

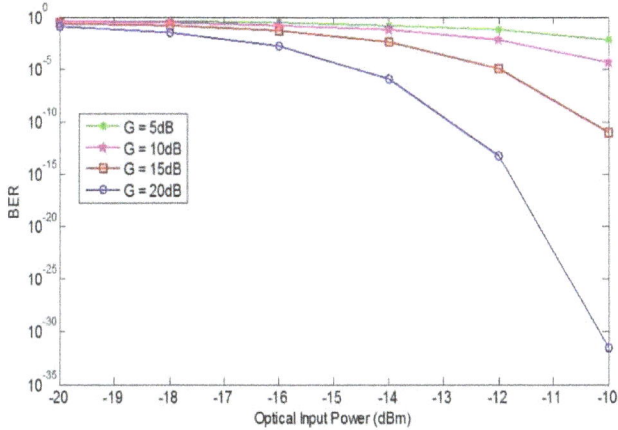

Figure 2. *BER vs optical input power varying amplifier gain (G=5dB, 10dB, 15dB, 20dB) for a fixed number of amplifier (N=10).*

Figure 2 demonstrates the variation of BER with an optical power for different amplifier gain of 5dB, 10dB, 15 dB and 20dB. The BER is decreases with the increase of optical input power. The BER is highest for the amplifier gain of 5dB. So the BER is decreases with the increase of amplifier gain.

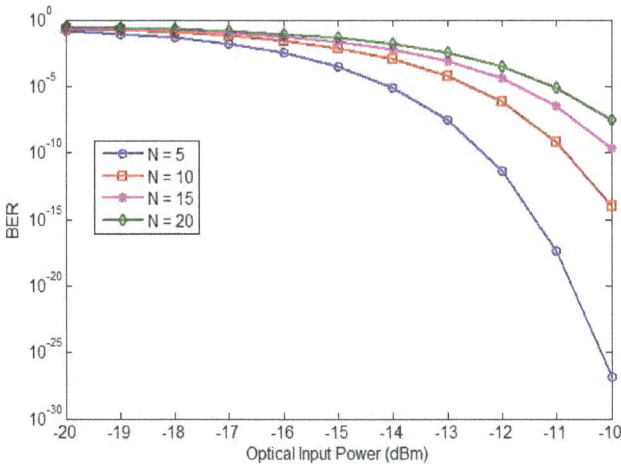

Figure 3. *BER vs optical input power varying different number of amplifiers (N=5, 10, 15, 20) for a fixed gain of G=10dB.*

In Figure 3, BER is plotted against optical input power varying different number of amplifiers for a fixed gain of 10dB. It shows that, the BER is lowest when number of amplifiers is 5. Thus the highest number of amplifiers produces highest BER.

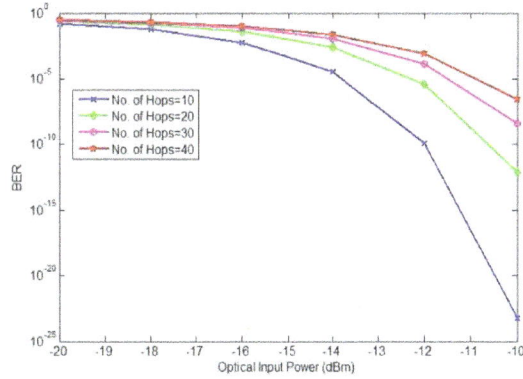

Figure 4. *BER vs optical input power for different number of hops (Nh=10, 20, 30, 40) with a fixed gain of G=10dB and bandwidth, B=5GHz.*

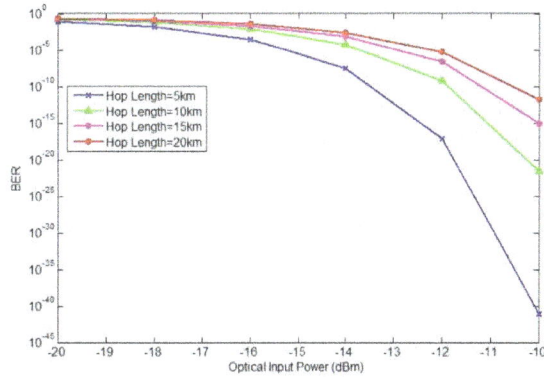

Figure 5. *BER vs optical input power varying hop length of 5km,10km,15km,20km for a fixed gain of G=10dB and bandwidth, B=5GHZ.*

In Figure 4, we present a graphical representation of BER vs received optical power varying different number of hops is 10, 20, 30, and 40. It shows, the BER is highest when the number of hops is 40. So it is clear that with the increase of number of hops the BER is also increases. In Figure 5, BER is plotted against optical received power for different hop length. The plot depicts that the BER is lowest for the hop length of 5km. So as the hop length is increases the BER is also increase. Comparing Figure 4 and Figure 5, for a fixed value of gain 10dB and a bandwidth of 5GHz, higher number of hops produces higher BER than high values of hop length. So it can be concluded that higher hop length is more preferable than higher number of hops.

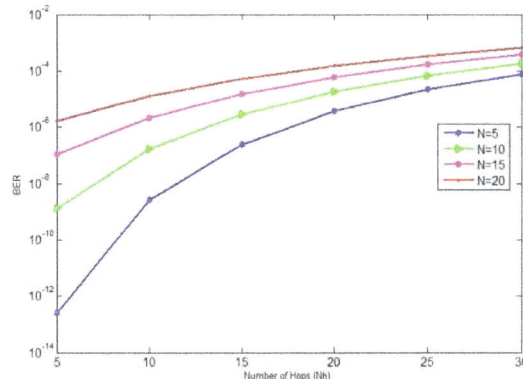

Figure 6. *BER vs number of hops varying number of amplifiers (N=5, 10, 15, 20) for a fixed gain of G=10dB and bandwidth B=5GHz.*

In Figure 6, BER vs number of hops has been plotted varying number of amplifiers. The BER increases with the increase of number of hops and also with the increase of number of optical amplifiers. Thus with the increase of the value of both parameters, BER performance is very inferior.

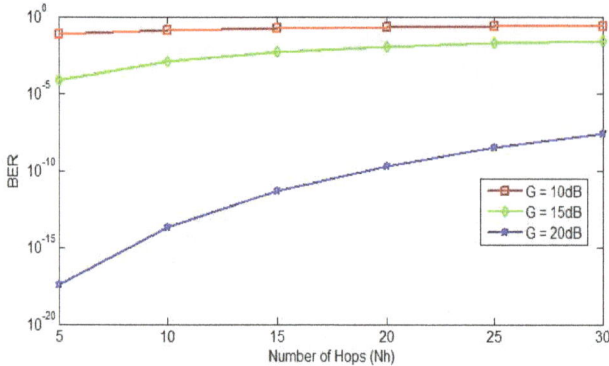

Figure 7. BER vs number of hops varying amplifier gain (G=10dB, 15dB, 20dB) for a fixed number of amplifiers of N=10, optical received power =-5dBm, bandwidth, B=5GHz.

In figure 7, BER vs number of hops varying amplifier gain for a fixed number of amplifiers of N=10, optical received power =-5dBm, bandwidth of B=5GHz has been plotted. Unlike Figure 6, here the BER is decreases with the increase of value of amplifier gain.

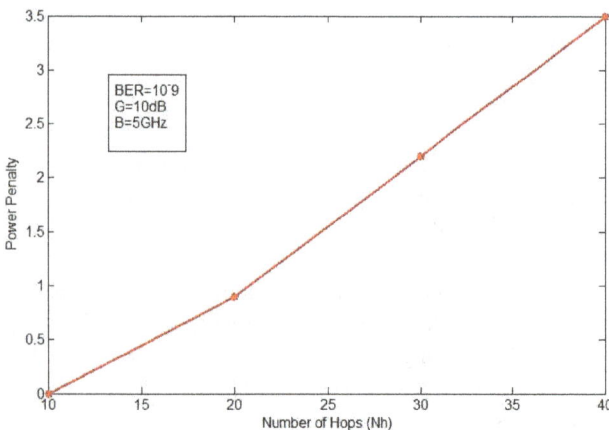

Figure 8. Power penalty versus number of hops for BER = 10^-9 with a fixed gain of G=10dB and bandwidth B=5GHz.

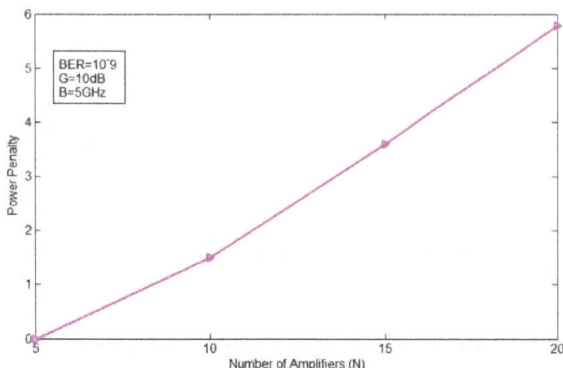

Figure 9. Power penalty vs No. of amplifiers for BER=10^-9 with a fixed gain of G=10dB and bandwidth B=5GHz.

In Figure 8, the power penalty vs no. of hops has been demonstrated with a fixed gain. The power penalty increases with the increase of number of hops. In Figure 9, the power penalty has been plotted against number of amplifiers. So combining Figure 8 and Figure 9 it can be concluded that, power penalty increases with the increase of both the value of number of amplifiers and number of hops.

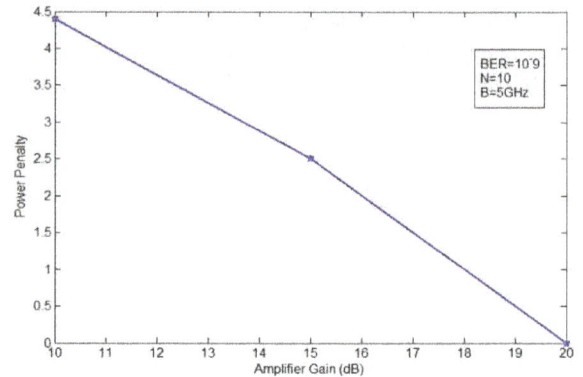

Figure 10. Power penalty vs amplifier gain for BER=10^-9 for N=10, B=5GHz.

In Figure 10, power penalty has been plotted against amplifier gain. It depicts that power penalty decreases with the increase of amplifier gain.

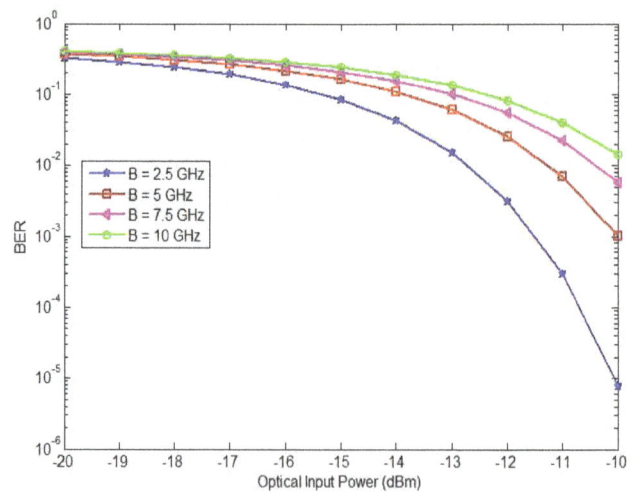

Figure 11. BER versus optical input power varying bandwidth for a fixed gain 10dB.

In Figure 11, BER has been demonstrated with an optical input power by varying different bandwidth. From the equation of (1, (2), (3) and (5) it can be seen that all the receiver noises and ASE noise are proportional to receiver bandwidth. So as the value of bandwidth is increases the SNR will be decreased and hence the BER will increase. In this figure BER is highest for 10GHz bandwidth. Therefore, a lower bandwidth produces a lower BER. It can be concluded that improved BER performance will be obtained when bandwidth is smaller. Also a small bandwidth corresponds to a smaller receiver size thus minimizes the cost. This figure is plotted for P-i-N receiver but this observation is valid for APD receiver also.

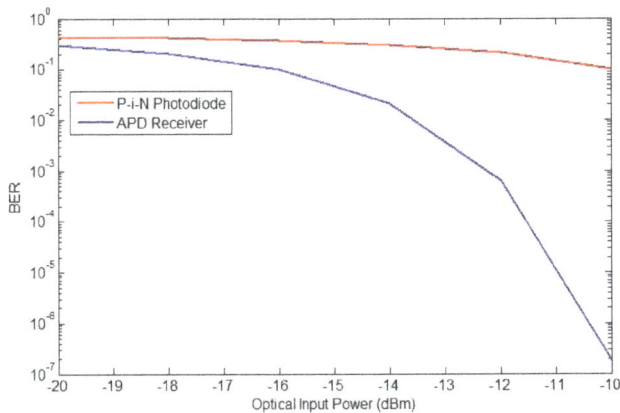

Figure 12. *BER vs optical input power for P-i-N and APD Receiver for a fixed bandwidth of B=5GHz.*

In Figure 12, BER is plotted against optical received power for P-i-N receiver and APD receiver. From Equation (8) and (9) it is seen that when the value of APD gain is 1, the SNR expression will be identical for both optical receiver. From this figure it is observed that BER is lower for APD receiver and it can be concluded that APD provides better performance compared to P-i-N optical receiver.

The performance of APD receiver depends on APD gain (M). After a certain value of APD gain (M=20dB), the improvement of DWDM system performance is not significant. We also identified the limitations of this article. From figure 11, for a given value of amplifier gain (10dB) the target BER 10^{-9} has not been achieved by varying receiver bandwidth. Therefore, in order to achieve desired threshold BER, the value of APD gain and receiver bandwidth should be optimum.

5. Conclusion

In this article, a general approach has been taken to evaluate the performance of DWDM system considering the effect of number of cascaded optical amplifiers in the presence of crosstalk. The SNR and BER have been analyzed by varying different parameters. Outcome of our investigation shows that, the system suffers from a power penalty which increases with the decrement in channel spacing and with the increment of optical amplifiers, and number of hops. The system performance can be substantially improved by using optimum number of optical amplifiers with higher gain and optimum receiver gain having moderate bandwidth. In future we expect to extend our analysis with intensity modulation coherent detection (IM/CD) and Raman amplifiers operating in a large wavelength as well as a wide bandwidth.

Acknowledgements

The author would like to express grateful thanks to his supervisor Dr. Satya Prasad Majumder, Professor, Dept. of EEE, BUET for his help, guidance and advice. The author also wishes to thank to Md. Mahfuzul Haque, Assistant Professor, BUBT, Bangladesh. The author's special thanks are sent to his father, mother and sister for their support and encouragement.

References

[1] Borella, M., Jue, J., Banerjee, D., Ramamurthy, B. and Mukherjee, B., "Optical components for WDM lightwave networks", Proceedings of the IEEE, Vol. 85 No. 8, August 1997, pp. 1274-1307.

[2] Bracket, C., "Dense wavelength division multiplexing networks: principles and applications", IEEE Journal Select. Areas Commun., August 1990, Vol. 8, No. 6, pp. 948-964.

[3] Jahid, A. and Alam, Z., "Performance analysis of DWDM system considering the effects of cascaded optical amplifiers with optimum receiver gain", American Journal of Engineering and Research, July 2015, Vol. 4, Issue 8, pp. 01-08.

[4] Tian, C. and Kinoshita, S., "Analysis and control of transient dynamics of EDFA pumped by 1480- and 980-nm lasers", IEEE/OSA Journal of Lightwave Technology, August 2003, vol. 21, No. 8, pp. 1728- 1734.

[5] Paraschis, L., Gerstel, O. and Frankel, M., "Metro networks: Services and technologies in optical fiber telecommunications, B: Systems and Networks", ed. Kaminow, I., Li, T. and Willner, A., Academic Press editors, EUA, 2008, ch. 12.

[6] E. Iannone, and R. Sabella, "Optical path technologies: A comparison among different cross-connect architectures," Journal of Lightwave Technology, Oct. 1996, vol. 14, no. 10, pp. 2184-2194.

[7] Elrefaie, A., Goldstein, E., Zaidi, S. and Jackman, N., "Fiber-amplifier cascades with gain equalization in multiwavelength unidirectional inter-office ring network," IEEE Photon. Technol. Lett., vol. 5, Sept. 1993, pp. 1026- 1031.

[8] Feuer, M., Kilper, D. and Woodward, S., "ROADMs and their system applications" in "Optical fiber telecommunications V, B: Systems and Networks", ed. Kaminow, I., T. and Willner, A., Academic press editors, EUA, 2008, ch.8.

[9] Dods S. D., Anderson T. B., "Calculation of bit-error rates and power penalties due to incoherent crosstalk in optical networks using Taylor series expansions," J. Lightwave Technol., vol. 23, April 2005, pp. 1828–1837.

[10] Maria Teresa Pinto, Ferreira Palma Ramalho, "Performance analysis of an optical link in DWDM sytems".

[11] Connely, M. J., "Semiconductor optical amplifiers", Kluwer Academic, 2002.

[12] Ramamurthiu, B., "Design of Optical WDM Networks, LAN, MAN, and WAN Architectures:, Kluvier, 2001.

[13] Ramaswami, R. and K. Sivarajan, "Optical Network: A Practical Perspective",3rd edition, Morgan Kaufmann, 2002.

[14] Keiser, G., "Optical Fiber Comunications", 3rd edition,McGraw-Hill, 2000.

[15] Kelvin B. A., Afa J. T., "Bit Error Rate Performance of Cascaded Optical Amplifiers Using Matlab Computation Software," European Scientific Journal, vol. 9, No. 3, January 2013.

[16] H. Takahashi, K. Oda, and H. Toba, "Impact of crosstalk in an arrayed waveguide multiplexer on NxN optical interconnection", Journal of Lightwave Technology, vol. 14, no. 6, June 1996, pp. 1120-1126.

[17] S. D. Dods, J. P. R. Lacey, and R. S. Tucker, "Homodyne crosstalk in WDM ring and bus networks", IEEE Photonic Technology Letter, vol. 10, no. 3, Mar. 1998 pp. 457-458.

[18] K. P. Ho, C.K. Chan, F. Tong, and L.K. Chen, "Exact analysis of homodyne crosstalk induced penalty in WDM networks", IEEE Photonic Technology Letter, vol. 9, no. 3,Sept. 1997, pp. 1285-1287.

[19] Y. Shen, K. Lu, and W. Gu, "Coherent and incoherent crosstalk in WDM optical networks", Journal of Lightwave Technology, vol. 17, no. 5, May 1999, pp. 759-764.

[20] Senior, J.M. ''Optical fibre communication principles and practice'' 3rd edition Prentice Hall 2009.

7

Study on Data Compression and Reduction of the Aviation Network Based on Multi-resolution Wavelet Analysis

Yao Hong Guang

School of Air Transportation / Flying, Shanghai University of Engineering and Science, Shanghai, China

Email address:
yhg1yhg@sina.com

Abstract: This paper proposes complex network data compression idea based on the multi-resolution wavelet decomposition theory, analyzes the concrete form of wavelet basis choice and wavelet decomposition, and then puts forward the determination method of network decomposition levels and parameters reduction method after decomposition. The empirical study shows that four-level wavelet decomposition for Chinese aviation network adjacency matrix is carried out by Haar wavelet basis, and the lowest frequency sub-band is 10×10 matrix. Moreover, the average degree, the average shortest path length and clustering coefficient of original network are restored in the lowest frequency sub-band after decomposition.

Keywords: Aviation Network, Multi-resolution Wavelet Decomposition, Complex Network, Data Compression and Reduction

1. Introduction

Aviation network can be regarded as the complex network composed of cities which the airports stand in and the air routes which link these cities, and navigable cities and air routes constitute the basic framework of aviation network. In the aviation network, one air route connects two navigable cities, which realizes the bidirectional air transport between the cities, so the network can be considered as undirected network. Aviation network has the characteristics of typical complex network [1]. It is the premise and foundation for the network analysis to measure the statistical characteristic parameters of aviation network. However, there appears considerable difficulty for the empirical study of aviation network characteristics due to the complexity and dynamics of aviation network node and air route. Wavelet multi-resolution decomposition is currently main technical means for the complex network. Marr. D proposed wavelet multi-resolution theory by simulating the human visual system for the first time [2]. Furthermore, the study from Fan J et al. showed that multi-resolution decomposition technology has been an effective tool for studying complex network structure properties.[3] The adjacency matrix $A(p)=a_{ij}$ should be firstly determined for the wavelet decomposition of a network. If there exists connection edge between i and j, then $a_{ij}=1$, if not, $a_{ij}=0$. The wavelet decomposition is carried out for

$A(p)$, and the appropriate wavelet basis is chosen. The high-pass filter H and low-pass filter L are used for filtering along the horizontal and vertical directions of network. And then four subsegments (HH, HL, LH, and LL) are output. The wavelet decomposition proceeds in the low frequency sub-band until the decomposition requirements are satisfied [4], as shown in Figure 1.

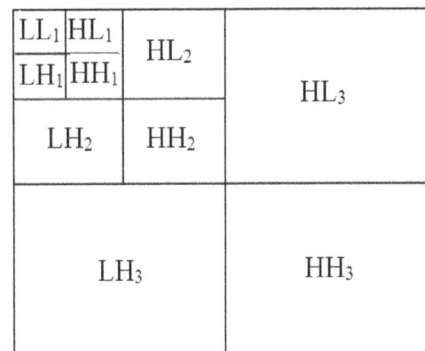

Fig. 1. Three-level wavelet decomposition diagram.

The study shows that multi-resolution wavelet decomposition has quite good energy tightness [5]. Most of the system energy concentrates on the low frequency sub-band, but the high frequency sub-band has little or no energy. The information contained by low frequency sub-band LL_1 can be regarded as the coarse graining description of network, other sub-bands include information with higher resolution, and different sub-bands can describe the network with different accuracy. This analysis thinking is similar to visual processing of human eyes [6]. Multi-resolution wavelet decomposition provides a natural way to analyze complex network. Namely, firstly, the network structure is analyzed by the coarse resolution, and then the analytical precision is gradually increased according to the requirements [7]. Hence, this method has excellent adaptation for processing fairly complex system of aviation network.

2. Major Parameters Analysis of Chinese Aviation Network

Network parameters intensively reflect the basic situation of the network, which is the foundation of network analysis.

2.1. Adjacency Matrix and Node Degree of Aviation Network

(1) Adjacency matrix of aviation network

The establishment of adjacency matrix is the basic work of complex network study [8]. The adjacency matrix of the aviation network stores the mutual connection information of the network in the form of matrix. If there is conjoint air routes between the navigable cities i and j in the aviation network, the element of adjacency matrix $a_{ij} = 1$, or else $a_{ij} = 0$.

(2) Degree

In aviation network, the degree k_i of node city i is defined as the total number of air routes connecting with the node city i. Namely, the greater the degree of one city is, the more the routes connecting with the city are [9].

2.2. Average Shortest Path of Aviation Network

The shortest path is the number of air routes included in the shortest routes between two node cities in the aviation network. And average shortest path length is the number of average shortest routes of all the node cities pairs in aviation network, which reflects the needed average least transit times for the passengers and goods from one node to another one[10]. The calculation method for the average shortest path length L of the aviation network is as shown in equation (1).

$$L = \frac{\sum L_{ij}}{\frac{1}{2}N(N+1)} \tag{1}$$

Where N is the number of nodes in the aviation network, L_{ij} is the number of air routes of the shortest path between the node cities i and j.

2.3. Clustering Coefficient

The clustering coefficient C_i of the node i can be used to describe the dense degree that the node i of aviation network connects with the other nodes; its value can be calculated by the equation (2).

$$C_i = \frac{\text{The number of triangles connecting with node i}}{\text{The number of triples connecting with node i}} \tag{2}$$

The clustering coefficient C of aviation network is the average value of clustering coefficient Ci of all the nodes i. The greater value reflects that the distribution of the network routes is more uneven, namely, local area density is greater, but some regional routes are sparse [11].

3. Wavelet Decomposition of Chinese Aviation Network

3.1. Wavelet Decomposition of Aviation Network Based on the Haar Wavelet Conversion

(1) Applicability analysis of Haar wavelet

Haar function is a group of function set, from which the Haar wavelet is derived, which is the simplest binary wavelet function with discontinuous time-domain, and it is defined as follows.

$$\psi(t) = \begin{cases} 1 & 0 \leq t < 1/2 \\ -1 & 1/2 \leq t < 1 \\ 0 & \text{other} \end{cases} \tag{3}$$

Due to the characteristic of Haar wavelet function, it has a good applicability for the decomposition and conversion of binary system or 0-1 matrix. The adjacency matrix is used to store the data information for the aviation network. The adjacency matrix is a 0-1 matrix, which can reflect the connection condition between i and j of any two points in the network. Therefore, the Haar wavelet is the ideal wavelet for aviation network decomposition.

(2) Haar wavelet decomposition of aviation network

Multi-resolution decomposition is carried out for the aviation network by using the Haar wavelet and adjacency matrix. The first level decomposition is that the original connection matrix is decomposed into a low frequency part (relative to the average) and three high frequency parts (corresponding to the details in the horizontal, vertical and diagonal locations). The process mentioned above is repeated to realize aviation network of the multi-layer wavelet decomposition.

It is assumed that the adjacent matrix of aviation network is A, and the r-level wavelet decomposition of A is denoted by $\tilde{A}^r$. N nodes in aviation network are divided into 2^r groups, and the ith group includes $N/2^r$ nodes in sequence[12], namely

$$2^r(r-1)+1, \ 2^r(r-1)+2, \ 2^r(r-1)+3, \ ; \ i=1, \ 2, \ .., \ N/2^r \tag{4}$$

The element value of adjacent matrix by r-level wavelet transform can be derived by the formula (4).

$$\tilde{a}_{ij}^r = \frac{1}{2^r}(\tilde{a}_{2i-1,2j-1}^r + \tilde{a}_{2i-1,2j}^r + \tilde{a}_{2i,2j-1}^r + \tilde{a}_{2i,2j}^r) = \frac{1}{2^r}\sum_{s=2^r(i-1)+1}^{2^r i}\sum_{t=2^r(j-1)+1}^{2^r j} a_{st} \quad i,\ j=1,\ 2,\ ,N/2^r \tag{5}$$

Hence, r-level wavelet transform of aviation network adjacent matrix A corresponds to the divided 2^r groups of network nodes, and each group includes $N/2^r$ nodes in sequence. The diagonal element of the lowest frequency sub-band LL_1 of wavelet transform signifies the normalized number of internal edges in each group, while the LL_1 non-diagonal element expresses the normalized number of edges connecting different groups, and the normalized constant is 2^r. This research shows that the elements in the lowest frequency sub-band LL_1 matrix after network decomposition contains most of the original network information [13]. Hence, the original adjacency matrix can be replaced by the lowest frequency sub-band LL_1 to simplify network.

3.2. Parameter Analysis after the Network Decomposition

3.2.1. Connection Property Analysis

The off-diagonal element $\tilde{a}_{ij}^r$ of the network low frequency sub-band LL_1 after wavelet decomposition depicts the connection property between the ith group and the jth group in the primary network. $\tilde{a}_{ij}^r \neq 0$ means that if and only if there is one edge at least between the two groups, and the greater value of $\tilde{a}_{ij}^r$ indicates that the number of edges is more between the two groups. In a similar way, the diagonal element $\tilde{a}_{ii}^r$ of LL_1 describes the connection property in the interior of the ith group. The greater value of $\tilde{a}_{ii}^r$ is, the more the number of the interior edges is[14].

3.2.2. Average Shortest Path Length Analysis

One node of new network after wavelet corresponds to one group of the original network. One edge between two different nodes i and j in new network is equivalent to one or more edges between the ith group and jth group in the original network. If the average shortest path length of new network after wavelet decomposition is L_{new}, the average shortest path length of original network L should meet equation (6).

$$L \le L_{new} + 2\max_{i,j} L_{ij} \tag{6}$$

3.2.3. Clustering Property Analysis

The clustering property of original network can be obtained by the lowest frequency sub-band LL_1 matrix. The value of $\tilde{a}_{ii}^r$ is much greater than that of $\tilde{a}_{ij}^r$ for all the i and j, which illustrates that the number of edges in each group is quite more than that of edges among different groups, and

means high clustering network, otherwise, the network clustering is low[15].

3.3. The Determination of the Number of Network Decomposition Levels r

For the global coupling network with an average path length of 1, there exists one edge at least between the arbitrary two groups in the new network after decomposition. It is supposed to remove all the edges between different groups, and each isolated group is regarded as a sub-network. Each group is a regular grid, and the formula (7) is expressed as follows. Where k is the mean value of each node degree[16].

$$\max_{i,j} L_{ij} \le 2^r/k \tag{7}$$

It can be derived that L_{ij} is the logarithmic growth function of the total number of network nodes N from the formula (7), namely

$$\max_{i,j} L_{ij} \le c\ln N \tag{8}$$

Network decomposition levels r can be obtained by formulas (7) and (8), which should meet the formula (9).

$$r \le \ln(ck) + \ln(\ln N) \tag{9}$$

4. Empirical Research on Multi-Resolution Wavelet Analysis of Chinese Aviation Network

4.1. Network Selection and Parameter Calculation of the Network

This paper chooses the Chinese aviation transport network with relatively less number of nodes as the empirical research object. The data are from the flight schedule of Ctrip website (http://flights.ctrip.com/schedule/ScheduleIndex.aspx). The number of collected navigable cities is 163, and the number of domestic routes is 2198. And 163 navigable cities are adopted as the rows and columns of the matrix, the number of air routes between two cities is taken as the weight, then forming 163×163 aviation network adjacency matrix A.

All the parameter values of Chinese aviation network are calculated by using Ucinet software and adjacency matrix A, as follows.

(1) Network connection situation: the number of nodes is 163, the number of edges in the network is 2198, and their connection probability is 0.086.

(2) The average shortest path length of network is $L = 2.16$, namely, starting from any navigable city, it only need one transfer to reach other navigable cities.

(3) The degrees of part node cities are shown in Table 1.

Tab. 1. *The degree values of part node cities.*

City	Beijing	Shanghai	Guangzhou	Xi'an	Shenzhen	Kunming	Chengdu	Chongqing	Hangzhou	Qingdao
Degree	118	95	96	62	66	67	68	63	41	40

(4) Cluster analysis of the network

Clustering coefficient is the average aggregation extent of network composed by the navigable cities and adjacent cities [9], which represents the breadth of air transport. In Chinese aviation network, the clustering coefficient is 0.828, which shows the stronger clustering.

4.2. Empirical Analysis of the Number of the Level of Network Decomposition

In the Chinese aviation network, the maximum path length between two nodes $\max L_{ij}$ equals 4, and the number of network nodes N equals 163. The formula $c \geq 0.786$ can be obtained by formula (10).

The average value of each node degree in the network is

13.779, which is calculated by the software Ucinet. The formula (9) can be used to determine the final decomposition levels r, that is $r \leq \ln(0.786*13.779) + \ln(\ln 163) = 4.010$. There should be four-level wavelet decomposition at least in Chinese aviation network.

4.3. Four-level Wavelet Decomposition of Chinese Aviation Network

The wavelet decomposition of network is carried out by formula (6), the 163 nodes of network are divided into 16 groups, and there are 10×10 elements in each group. The four-level wavelet decomposition of network is shown in Figure 3.

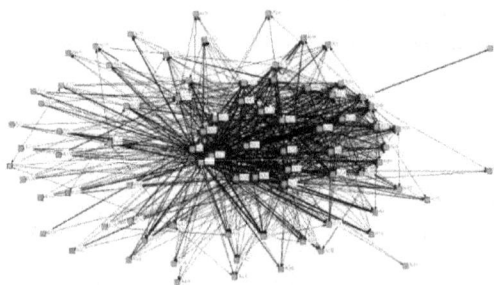

(a) The aviation network after one-level wavelet decomposition

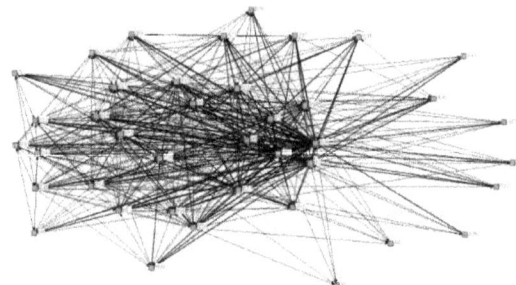

(b) The aviation network after two-level wavelet decomposition

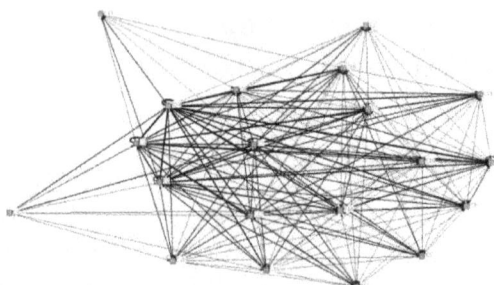

(c) The aviation network after three-level wavelet decomposition

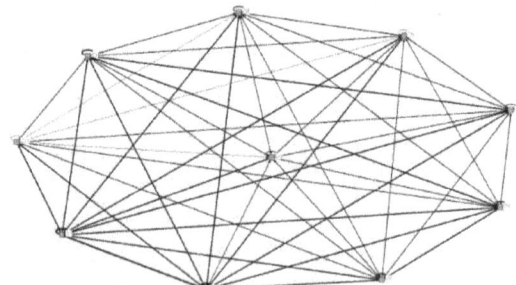

(d) The aviation network after four-level wavelet decomposition

Fig. 2. *Four-level wavelet decomposition comparison chart.*

Each element value of the lowest frequency sub-band LL_1 matrix of the network after decomposition can be obtained by the formula (7), as shown in Table 2.

Tab. 2. *The adjacency matrix of Chinese aviation network after four-level wavelet decomposition.*

	D1	D2	D3	D4	D5	D6	D7	D8	D9	D10
D1	1.500	1.312	1.813	1.438	1.188	1.125	1.875	1.125	1.688	1.313
D2	1.375	1.750	1.500	1.313	1.375	1.063	2.250	1.125	1.625	0.813
D3	1.813	1.625	2.313	2.000	1.063	1.625	2.813	2.063	2.188	1.500
D4	1.438	1.313	2.000	0.875	0.750	0.938	2.125	1.250	1.313	0.688
D5	1.188	1.375	1.063	0.750	0.750	0.688	1.688	0.688	1.313	0.500
D6	1.125	1.063	1.625	0.938	0.688	0.500	2.188	0.750	0.750	0.500
D7	1.875	2.250	2.813	2.125	1.688	2.188	3.500	2.125	2.188	1.563
D8	1.125	1.125	2.063	1.250	0.688	0.750	2.125	0.875	1.250	0.500
D9	1.688	1.625	2.188	1.313	1.313	0.750	2.188	1.250	1.000	0.875
D10	1.313	0.813	1.500	0.688	0.500	0.500	1.563	0.500	0.938	0.188

4.4. Reduction and Contrast Analysis for the Decomposition Results of Multiresolution Wavelet

Because the wavelet transform has quite good energy tightness, the system energy mostly focuses on the low frequency sub-band, while the high frequency sub-band has a little or no energy. The information contained in the low frequency sub-band can be regarded as the rough description of the network. Therefore, the lowest frequency sub-band LL_1 matrix of aviation network after four-level wavelet decomposition contains most of the original network information, and the original network can be restored by the data in Table 2.

4.4.1. Reduction and Contrast of Network Connectivity

The nondiagonal element in Table 2 describes the connection property between *ith* group and *jth* group in original network, the greater the value of the nondiagonal element is, the larger the number of the edges between the two groups is, and the number of the original network edges equals $a_{ij}^4 \times 2^4$. For example, the corresponding element value of D2-D5 is 1.375, then the number of the edges between *2nd* group and *5th* group is 22 (1.375×16). Similarly, the diagonal element LL_1 describes the internal connection property of the *ith* group. The number of internal edges in the *ith* group is $a_{ii}^4 \times 2^4$, the greater the value is, the more the number of edges in the group interior is. For instance, the corresponding element value of D2-D2 is 1.750, and then the number of edges in the *2nd* group interior is 28 (1.750×16). All the elements in Table 2 are added up, multiplied by the standardized coefficients 2^4, and the number of all the edges in the original network can be obtained. The sum of all the elements is 137.123, then is $2193.968 (137.123 \times 2^4)$ which is close to the 2198 edges in the original network. The cause of error is that there are a total of 163 nodes in the original network, cannot be evenly divisible by 2^4. When the wavelet decomposes, three nodes are ignored.

4.4.2. Reduction and Contrast of Average Shortest Path Length

The lowest frequency sub-band LL_1 is the entire connection network, and its average shortest path length is 1, which can be obtained by formulas (6) and (7).

$$L \le L_{new} + 2\max_{i,j} L_{ij} \le 1 + 2 \times 2^r / k = 1 + 2 \times 16 / 13.779 = 3.32 \quad (10)$$

The average shortest path length of original network is 2.16, so the formula (10) is workable. The lowest frequency sub-band LL_1 can determine that the average shortest path value of the original network is less than 3.32.

4.4.3. Reduction and Contrast of Average Degree and Clustering Condition

the average degree and clustering situation information of the network can be restored by using the lowest frequency sub-band LL_1 of network after decomposition. The data in Table 2 can restore all the edges of the original network, a total of 2194 edges. And then the average degree of original network $k' = 2194 / 163 = 13.460$, which is close to the average value of original network degree 13.779. The difference between the minimum diagonal element 2.813 and maximum off-diagonal element 0.188 is 2.625, which is far greater than the average value 1.371 in Table 2. This indicates that the original network has high integration.

5. Conclusions

The study shows as follows.

(1) Chinese aviation network should be carried out four-level wavelet decomposition. The data in Table 2 cover the main information of Chinese aviation network characteristic.

(2) The lowest frequency sub-band LL_1 of 4 layers wavelet decomposition of Chinese aviation network includes most of the information of the original network. And the relevant information of original network can be accurately restored by adopting corresponding methods.

This paper provides an effective simplified technology for studying the quite complex network with the giant data. The subsequent study can focus on the fields such as network grouping methods, the influence of different grouping on the network information, and so on.

Acknowledgment

This study is funded by "Key Projects of Scientific Research Innovation of Shanghai Education Committee, 2013 (Project Grant Number: 13ZS127)", "Foundation Project for Youths of Humanities and Social Science Research of the Ministry of Education, 2014 (Project Grant Number: 14YJCZH183)".

Reference

[1] Liu Hongkun, Zhou Tao. Research on aviation network [J]. progress in natural science, 2008(6): 601-608.

[2] Marr D. Vision[M]. New York: Freeman Publishers，1982.

[3] Fan J，Wang X F. A wavelet view of small-world networks[J]. IEEE Trans. Circuits & Systems-II，2005,52(5):238-242.

[4] Wang Xiaofan, Li Xiang, Chen Guanrong. The complex networks theory and application [M]. Beijing: Tsinghua University press, 2006

[5] Dai Houping. A wavelet transform optimization method of complex networks [J]. Journal of Chongqing University of Science and Technology (Natural Science Edition). 2008(6):84-86.

[6] Zhong Zhili. The study of fabric pilling objective evaluation Based on wavelet analysis [D]. Tianjin: Tianjin University of Technology, 2006

[7] Zhao Hailong, Mu Zhichun, , Ding Wenkui, Zhang Xia. Ear Recognition Based on Wavelet Transform and Block DCT[J]. Acta Scientiarum Naturalium Universitatis Pekinensis, 2009 (3):243 -247.

[8] Liu Hongkun. The structure of Chinese aviation network and analysis of its influencing factor [D]. Chengdu:Southwest Jiao Tong University, 2007

[9] Chunhua Gao. Airline Integrated Planning and Operations [D].Georgia Institute of Technology，2007.

[10] Ye Wu, Ping Li, Maoyin Chen. Response of Scale-free Networks with Community Structure to External Stimuli[J].Physics A, 2009, 388 (14):2987-2994.

[11] Albert R, Barabásia L. Statistic mechanics of complex networks [J].Review of Modern Physics, 2002 (74):47-97.

[12] Kurant M, Thiran P. Extraction and analysis of traffic and topologies of transportation networks. Phys Rev E. 2006, 74:036114.

[13] Bagler G. Analysis of the airport network of India as a complex weighted network[J]. Physica A, 2008, 387: 2972-2980.

[14] Liu HK, Zhou T. Topological properties of Chinese city airline network. Dynamics of Continuous, Discrete and Impulsive Systems B, 2007, 14: 135-138.

[15] Guimerà R, Mossa S, Turtschi A. The world-wide air transportation network: Anomalous centrality, community structure, and cities'global roles. Proc Natl Acad Sci USA, 2005, 102(22): 7794-7799.

[16] Yao Hongguang, Zhu Liping. Research on Robustness of China's Aviation Network Based on Simulation Analysis[J]. Journal of Wuhan University of Technology (Transportation Science & Engineering),2012(1):42-46.

8

Review of Performance of Impedance Based and Travelling Wave Based Fault Location Algorithms in Double Circuit Transmission Lines

Ankamma Rao Jonnalagadda, Gebreegziabher Hagos

Department of Electrical & Computer Engineering, School of Engineering & Technology, Samara University, Semera, Afar Region, Ethiopia

Email address:
jaraoeee04@gmail.com (A. R. Jonnalagadda), safehagos@gmail.com (G. Hagos)

Abstract: Parallel transmission lines or Double circuit transmission lines have been extensively utilized in modern power systems to enhance the reliability and security for transmission of electrical energy. This paper presents two fundamental algorithms: Impedance based, Travelling wave (TW) based algorithms for 100km, 400KV Double circuit transmission lines. MATLAB/ Simulink software was used to implement these algorithms. The accuracy of fault location on power transmission line are reviewed for these two methods by varying various parameters like fault type, fault location on a given power system model.

Keywords: Fault Location, MATLAB, Impedance Based, Travelling Wave Based, Accuracy of Fault Location

1. Introduction

Location of faults in power transmission lines is one of main concerns for all electric utilities as the accurate fault location can help to restore the power supply in shortest possible time. Fault location methods are broadly classified as impedance based method which uses the steady state fundamental component of voltage and current values [1-6], Travelling wave(TW) based method which uses the incident and reflected TWs observed at measuring ends of the line[7-10],and knowledge based method which uses artificial neural network and/or pattern recognition techniques[11]. Conventional fault detection algorithms are designed based On current or voltage magnitude measurements .When a fault occurs on a transmission line it causes a sudden change in the current and voltage signals as well as measured impedances at the relay location. Increase of current magnitude **or** decrease of voltage/impedance magnitude could be considered as a measure to detect a system fault; these algorithms are dependent on various factors such as fault resistance and power system short circuit capacity. This paper describes two fundamental algorithms: Impedance based, Travelling wave (TW) based algorithms are implemented in 100km,400KV Double circuit transmission lines; it compares two algorithms by varying various parameters like fault type; fault location etc.

2. Theory of Impedance Based Fault Location Algorithm

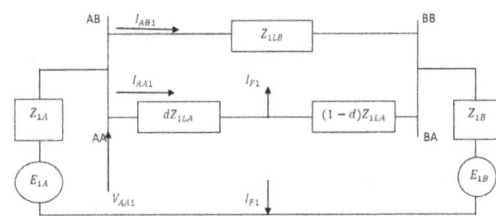

Fig. 1. Equivalent Positive sequence circuit diagram for double circuit transmission lines.

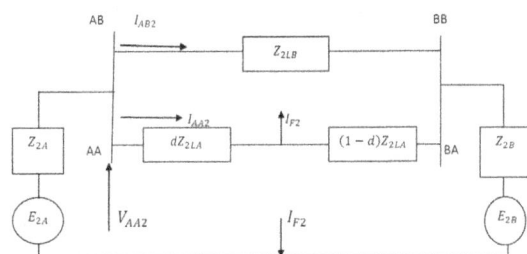

Fig. 2. Equivalent negative sequence circuit diagram for Double circuit transmission lines.

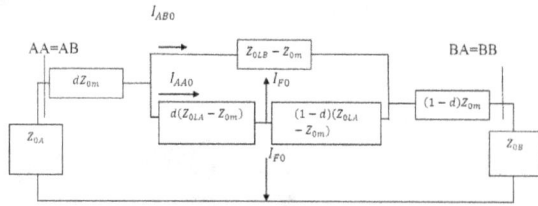

Fig. 3. *Equivalent zero sequence circuit diagram for double circuit transmission lines.*

To derive the Fault location algorithm, the fault loop composed according to the fault classified type is considered. This loop contains the faulted line segment (between points AA and F) and the fault path itself. A generalized model for the fault loop is stated as fallows

$$V_{AA\_P} - dZ_{1LA} * I_{AA\_P} - I_F * R_F = 0 \qquad (1)$$

Where

$$I_F = a_{F1} * I_{F1} + a_{F2} * I_{F2} + a_{F0} * I_{F0} \qquad (2)$$

Fault loop voltages and current can be expressed interns of the local measurements and with using coefficients gathered in Table 1.

$$V_{AA\_P} = a_1 V_{AA1} + a_2 V_{AA2} + a_0 V_{AA0} \qquad (3)$$

$$I_{AA\_P} = a_1 I_{AA1} + a_2 I_{AA2} + a_0 \frac{Z_{0LA}}{Z_{1LA}} I_{AA0} + a_0 \frac{Z_{0m}}{Z_{1LA}} I_{AB0} \qquad (4)$$

Table 1. *Coefficients for determining signals defined in Equations (2) and (3).*

Fault Type	a_1	a_2	a_0
AG	1	1	1
BG	a^2	a	1
CG	a	a^2	1
AB, ABG, ABC, ABCG	$1-a^2$	$1-a$	0
BC, BCG	$a^2 - a$	$a - a^2$	0
CA, CAG	$a-1$	a^2-1	0
	$a = exp(j2\pi/3)$		

Voltage drop across the fault path (as shown in the third term in Equation (1)) is expressed using sequence components of total fault current (I_{F0}, I_{F1}, I_{F2}). Determining this voltage drop requires establishing the weighting coefficients. These coefficients can accordingly be determined by taking the boundary conditions for particular fault type. However, there is some freedom for that. Thus, it is proposed firstly to utilize this freedom for avoiding zero sequence quantities. This is well known that the zero sequence impedance of a line is considered as unreliable parameter. This is so due to dependence of this impedance upon the resistivity of a soil, which is changeable and influenced by weather conditions. Moreover, as a result of influence of overhead ground wires the zero sequence impedance is not constant along the line length. Thus, it is highly desirable to avoid completely the usage of zero sequence quantities when determining the voltage drop across the fault path. This can be accomplished by setting $I_{F0} = 0$ as shown in Table 2, where the alternative sets of the weighting coefficients are gathered. Secondly, the freedom in establishing the weighting coefficients can be utilized for determining the preference for using particular

quantities. The negative sequence (Table 2) or the positive sequence (Table 2) can be preferred.

For example, considering AG fault one has:

$$\begin{vmatrix} I_{F0} \\ I_{F1} \\ I_{F2} \end{vmatrix} = \frac{1}{3} * \begin{vmatrix} 1 & 1 & 1 \\ 1 & a & a^2 \\ 1 & a^2 & a \end{vmatrix} * \begin{vmatrix} I_{FA} \\ 0 \\ 0 \end{vmatrix} \qquad (5)$$

Thus, symmetrical components of a fault current are:

$$I_{F0} = I_{F1} = I_{F2} = \frac{1}{3} * I_{FA} = I_F \qquad (6)$$

It follows from Equation (6) that the total faults current ($I_F = I_{Fa}$) can be expressed in the following alternative ways, depending on which symmetrical component is preferred:

$$I_F = 3 * I_{F1} \qquad (7)$$

$$I_F = 3 * I_{F2} \qquad (8)$$

$$I_F = 3 * I_{F0} \qquad (9)$$

$$I_F = 1.5 * I_{F1} + 1.5 * I_{F2} \qquad (10)$$

Application of Equation (1) for fault location requires determining the positive and the negative sequence components of the fault path current. Considering the two different paths in the circuits of Fig.1 and Fig.2:- the faulted line segment adjacent to the local substation, the healthy line together with the remote segment of the faulted line, one obtains:

$$I_{F1} = \frac{I_{AA1} - \frac{Z_{1LB}}{Z_{1LA}} * I_{AB1}}{1-d} \qquad (11)$$

$$I_{F2} = \frac{I_{AA2} - \frac{Z_{2LB}}{Z_{2LA}} * I_{AB2}}{1-d} \qquad (12)$$

Substituting Equation (11) and (12) in Equation (2)

$$I_F = a_{F1} * \frac{I_{AA1} - \frac{Z_{1LB}}{Z_{1LA}} * I_{AB1}}{1-d} + a_{F2} * \frac{I_{AA2} - \frac{Z_{2LB}}{Z_{2LA}} * I_{AB2}}{1-d} \qquad (13)$$

$$I_F = \frac{a_{F1} * \left(I_{AA1} - \frac{Z_{1LB}}{Z_{1LA}} * I_{AB1} \right) + a_{F2} * \left(I_{AA2} - \frac{Z_{2LB}}{Z_{2LA}} * I_{AB2} \right)}{1-d} \qquad (14)$$

Let

$$N_{12} = a_{F1} * \left(I_{AA1} - \frac{Z_{1LB}}{Z_{1LA}} * I_{AB1} \right) + a_{F2} * \frac{I_{AA2} - \frac{Z_{2LB}}{Z_{2LA}} * I_{AB2}}{1-d} \qquad (15)$$

Therefore

$$I_F = \frac{N_{12}}{1-d} \qquad (16)$$

Substitute Equation (16) in Equation (1)

$$V_{AA\_P} - dZ_{1LA} * I_{AA\_P} - \frac{N_{12}}{1-d} * R_F = 0 \qquad (17)$$

Resolving (17) into real and imaginary parts gives:

$$Re(V_{AA\_P}) - d * Re(Z_{1LA} * I_{AA\_P}) - \frac{R_F}{1-d} * Re(N_{12}) = 0 \quad (18)$$

$$Im(V_{AA\_P}) - d * Im(Z_{1LA} * I_{AA\_P}) - \frac{R_F}{1-d} * Im(N_{12}) = 0 \quad (19)$$

Elimination of the agent ($R_F/(1-d)$) yields the following formula for a sought distance to fault:

$$\begin{vmatrix} Re(Z_{1LA} * I_{AA\_P}) & Re(N_{12}) \\ Im(Z_{1LA} * I_{AA\_P}) & Im(N_{12}) \end{vmatrix} * \begin{vmatrix} d \\ \frac{R_F}{1-d} \end{vmatrix} = \begin{vmatrix} Re(V_{AA\_P}) \\ Im(V_{AA\_P}) \end{vmatrix} \quad (20)$$

$$d = \frac{\begin{vmatrix} Re(V_{AA\_P}) & Re(N_{12}) \\ Im(V_{AA\_P}) & Im(N_{12}) \end{vmatrix}}{\begin{vmatrix} Re(Z_{1LA} * I_{AA\_P}) & Re(N_{12}) \\ Im(Z_{1LA} * I_{AA\_P}) & Im(N_{12}) \end{vmatrix}} \quad (21)$$

$$d = \frac{Re(V_{AA\_P}) * Im(N_{12}) - Im(V_{AA\_P}) * Re(N_{12})}{Re(Z_{1LA} * I_{AA\_P}) * Im(N_{12}) - Im(Z_{1LA} * I_{AA\_P}) * Re(N_{12})} \quad (22)$$

$$d = \frac{Im(V_{AA\_P}) * Re(N_{12}) - Re(V_{AA\_P}) * Im(N_{12})}{Im(Z_{1LA} * I_{AA\_P}) * Re(N_{12}) - Im(N_{12}) * Re(Z_{1LA} * I_{AA\_P})} \quad (23)$$

The formula (23) can be written down in a more even compact alternative form:

$$d = \frac{Im(V_{AA\_P} * N_{12}{}^*)}{Im(Z_{1LA} * I_{AA\_P} * N_{12}{}^*)} \quad (24)$$

Table 2. *Alternative sets of weighting coefficients.*

Fault type	Set I			Set II		
	a_{F1}	a_{F2}	a_{F0}	a_{F1}	a_{F2}	a_{F0}
AG	0	3	0	3	0	0
BG	0	$-1.5 + j1.5\sqrt{3}$	0	$-1.5 - j1.5\sqrt{3}$	0	0
CG	0	$-1.5 - j1.5\sqrt{3}$	0	$-1.5 + j1.5\sqrt{3}$	0	0
AB	0	$1.5 - j0.5\sqrt{3}$	0	$1.5 + j0.5\sqrt{3}$	0	0
BC	0	$j\sqrt{3}$	0	$-j\sqrt{3}$	0	0
CA	0	$-1.5 - j0.5\sqrt{3}$	0	$-1.5 + 0.5\sqrt{3}$	0	0
ABG	$1.5 + j0.5\sqrt{3}$	$1.5 - j0.5\sqrt{3}$	0	$1.5 + j0.5\sqrt{3}$	$1.5 - j0.5\sqrt{3}$	0
BCG	$-j\sqrt{3}$	$j\sqrt{3}$	0	$-j\sqrt{3}$	$j\sqrt{3}$	0
CAG	$1.5 - j0.5\sqrt{3}$	$1.5 + j0.5\sqrt{3}$	0	$1.5 - j0.5\sqrt{3}$	$1.5 + j0.5\sqrt{3}$	0
ABC, ABCG	$1.5 + j0.5\sqrt{3}$	$1.5 - j0.5\sqrt{3}$	0	$1.5 + j0.5\sqrt{3}$	$1.5 - j0.5\sqrt{3}$	0

3. Traveling Wave Based Fault Location Algorithm

The proposed fault location algorithm using Wavelet Transform is show in the following steps:

1. Get the signals from transducer output.
2. Transform the signals into modal domain.
3. Apply Discrete Wavelet Transform and obtain the Wavelet Transform Coefficients (W_{mm}).
4. If the mode 0 (W_{mm0}) is zero, then the fault is identified as an ungrounded fault and the fault distance is given by the equation :

$$d = (v \times t_d)/2 \quad (25)$$

where d is the fault location from source A, v is the wave velocity of mode 1 having magnitude slightly less than velocity of light, and t_d is the time gap between first two peaks of WTC of mode 1.

5. If the mode 0(W_{mm0}) is nonzero, then the fault is identified as a grounded fault and the calculate the time gap t_{dm} between the first peaks of mode 0 and mode 1.
If tdm > tl/2, then

$$t_d^l = (2l/v) - t_x \quad (26)$$

$$d = (v \times t_d^l)/2 \quad (27)$$

where $tl/2$ is the travel time delay between mode 0 and mode 1 if the fault is located at the center of the line, x is the distance to the fault, v is the wave velocity of mode 1, and tx is the time delay between two consecutive peaks of the WTC mode 1.

Else, the fault distance using (Fault is in second half section of line).

$$d = (v \times t_d)/2 \quad (28)$$

4. Power System Model

The SimPowerSystem which is an extension to the simulink of MATLAB software was used to simulate the double end fed power system. The 100 km, 400 kV Double circuit transmission line was modeled using distributed parameter model as shown in Fig.4

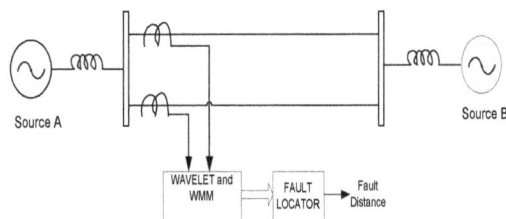

Fig. 4. *Power System model.*

The transmission line parameters are as follows:
Positive Sequence Resistance, R_1 : 0.0275 Ω / km
Zero Sequence Resistance, R_0 : 0. 275 Ω/km
Zero Sequence Mutual Resistance, R_{0m} : 0.21 Ω/km
Positive Sequence Inductance, L_1 : 0.00102 H/km
Zero Sequence Inductance, L_0 : 0.003268 H/km
Zero Sequence Mutual Inductance, L_{0m} : 0.0020 H/km
Positive Sequence Capacitance, C_1 : 13 $e^{-0.009}$ F/km
Zero Sequence Capacitance, C_0 : 8.5 $e^{-0.009}$ F/km
Zero Sequence mutual Capacitance, C_{om} : -5e^{-009} F/km

5. Simulation Results

The simulation is carried out for these algorithms by varying various fault parameters like fault type, fault location etc. The accuracy of fault location of these three algorithms are compared and shown in Table.3.

Table 3. Results of two algorithms.

Fault Type	Actual fault location	Impedance based Method		Travelling wave based Method	
		d_{esti}	Error%	d_{esti}	Error%
A1G	10	9.6896	0.3104%	9.911	0.089%
B1G	20	19.7514	0.2486%	19.944	0.056%
C2G	30	29.3602	0.6398%	29.96	0.040%
A2B2	40	39.938	0.062%	39.997	0.003%
B2C2	50	49.3868	0.6132%	49.944	0.056%
C1A1	60	59.248	0.752%	59.039	0.961%
A2B2G	70	69.9129	0.0871%	69.901	0.099%
B1C1G	80	78.9692	1.0308%	79.879	0.121%
C1A1G	85	84.9617	0.0383%	84.963	0.037%
A1B1C1,A1B1C1G	90	89.94	0.06%	89.928	0.072%

The fault location error is calculated as

$$Error(\%) = \frac{|Calculated\ Fault\ Location - Actual\ Fault\ Location|}{Total\ Line\ Length} * 100 \qquad (29)$$

6. Conclusion

The use of double circuit lines are becoming common when constructing and updating newlines. In this paper, two fundamental algorithms: Impedance based, Travelling wave(TW) algorithms for 100km, 400KV Double circuit transmission lines are implemented using Matlab Simulink and programing. The performance of these two algorithms are reviewed by varying various parameters like fault type, fault location etc. The simulation results show that all ten types of faults are correctly located and travelling wave (TW) based algorithm locates faults with accuracy less than 0.5% and Impedance based algorithms locates faults with accuracy less than 2%.

Nomenclature

D: Estimated distance to the fault (units: p.u)

V_{AA_P}: Fault loop voltage composed according to fault, Type for double circuit line

I_{AA_P}:Fault loop current composed according to fault, type for double Circuit line.

I_F :Total fault current

a_{F0}, a_{F1}, a_{F2} :Weighting coefficients (complex numbers), dependent on fault type and the assumed priority for using particular symmetrical components

I_{F0}, I_{F1}, I_{F2}: Zero, positive and negative sequence components of total fault current, which are to be calculated or estimated.

Z_{1A}, Z_{1B}: Positive sequence source impedances at terminals A and B respectively.

Z_{2A} , Z_{2B} : Negative sequence source impedances at terminals A and B respectively .

E_{1A}, E_{1B}: Positive sequence source voltages at terminals A and B respectively.

Z_{1LA}: Positive sequence impedance of the faulted line AA

Z_{0LA} :Zero sequence impedance the of faulted line AA

I_{AA1}, I_{AA2}, I_{AA0} : Total sequence currents from faulted line (AA)

V_{AA1}, V_{AA2}, V_{AA0} : Total sequence voltages from faulted line (AA)

I_{AB1}, I_{AB2}, I_{AB0} : Total sequence currents from healthy line (AB)

References

[1]   L. Eriksson, M. M. Saha, and G. D. Rockefeller, ``An accurate fault locator with compensation for apparent reactance in the fault resistance resulting from remote-end infeed," IEEE Trans. Power App. Syst., vol. PAS-104,no. 2, pp. 423_436, Feb. 1985.

[2]   Izykowski J, Rosolowski E, Saha MM (2004) "Locating faults in parallel transmission lines under availability of complete measurements at one end". IEE Proc – Gener Transm Distrib 151(2):268–273.

[3]   M. S. Sachdev and R. Agarwal, "A technique for estimating transmission line fault locations from digital impedance relay measurements," IEEE Trans. Power Del., vol. 3, no. 1, pp. 121–129, Jan. 1988.

[4]   Saha MM, Wikstrom K, Izykowski J, Rosolowski E (2001) "New fault location algorithm for parallel lines". In: Proc of 7th Int Conf on Developments in Power System Protection – DPSP, IEE CP476 pp 407–410.

[5]   Wiszniewski A. "Accurate fault impedance locating algorithm". IEE Proc C1983:130(6):311-5.

[6]   Izykowski J, Kawecki R, Rosolowski E (2002) "Accurate location of faults in parallel transmission lines under availability of measurements from one circuit only".In: Proc of Power Systems Computation Conference –PSCC'02 (CD ROM), Sevilla,paper 6.

[7] Magnago FH, Abur A (1998) "Fault location using wavelets". IEEE Trans on Power Deliv 13(4):1475–1480.

[8] Abur A, Magnago FH (2000) Use of time delays between modal components in wavelet based fault location. Int J Electr Power and Energy Syst 22(6):397–403.

[9] V.S.Kale, S.R.Bhide, P.P.Bedekar, Faulted Phase Selection Based on Wavelet Analysis of Traveling Waves", International Journal of Computer and Electrical Engineering, Vol. 3, No. 3, June 2011.

[10] AnkammaRao J, BizuayehuBogale ."Double Circuit Transmission Line Fault Distance Location using Wavelet Transform and WMM Technique ",International Journal of Science and Research (IJSR), Vol.4, Issue.1,January-2015.

[11] Anamika Jain, Kale VS, Thoke AS. Application of artificial neural networks to transmission line faulty phase selection and fault distance location. In: IASTED, Chiang Mai, Thailand; 29–31 March 2006. p. 262–7.

[12] MATLAB user's guide, The Math Works Inc., Natick, MA.

9

Unrepeatered OTDM Data Transmission over Long Legacy Fiber Span Using Unidirectional Backward Raman Amplification

Mousaab M. Nahas

Electrical and Computer Engineering Department, Faculty of Engineering, University of Jeddah, Jeddah, Saudi Arabia

Email address:
mnahas1@uj.edu.sa

Abstract: This paper presents experimental results for transmitting 40 Gb/s OTDM signal over in-line long fiber span using unidirectional backward Raman amplification. The investigation uses legacy dispersion-managed SMF-DCF configuration where remote Erbium amplification is used to compensate for the DCF spans losses. It is practically shown that the system performance improves significantly with more Raman pump power if we use an appropriate signal wavelength, Raman pump power and Erbium gain. As a result, successful unrepeatered transmission over 206 km SMF is achieved using 1545 nm signal wavelength, 1.58 W Raman power and unsaturated EDFA gains into the DCF spans. We believe that the results of such investigation can be useful for enhancing systems that still use legacy cables without the need for substantial alteration.

Keywords: Fiber-Optic Communications, OTDM, Raman Amplification

1. Introduction

It is well known in the optical telecommunications arena that Raman amplification is referred to the Stimulated Raman Scattering (SRS) phenomenon where signal amplification is attained via nonlinear power transfer from an intense pump beam propagating simultaneously through the fiber [1]. It is also traditionally recognized that Raman amplification can be applied in either forward or backward direction. In forward Raman pumping, the energy is transferred from the pump beam to the data signal as the two beams co-propagate inside the fiber. In backward Raman, the pump and data signal beams counter-propagate in the fiber, and this type is commonly used in practice due to better amplification results [2]. In reality, most of the long haul optical transmission systems ($\geq$ 200 km) use Raman amplification since the early part of the 20th century [3]. Moreover, plenty of researches have already demonstrated successful transmission using unrepeatered Raman amplification. Many of these researches use multiple 10 Gb/s WDM signals [4-7], while many others use higher bit rates through OTDM as we demonstrate in this research. However, some of these OTDM projects use all-Raman amplification that is split between the SMF and DCF spans, where no EDFA amplification is applied [8]. In other projects, unconventional large effective area fiber (LEAF) is used in which the nonlinear penalty is reduced including SRS, thus high power of Raman pump is required to increase amplification [9-11]. Other unconventional fibers are also used like DSF and NZ-DSF [12-13] that we are not interested in hereby. In later project [14], conventional SMF is used but the authors apply bidirectional Raman pumping scheme. Recently, some projects use conventional SMF in all-distributed Raman configuration [15-16] where multiple Raman modules are distributed around short or medium SMF spans so that transmission over unrepeatered long spans is not demonstrated.

In this paper, we investigate the application of unidirectional backward Raman amplification over unrepeatered long conventional SMF using 40 Gb/s OTDM signal with 2^{31}-1 data length. This signal is being encoded by a simple traditional coding scheme, which is RZ-IMDD, such that the complexity of the transmitter and receiver is minimized. This effectively opens the door for upgrading already installed systems or legacy parts of a network without the need for adding any complexity. However, there have been results of unidirectional Raman over long conventional fibers [17], but since the experiments are too old they use 2^7-1 word length which is much shorter that what we have in this project.

2. Experimental Setup

Fig. 1 shows the experimental setup for our investigation. An RF generator produces 10 GHz electrical signal that is used to drive both a laser source and a pulse pattern generator (PPG). The PPG produces a PRBS data signal with 2^{31}-1 length which is used to modulate the laser signal at a LiNbO$_3$ intensity modulator. The output of this process is basically a 10 Gb/s RZ-IM data signal. This resulting bit rate is experimentally increased to 40 Gb/s through Mach Zehnder optical time division multiplexing (MZ-OTDM) that is shown in Fig. 2, where the output pulses produced by the laser mentioned above are sufficiently narrow. In MZ-OTDM, a 10 Gb/s signal whose pulse time window is 100 ps is split into two channels; one is delayed by 50 ps and then couples back with the other channel. The resultant is 20 Gb/s signal. If this stage is repeated with 25 ps delay, the output signal is then 40 Gb/s. In practice, such delays are insufficient for adequate mixing of bits as they would result in sequential bit repetition within the combined 40 Gb/s random data. Therefore, we use additional 100 ps (one time widow) delay to avoid this effect [18]. Polarization controller is used in one Mach Zehnder arm to equalize the polarization of the two arms. The MZ-OTDM input 10 Gb/s and output 40 Gb/s signals are shown in Fig. 3.

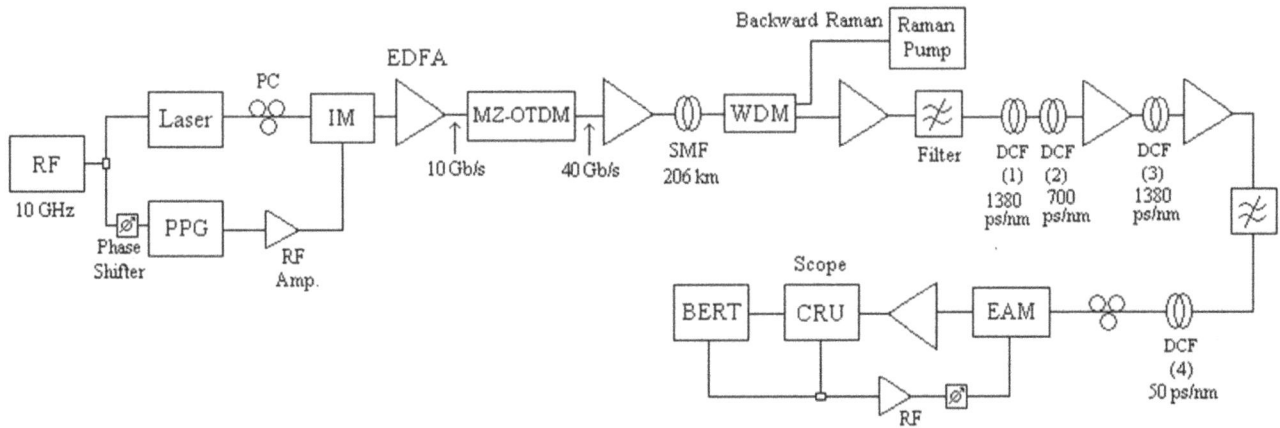

Figure 1. Experimental setup.

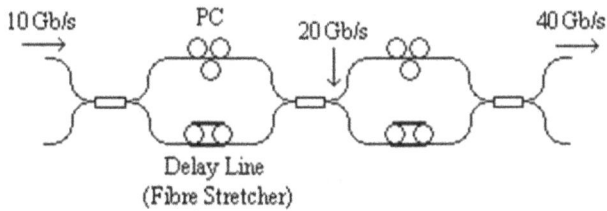

Figure 2. MZ-OTDM.

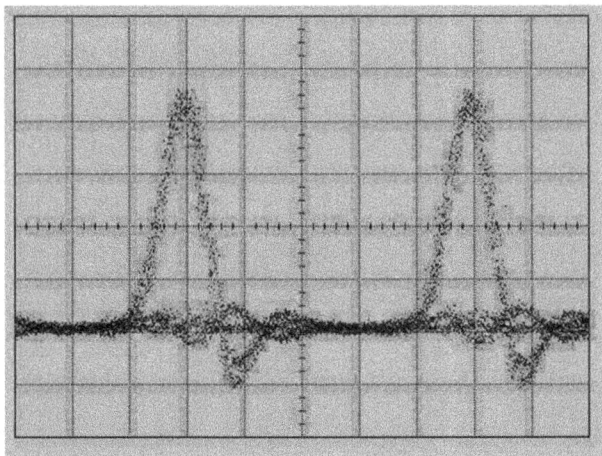

(a)

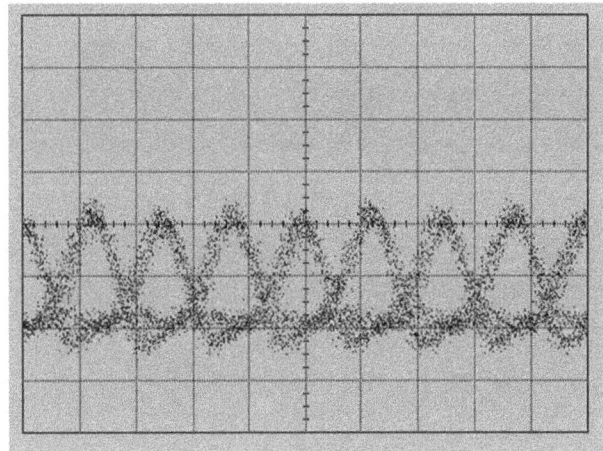

(b)

Figure 3. (a) MZ-OTDM input 10 Gb/s signal; (b) MZ-OTDM output 40 Gb/s signal.

The resulting 40 Gb/s signal is boosted by an EDFA and then transmitted over a 206 km SMF span. The average fiber attenuation is measured to be 0.2 dB/km at 1550 nm, while the total dispersion is measured to be 3510 ps/nm. This dispersion is compensated by the DCF modules shown in the setup. The loss of the SMF span is to be compensated by backward Raman as seen, while the cascaded EDFAs are used to compensate for losses in the DCF spans and in other components such as filters. The filters are used to eliminate the accumulative ASE noise along the system.

At the receiver, the individual 10 Gb/s signal must be extracted from the 40 Gb/s OTDM bit stream for measurements. Since each pulse occupies a 25 ps window in the 40 Gb/s signal, it is required to create a 25 ps switching window every 100 ps to extract a 10 Gb/s channel. To achieve this, an electro-absorption modulator (EAM) is used to absorb the unwanted three channels and leave only one channel in the time window. The EAM is initially driven by a 10 GHz electrical signal generated by clock recovery unit (CRU) to enable modulation of the first arrived bits, and then it is driven through feedback clock recovery for the next coming bits. The phase of the 10 GHz signal can be adjusted using a phase shifter which enables sliding the switching window in the time domain, giving the ability to select which of the four OTDM channels to be detected. The output 10 Gb/s signal is isolated and pre-amplified before being detected and analyzed.

3. Results and Discussion

Fig. 4 shows the back-to-back demultiplexed RZ data signal obtained from the above setup (without transmission) using -1 dBm EAM input power. It is obvious that the receiver demonstrated above can successfully extract a single 10 Gb/s channel out of the entire OTDM signal, where the original pluses are recovered properly and no errors are counted at the BERT.

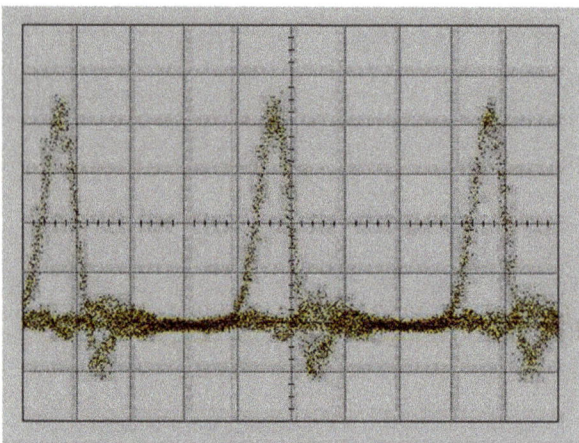

Figure 4. Back-to-back 10 Gb/s signal.

In transmission over 206 km, it is initially observed that the signal performance is varying with the wavelength, which is common with most laser sources that produce narrow pulses due to chirp variation. Therefore, it is significant to identify the best operating wavelength for this system before applying Raman amplification. Fig. 5 shows the signal BER as a function of wavelength using high launched power (17 dBm) and EDFA amplification only. As a result, the best performance is found to be around 1545 nm, thus this is the operating wavelength for our data signal from now on. Obviously, the BER is not aimed to be optimized at this stage as it is used here just for comparison purpose.

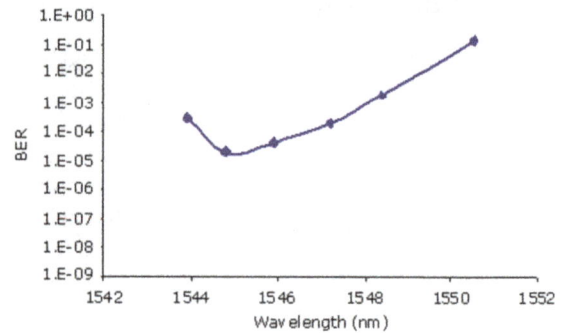

Figure 5. BER vs operating wavelength.

To apply Raman amplification, the Erbium gains are set to mainly compensate for the DCF losses which are ~0.5 dB/km in all DCF spans. This Erbium gain should also cover the filters losses as stated before. However, the signal launched power used at this stage is chosen to be 13 dBm, and this is based on the existing EDFAs saturation levels. This would imply that all Erbium amplifiers are allowed to work near saturation. By applying backward Raman pump signal at 1455 nm to the 206 km SMF span, the received signal counts considerable number of errors and is significantly distorted. This can be understood from Fig.6 (a) that shows the output spectra from the first amplifier, i.e. after the SMF span, using different pump powers in the range 40-100% according to the existing pump module. The corresponding values of the percentages mentioned above (and seen in the figure) are 0.06, 0.353, 0.624, 0.894, 1.13, 1.37 and 1.58 W, respectively. In (b), the OSNR measurements at the same point are presented versus Raman pump for higher and lower wavelengths. It is clear from both (a) and (b) that, the lower the Raman pump, the higher the ASE peak at 1530 nm, while the higher the pump, the higher the ASE noise around 1555 nm. However, the received BER against Raman pump is shown in Fig. 7. It is noticeable that the Raman pump increases the signal BER so far. Fig. 8 shows the eye closure due to increased Raman pump, where the worst eye is observed at 100% pump power. Practically, this degradation is mainly caused due to the interaction between Raman signal and the ASE noise signal caused by the lumped Erbium amplification, where this interaction is wavelength dependent.

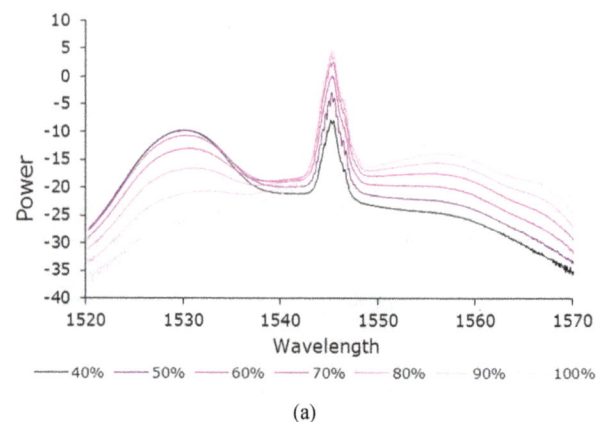

(a)

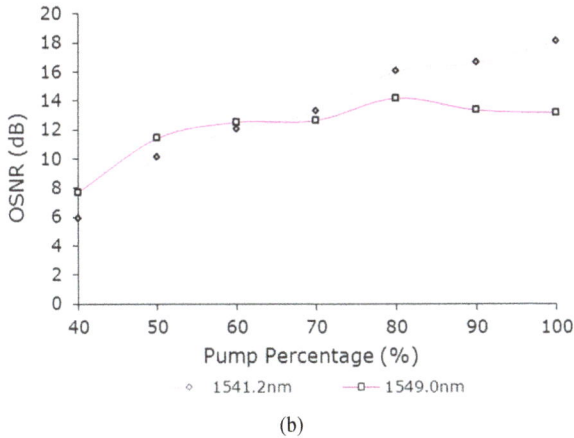

(b)

Figure 6. (a) SMF output spectra for different Raman pumps. (b) SMF output OSNR vs Raman pump.

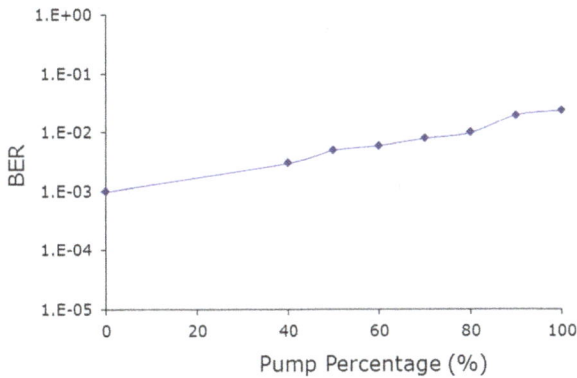

Figure 7. BER vs Raman pump.

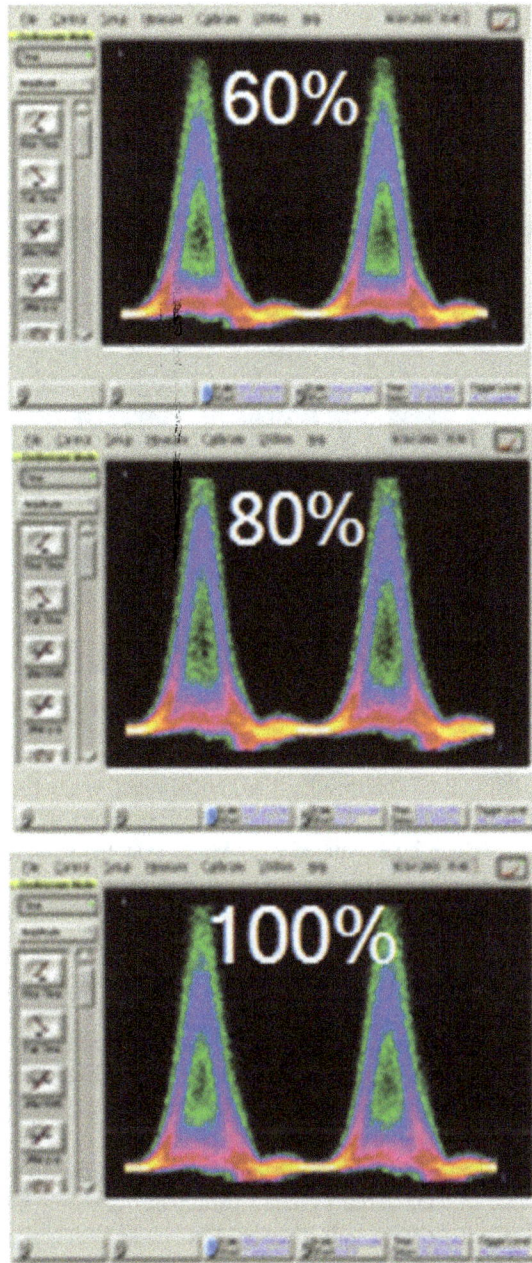

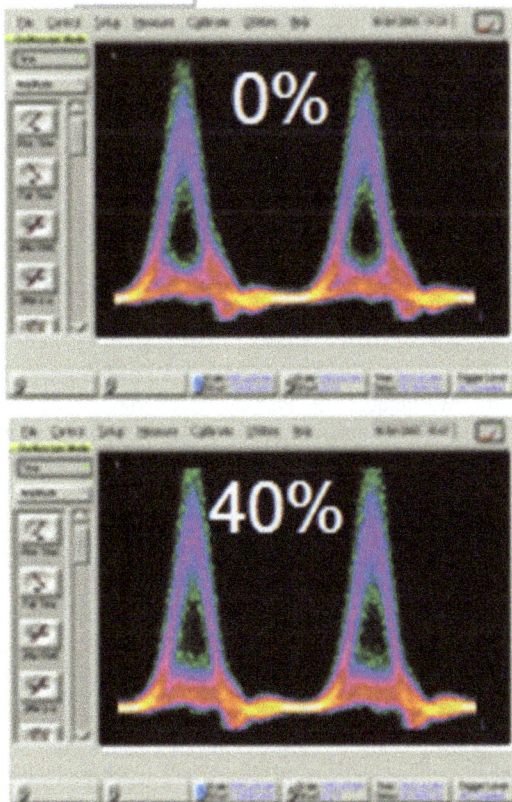

Figure 8. Received eye diagrams with different Raman pumps.

To solve the above problem, the lumped EDFA gains are empirically reduced such that the accumulated ASE noise is improved and the entire spectrum is nearly flattened. By doing this, all EDFA gains in our system (excluding the booster and pre-amplifier) are dropped by about 2 dB and no longer operate at saturation. This essential adjustment allows using higher Raman powers comfortably where it is noticed that the signal performance is now improved with Raman rather than degraded as before. In this case, the Raman amplification effectively compensates for the drops in the Erbium gain thus the data signal power throughout the system is balanced on average. As a result, successful transmission of the 40 Gb/s data is achieved where the best signal is obtained by using the highest Raman percentage (100%) which is corresponding to 1.58 W. The eye diagram of the received 10 Gb/s signal is presented in Fig. 9 where it measures ~10^{-9} BER. We believe that this result is satisfactory for backward Raman amplification, and there is no need to afford higher Raman power as long as the intended OTDM data signal is successfully transmitted and received via our system.

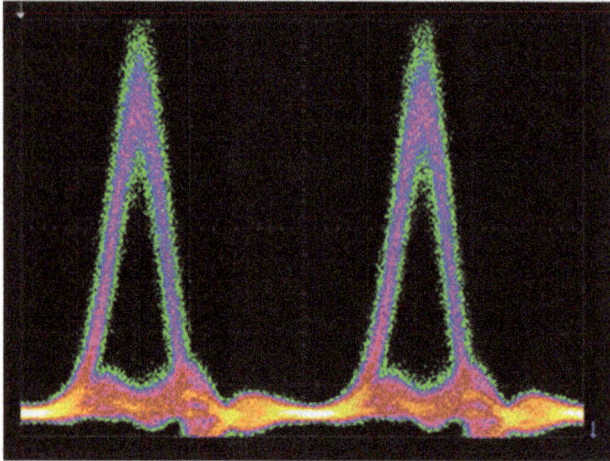

Figure 9. Received signal obtained by 1.58 W Raman pump and reduced Erbium gain.

In fact, this kind of results can encourage upgrading systems that use traditional fiber types/configurations where the above study can be expanded over larger systems such as long and ultra-long haul transmission systems, whether having real or recirculating-loop-based setup [19-20]. These large systems are composed of multiple unrepeatered fiber spans that are typically << 200 km and are conventionally amplified by EDFAs. To make use of the above results without major alteration on the basic configuration, it is possible to insert multiple Raman pumps where each one can serve single unrepeatered 200 km SMF span. In this case, each section of the entire system can be simulated by our single in-line long fiber span presented in this paper. The main purpose in such large systems would be reducing the required lumped Erbium gain (hence ASE noise) such that the total number of EDFAs can be reduced significantly while the overall performance is improved. Additional work can also be done where the investigation can be extended to be applied with complex modulation formats such as QPSK, DPSK, etc.

4. Conclusions

In this paper, we demonstrate unrepeatered 40 Gb/s OTDM data transmission over 206 km conventional SMF span using backward Raman amplification. The experiment uses dispersion-managed SMF-DCF configuration so that Raman amplification is applied to compensate for the SMF span loss while remote EDFAs are used for the DCF spans. The system is optimized with respect to the operating wavelength, Erbium gains/spectra and Raman pump power. As a result, successful transmission of the intended OTDM signal is attained using 1545 nm signal wavelength, 1.58 W

Raman power and unsaturated gains in the cascaded EDFAs.

Acknowledgments

The author thanks Aston Photonics Research Group and Azea Networks for their assistance in this experiment.

References

[1] R. W. Boyd, Nonlinear Optics, San Diego, 2008.

[2] G. P. Agrawal, Nonlinear Fiber Optics, San Diego, 2007.

[3] M. Islam, "Raman amplifiers for telecommunications", IEEE Journal of Selected Topics In Quantum Electronics, vol. 8, no. 3, pp. 548-559, 2002.

[4] Z. Xu, J. Seoane, A. Siahlo, L. Oxenlewe, A. Clausen, and P. Jeppesen, "Experimental characterization of dispersion maps with Raman gain in 160 Gb/s transmission systems", Conference on Lasers and Electro-Optics (CLEO), pp. 2, 2004.

[5] M. Haris, J. Yu, and G. K. Chang, "Repeaterless transmission of 10 Gbit/s MD-RZ signal over 300 km SMF-28 by using Raman amplification", IEEE Lasers and Electro-Optics Society (LEOS), WE1, pp. 479-480, 2005.

[6] H. Maeda, G. Funatsu, and A. Naka, "Ultra-long-span 500 km 16 × 10 Gbit/s WDM unrepeatered transmission using RZ-DPSK format", Electronics Letters, vol. 41, no. 1, pp. 34-35, 2005.

[7] R. Jee, and S. Chandra, "Single-span transmission of WDM RZ-DPSK signal over 310 km standard SMF without using FEC and remote-pumping", International Conference on Advances in Computing, Communications and Informatics (ICACCI), pp. 172-177, 2015.

[8] Z. Huang, A. Gray, Y. Lee, I. Khrushchev, and I. Bennion, "All-Raman amplified transmission at 40 Gbit/s in standard single mode fiber", Conference on Lasers and Electro-Optics (CLEO), pp. 531, 2003.

[9] G. Charlet, M. Salsi, P. Tran, M. Bertolini, H. Mardoyan, J. Renaudier, O. Bertran-Pardo, and S. Bigo, "72 × 100 Gb/s transmission over transoceanic distance, using large effective area fiber, hybrid Raman-erbium amplification and coherent detection", Optical Fiber Communication Conference, p. PDPB6, 2009.

[10] D. Chang, W. Pelouch, P. Perrier, H. Fevrier, S. Ten, C. Towery, and S. Makovejs, "150 × 120 Gb/s unrepeatered transmission over 409.6 km of large effective area fiber with commercial Raman DWDM system", Optics Express, vol. 22, no. 25, pp. 31057-31062, 2014.

[11] H. Bissessur, C. Bastide, S. Dubost, and S. Etienne, "80 × 200 Gb/s 16-QAM unrepeatered transmission over 321 km with third order Raman amplification", Optical Fiber Communications Conference and Exhibition (OFC), pp. 1-3, 2015.

[12] H. Masuda, H. Kawakami, S. Kuwahara, A. Hirano, K. Sato, and Y. Miyamoto, "1.28 Tbit/s (32 × 43 Gbit/s) field trial over 528 km (6 × 88 km) DSF using L-band remotely-pumped EDF/distributed Raman hybrid inline amplifiers", Electronics Letters, vol. 39, no. 23, pp. 1668-1670, 2003.

[13] D. Rafique, T. Rahman, A. Napoli, R. Palmer, J. Slovak, E. Man, S. Fedderwitz, M. Kuschnerov, U. Feiste, B. Spinnler, B. Sommerkorn-Krombholz, and M. Bohn, "9.6 Tb/s CP-QPSK transmission over 6500 km of NZ-DSF with commercial hybrid amplifiers", IEEE Photonics Technology Letters, vol. 27 , no. 18, pp. 1911-1914, 2015.

[14] P. Rosa, N. Murray, R. Bhamber, J. Ania-Castañón, and P. Harper, "Unrepeatered 8 × 40 Gb/s transmission over 320 km SMF-28 using ultra-long Raman fiber laser based amplification", European Conference on Optical Communications (ECOC), pp. 1-3, 2012.

[15] D. Chang, S. Burtsev, W. Pelouch, E. Zak, H. Pedro, W. Szeto, H. Fevrier, T. Xia, and G. Wellbrock, "150 × 120 Gb/s field trial over 1,504 km using all-distributed Raman amplification", Optical Fiber Communications Conference and Exhibition (OFC), pp. 1-3, 2014.

[16] S. Burtsev, H. Perdo, D. Chang, W. Pelouch, H. Fevrier, S. Ten, S. Makovejs, and C. Towery, "150 × 120 Gb/s transmission over 3,780 km of G.652 fiber using all-distributed Raman amplification", Optical Fiber Communications Conference and Exhibition (OFC), pp. 1-3, 2015.

[17] M. Gunkel, F. Kiippers, J. Berger, U. Feiste, R. Ludwig, C. Schubert, C. Schmidt, and H. Weber, "Unrepeatered 40 Gbit/s RZ transmission over 252 km SMF using Raman amplification", Electronics Letters, vol. 37, no. 10, pp. 646-648, 2001.

[18] M. Nahas, "Investigation of different techniques to upgrade legacy WDM communication systems", PhD Thesis, Aston University, 2006.

[19] M. Nahas, and K. Blow, "Monitoring long distance WDM communication lines using a high-loss loopback supervisory system", Optics Communications, vol. 285, no. 10, pp. 2620-2626, 2012.

[20] M. Nahas, and K. Blow, "Investigation of the effect of an increased supervisory signal power in a high-loss loopback monitoring system", IET Communications, vol. 8, no. 6, pp. 800-804, 2014.

10

Harmonic Aggregation Techniques

Mohammad Mahdi Share Pasand

Department of Electrical and Electronics Engineering, Standard Research Institute - SRI, Alborz, Iran

Email address:

sharepasand@standard.ac.ir

Abstract: Different harmonic aggregation techniques for assessment of current harmonic levels at points of measurement are investigated and compared to each other. The interaction effect in aggregation of harmonic currents is investigated. Several approaches including IEC 61000-3-6 recommended method are utilized and compared.

Keywords: Electromagnetic Compatibility, Harmonic Distortion, Harmonic Aggregation, Power Quality, Current Harmonics, Interaction Effect

1. Introduction

The importance of harmonic emission as a measurefor electromagnetic compatibility is well known. [1] Harmonic pollution, a conductive form of electromagnetic emission, may cause distribution losses, damage to several components such as transformers, power switches and electric motors, accidental operation of remotely controlled switches and breakers(false tripping), excess heating of cablings, metering errors in power distribution, penalties on monthly bill units, generator failures etc. Acceptable levels of harmonic pollution in IEC and IEEE international standards are provided as static indices measured at PCCs [1][3, 4]. Equipment standards determine level of current harmonic emissions from the equipment while distribution standards determine the acceptable level of voltage harmonics at the bus. In order to check whether a set of loads require harmonic filters to fulfill distribution system harmonic requirements, one has to estimate the aggregated sum of current harmonics and the voltage harmonic pollution caused by that current. Approximate methods are recommended in IEC 61000 series for evaluating the total harmonic current based on single component current [3]. The main challenge in that approximation of aggregated harmonic current is that phase behaviors of equipments are significantly different. The phase of a harmonic current depends on the equipment type (e.g. rectifier, inverter, etc.), switching/ triggering behavior (i.e. time indices at which the equipment switches on and off either by user or automatic switching mechanisms) and its parameters (e.g. threshold voltages). A significantly contributing factor is time. It is known that current pollution increases at the early night in urban and industrial areas as a result of TVs and lighting loads.

It is desirable to have an estimation of harmonic level before actually adding a new load, a harmonic filter or restructuring the network. Also aggregation is important as it makes the designer decide where to install new filters and how to distribute reactive load between different capacitive banks.

Methods for passive filter design to compensate for harmonics are surveyed in [18]. Active filters are also of growing interest as in [19]; however active filters are expensive and complicated to repair.

Street lamps and their aggregated harmonic current are studied in Aggregation of current harmonics in wind farms which is due to the time-varying behavior of wind causing varying currents and inverter voltage harmonics is studied in [20,21].

Two major phenomena may impact aggregation of harmonic currents. First is the magnitude aggregation error which is caused by the phase differences between several harmonic components. Another complication is the interaction effect. Interaction is the effect of adding new loads on the harmonic components of previously installed loads. This phenomenon is caused because of line impedance which results in harmonic voltages due to harmonic currents. As a result, actual harmonic pollution of an individual load may change when installed in parallel with other loads.

---

1 Point of Common Coupling

2. Techniques for Computing Aggregated Current Harmonic

If one sums up harmonic current magnitudes of different orders, a conservative approximation is resulted. [3] This approximated value is too large compared to actual aggregated current measured from the network. [4] In other words, actual harmonics cancels each other due to phase differences, resulting in a very smaller aggregated harmonic level than arithmetic sum. (See [6] and references therein.)

Another method is to vector-sum the harmonic currents which lead to an exact value for the aggregated value. However the actual values of harmonic phases are unknown and due to changes as a result of switching. [6] In addition, measurement of phases requires more accurate and more expensive measurement tools. An approximate method is to use statistical methods as [6] on the basis of a probability density function for harmonic phases.

The third method is to establish formulae based on empirical data. IEC 61000-3-6 recommends to add up harmonics based on table 2 and equation 2, which is arithmetic summation for harmonic orders below 5 and a root of powered sums for higher orders.(2)

$$I_{hTOT} = \sum_{j=1}^{n} I_{hj} \qquad (1)$$

In which the following notation is used:

I_{hTOT}: Aggregated harmonic current (order h) for all of components

I_{hj}: Harmonic current (order h) for j^{th} component

$$I_{hTOT} = \sqrt[\alpha]{\sum_{j=1}^{n} I_{hj}^{\alpha}} \qquad (2)$$

In which:

Table 1. *IEC 61000-3-6 method for harmonic aggregation [5].*

α	Harmonic Order
1	h < 5
1.4	5 < h < 10
2	h > 10

Other aggregation techniques include summation of vector components presuming random phase differences driven by probability densities of different types. These methods require experimental data to estimate phase angle difference PDFs, from the experimental histograms. Statistical approaches are more logical to be applied especially when an appropriate density function is available. Actual PDFs can be modeled only through complicated analytical functions. [6] In addition, because of fixed limits recommended for THD levels [4, 9], it is more practical to derive conservative deterministic models based on statistical analysis. One approach is to define a percentile in the statistical model, not to be violated by the system. For instance [10] sets 5% limit for the THD (i.e. The THD level may not exceed 95% of the recommended limit) which is recommended by IEC as well. [3]A statistical analysis of harmonic phases is performed in [7], in which probability density functions are estimated on the basis of

simplifying assumptions. The PDF of harmonic phases is considered to be uniform in [7], [6], which are too optimistic. In fact, the statistical analysis of harmonics is valid only if experimental data for individual appliance harmonics is used to model phase behaviors. In this study, experimental data are drawn to model the phase behavior, examine the current aggregation method, evaluate the statistical assumptions of [7] and to derive statistical indices for THD levels which is optimistically considered to be in steady state in standards IEEE 519, IEC 61000.

The following figures depict the approximations for aggregated harmonics. [4] RSS is suitable when phase and magnitude differences are significant. When phases are significantly different but magnitudes are the same, RSSis not appropriate. The proposed method in [4] is an appropriate choice when estimations of statistical measures of phase and magnitude are present. For identical loads with similar magnitude and phases, only linear summation is satisfactory. The practical behavior of phase and magnitudes remains as an important question which helps choosing appropriate approximation methods.

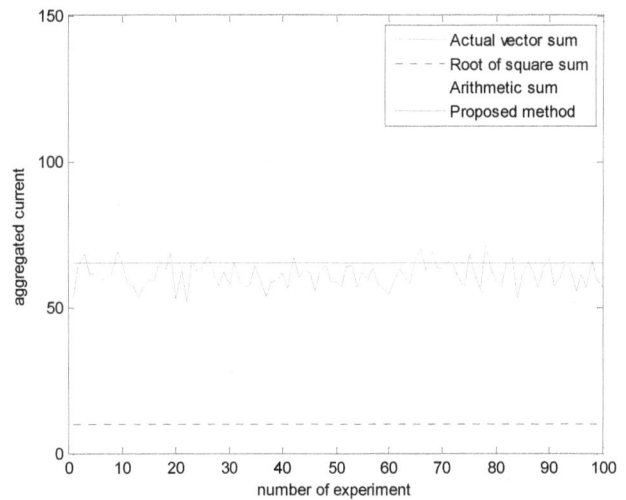

Figure 1. *Comparison of different estimates: Identical loads.*

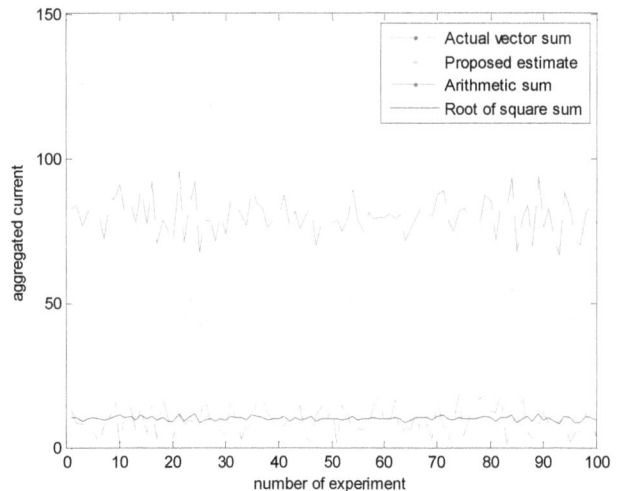

Figure 2. *Comparison of different estimates: Different loads.*

It should be noted that in most applications a few number of harmonic orders are of concern. Namely 3rd, 5th and 7th harmonics are the most important ones when current harmonic is considered. Even harmonics are usually insignificant as they usually cancel each other in the aggregation [15]. This holds also for 3rd order harmonics in 3phase symmetric loads. Therefore only 5th and 3rd harmonics are considered here to represent a measure for all harmonic orders. Existing approaches only consider aggregations of a single harmonic order and provide a total measure (e.g. THD) for all harmonics. As a result, there is no need to consider all harmonic orders in the analysis.

3. Interaction Effects on Individual Harmonic Currents

Interaction effect is caused because of the voltage drop between parallel loads due to resistance/impedance existing in wirings. It may also model the change in harmonic distortion due to current reduction in each load being paralleled with other loads. [22] Used simulation studies of identical and semi-identical loads of single phase rectifiers to show the effect of interaction. It is shown that interaction significantly disturbs aggregation approximations when line impedance is noticeable and harmonic currents of high orders exist in lines. This is similar to capacitive bank situation in which reactive power is preferred to be injected in lines from different places in order to reduce the distance in which reactive current is flowing.

[22] Recommends a computational method to compensate for interaction effect and improve aggregation formulae. Compared to phase effect on the aggregation, interaction is of much less effect and could be neglected unless for situations in which line impedance and THD levels are extremely high.

4. Conclusion

Different aggregation techniques are reviewed and the interaction effect which may adversely impact the accuracy of harmonic aggregation computations is studied. Also it is shown that phase behavior of loads has a decisive effect on the aggregated current and practical data should be gathered to extract realistic distributions for phase behavior. Effect of phase is much more severe than interaction and it could be neglected when line impedance is low.

References

[1] Allen, George W, "Design of Power-Line Monitoring Equipment", IEEE Transactions on Power Apparatus and Systems, Vol. PAS-90, Issue: 6, Nov. 1971.

[2] S Chattopadhyay, M Mitra, S Sengupta, "Electric power quality", Springer, 2011.

[3] G. D. Castro, A., M. Bollen and A. Moreno-Monuz, "Street Lamps Aggregation Analysis Through IEC 61000-3-6 Approach", International Conference on Electricity Distribution, Stockholm, June 2013.

[4] Pasand, Mohammad Mahdy Share, and Zahra Rahmatian. "Harmonic aggregation techniques for power quality assessment a standard framework." International Journal of Engineering & Technology 3.3 (2014): 365-371.

[5] Pasand, MM Share. "Harmonic Aggregation Techniques for Power Quality Assesment A review of different methods." International Journal of Engineering Science and Technology 6.7 (2014): 423.

[6] IEEE std. 519, "Recommended Practices and Requirements for Harmonic Control in Electrical Power Systems", 1992.

[7] IEC 61000, "Electromagnetic compatibility (EMC) – Part 3-6: Limits – Assessment of emission limits for the connection of distorting installations to MV, HV and EHV power systems", Basic EMC publication, ed.2, 2008. (IEC/TR 61000-3-6).

[8] Y. Baghzouz, et al., "Time-Varying Harmonics: Part II—Harmonic Summation and Propagation", IEEE Transactions On Power Systems, Vol. 17, No. 1, 2002.

[9] Y. Baghzouz, Task Force on Probabilistic Aspects of Harmonics: "Time-varying harmonics: Part I—Characterizing measured data," IEEE Trans. Power Delivery, vol. 13, pp. 938–944, 1998.

[10] A. Cavallini, R. Langella, et al., "Gaussian modeling of harmonic vectors in power systems," in IEEE 8th Int. Conf. Harmonics and Quality of Power, Athens, Greece, Oct. 14–16, 1998, pp. 1010–1017.

[11] IEC 61000 - 3 - 2: 2009, Electromagnetic Compatibility (EMC) - Part 3-2: Limits-Limits for harmonic current emissions (Equipment input current ≤ 16A per phase) (IEC 61000-3-2).

[12] CIGRE WG 36-05, "Assessing voltage quality with relation to harmonics, flicker and unbalance", France, CIGRE Rep. 36-203, 34th Session, 1992.

[13] Medeiros, F.; Brasil, D. C.; Ribeiro, P. F.; Marques, C. A. G.; Duque, C. A. "A new approach for harmonic summation using the methodology of IEC 61400-21", 14th International Conference on Harmonics and Quality of Power (ICHQP), 2010.

[14] Hsiao, Y. "Design of Filters for Reducing Harmonic Distortion and Correcting Power Factor in Industrial Distribution Systems." Tamkang Journal of Science and Engineering, 4(3), 193-199, 2001.

[15] Wakileh, G. J. "Power systems harmonics: fundamentals, analysis, and filter design", Berlin: Springer, 2001.

[16] Persson, Jonas,"Comparing Harmonics Mitigation Techniques." (2014).

[17] Buccella, Concettina, et al. "Harmonic mitigation technique for multilevel inverters in power systems." Power Electronics, Electrical Drives, Automation and Motion (SPEEDAM), 2014 International Symposium on. IEEE, 2014.

[18] Aye, Thet Mon, and Soewin Naing. "Analysis of Harmonic Reduction by Using Passive Harmonic Filters." (2014).

[19] Sharma, Vandana, and Anurag Singh Tomer. "Comparative Analysis on Control Methods of Shunt Active Power Filter for Harmonics Mitigation." International Journal of Science and Research 3 (2014).

[20] Yang, Kai, Math HJ Bollen, and E. O. Anders Larsson. "Aggregation and amplification of wind-turbine harmonic emission in a wind park." Power Delivery, IEEE Transactions on 30.2 (2015): 791-799.

[21] Yang, Kai, Math HJ Bollen, and E. O. Larsson. "Wind power harmonic aggregation of multiple turbines in power bins."

Harmonics and Quality of Power (ICHQP), 2014 IEEE 16th International Conference on. IEEE, 2014.

[22] Mohammad Mahdi Share Pasand, "Harmonic Aggregation Techniques: Methods to Compensate for Interaction Effects." American Journal of Electrical and Electronic Engineering, vol. 3, no. 3 (2015): 83-87. DOI: 10.12691/ajeee-3-3-4.

11

A Wideband Directional Microstrip Slot Antenna for On-Body Applications

Mehdi Hamidkhani*, **Behdad Arandian**

Department of Electrical Engineering, Dolatabad Branch, Islamic Azad University, Isfahan, Iran

Email address:

mehdi.hamidkhani@gmail.com (M. Hamidkhani), b.arandian@yahoo.com (B. Arandian)

Abstract: Antennas are one of the main components of every wireless telecommunication system. With wideband antennas for on-body applications, there are some additional features that should be considered including radiation and physical size. In this paper, we presented a wideband omnidirectional slot antenna with a reflector element improvised under the feed line in order to minimize the impact of body on the antenna. Curiously, this reflector element allows for directionality of the antenna. In the following sections of the paper we will show that less power can penetrate body tissues in the presence of reflector elements; thus, wideband directional antennas are influenced less by human body than omindirectional antenna.

Keywords: Directivity, On-Body Application, Slot Antenna, Reflector Element, Wideband

1. Introduction

There may not be direct transmission of microwave signals in the human body, because they encounter significant losses there. However, wideband signals may be used for the transmission of narrow pulses with little multipath fading and interference, which are present in signal frequency communication channels. One of the requirements of the on-body installations of communication systems is their low transmission power, so that no adverse radiation effects are incurred on the human body. Consequently, there is a need for the development of wideband antennas with appropriate near and far field radiation characteristics for the on-body application in body-centric wireless communication systems [1-4].

For the installation of lower power transmitter and receiver equipment on the human body or clothing, the near field radiation of antennas inside or in the vicinity of human body should be considered. Also the dissipation of RF power inside the human body should be minimized. Furthermore, the antenna far field radiation should be considered for the communication among the on-body sensors and among these sensors and among these sensors and some nearby equipment (such as a mobile phone or a laptop personal computer). Consequently, the wideband antennas should be designed according to the specified characteristics [2,5,6].

In this paper, we present a novel wideband slot antenna [7-10] for on-body installation, which may be applied in low power communication systems to provide a very low radiation power towards the human body and an appropriate directional for field radiation pattern. The criterion for the wideband impedance bandwidth is SWR<2.5 for the frequency band 6-10 GHz. An appropriate circuit is devised for the impedance matching at the antenna feed. Due to the complex interaction between an on-body antenna and the human body, we use full-wave numerical electromagnetic solvers, such as HFSS and FEKO.

2. The Proposed Antenna Structure

We use the substrate Rogers 5880 with dielectric constant ϵ_r=2.2, height h=1.6m and loss tangent tanδ=0.0009. Two U-shaped slots are made on the ground plane on one side of the substrate with the dimensions shown in figure 1. The feed line (with characteristic impedance 50Ω) is branched into two symmetrical strips (of impedance 100Ω) on its other side. This feed configuration provides wideband matching for slotted and stacked patches. Such a slot antenna provides an omnidirectional pattern in the XZ plane. In order to reduce the radiation on the feed line side of the antenna, a narrow rectangular conducting plane as a reflector is placed at a distance h from it.

(a)

(b)

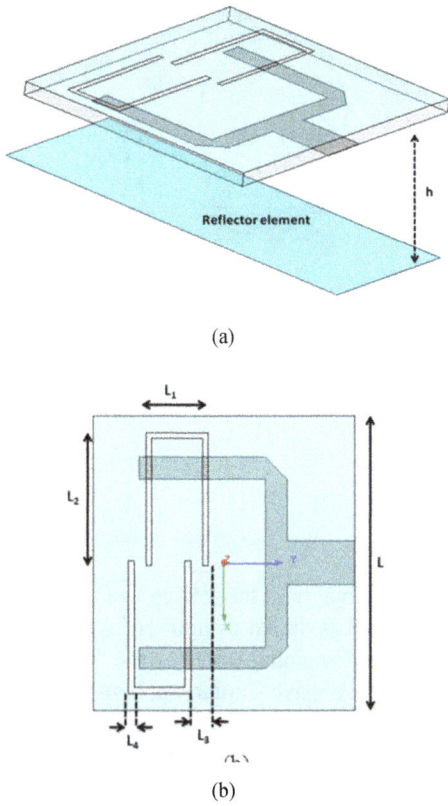

Figure 1. *Proposed antenna configuration; (a) 3d view, (b) top view.*

3. Computer Simulation of Isolated Antenna

The proposed antenna is designed and optimized by the full-wave simulation software HFSS. The simulation results and measurement data for SWR of the slot antenna without the reflector plane are shown versus frequency in figure 2. The frequency band for SWR≤2.5 is 5.2-10GHz. Now consider a reflector plane as a rectangular conductor plate of dimensions 16mm*44mm placed under the feed line at a distance h=3mm. Its SWR is drawn in figure 3, where the frequency band is 5.6-10.3GHz for SWR≤2.5. The radiation patterns of the antenna without and with the reflector plate are drawn in

figure 4, at three frequencies 6, 8 and 10GHz, respectively. The antenna gains with and without reflector plate are drawn in figure 5. It can be seen that the gain of the antenna with the reflector plane is relatively higher, which makes it a directional antenna.

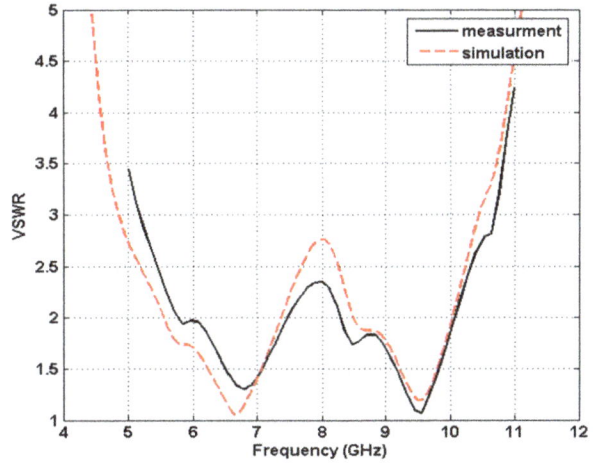

Figure 2. *Simulated and measured VSWRs of proposed antenna without the reflector plane. The geometrical parameters are L=32mm, L1=6.9mm, L2=14.4mm, L3=2mm, L4=0.6mm.*

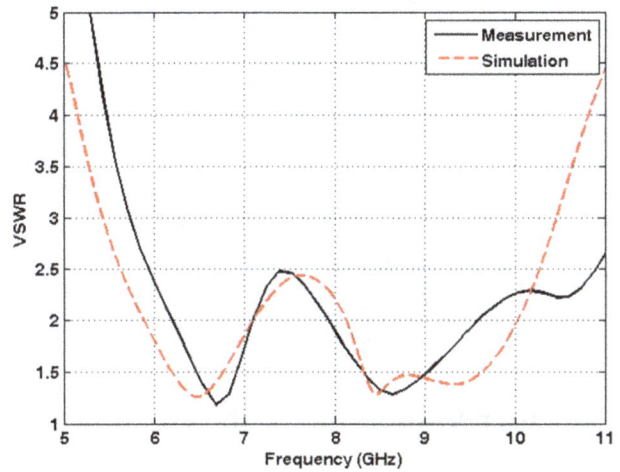

Figure 3. *Simulated and measured VSWRs of proposed antenna with the reflector plane.*

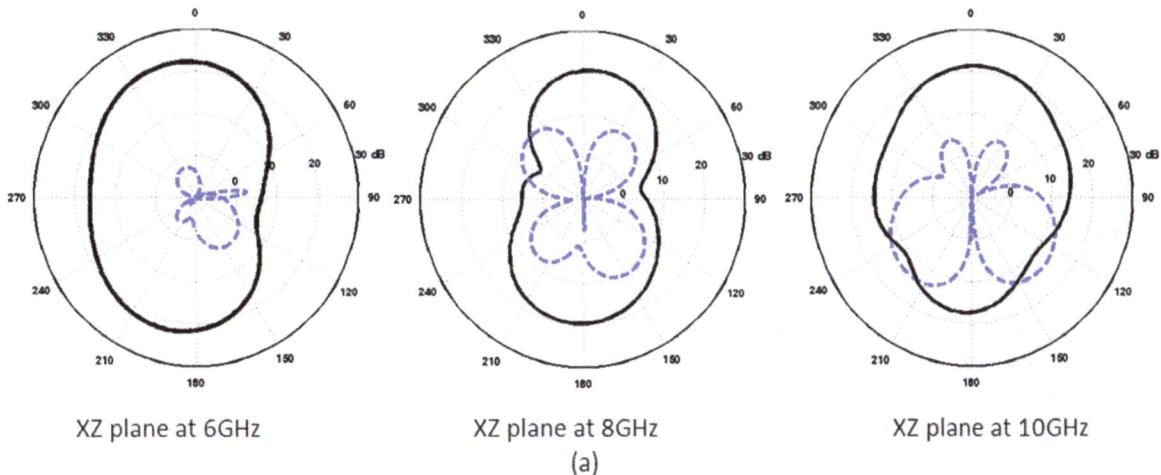

XZ plane at 6GHz   XZ plane at 8GHz   XZ plane at 10GHz

(a)

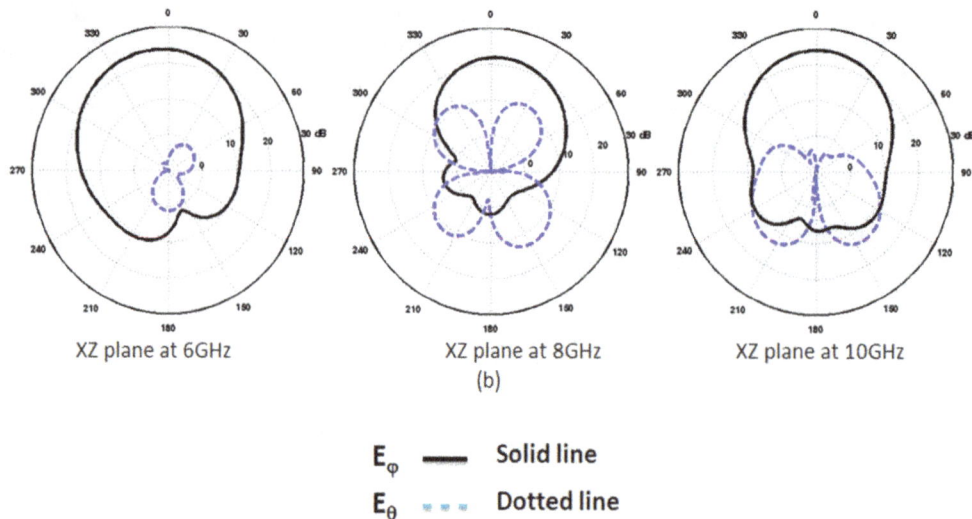

| XZ plane at 6GHz | XZ plane at 8GHz | XZ plane at 10GHz |

(b)

E_φ —— **Solid line**

E_θ - - - **Dotted line**

Figure 4. Measured radiation patterns of the proposed antenna in the XZ plane; (a) without the reflector plate, and (b) with the reflector plate.

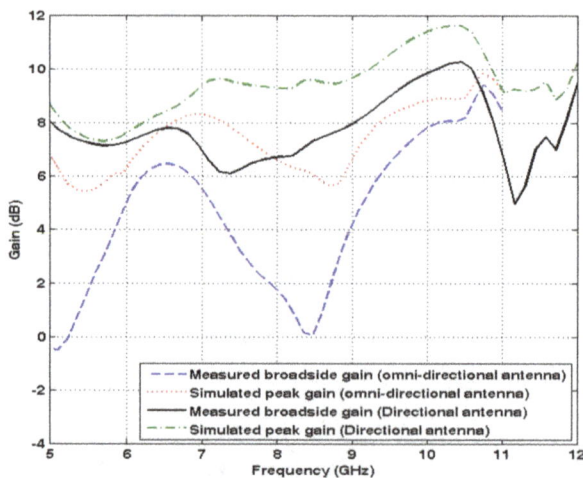

Figure 5. Simulated and measured gain of the proposed antenna.

4. Computer Simulation of the On-Body Antenna

We may consider two simplified models for the operation of the on-body antenna as given in [2]:

1- A three tissue model consisting of three layers of skin layer (with thickness 1mm), fat layer (with thickness 3mm) and muscle layer (with thickness 40mm);

2- A homogenous model composed of a muscle layer (of thickness 44mm) with dielectric constant ϵ_r =52.7, conductivity σ=1.773 and loss tangent tanδ=0.242.

The dimensions of both models are $120 \times 110 \times 44 mm^3$. The size of this simplified model is obtained by computer simulation of on-body antennas. For simplicity we use the second model. First, place the antenna feed line (without the reflector plate) at a distance of 3mm from the body. In figure 6(a) performances of the isolated antenna and on-body antenna is compared based on their SWR versus frequency. We can note that the bandwidth for SWR≤2.5 of the on-body antenna is quite narrower than that of the isolated antenna.

Next, place the antenna with the reflector plate on the body muscle tissue. Its SWR is drawn in figure 6(b) and compared with that of the isolated antenna. Based on the figure the SWR versus frequency curve behave similarly at higher frequencies, but are somewhat different at lower frequencies. Curiously, these lower frequencies lie outside the working frequency of the antenna. Consequently, the reflector plate behaves as an effective shield to protect the body against exposure to the antenna radiations.

This is because less power can penetrate body tissues in the presence of the reflector elements. In general, wideband directional antenna is influenced less by human body than omnidirectional antenna. In addition, changes observed for omnidirectional antenna during proximity to human body is dependent upon its distance from the body. More specifically, when small-size wideband antennas are studied in proximity to body models (reactive near field), the results rely the size of the antenna and its distance from the body. However, orientation is not necessarily a good solution for on-body antennas. The reason is that orientation of the antenna is measured in far field, while near fields of the antenna is not accounted for. Even if a directional antenna is placed in a specific angle, the near fields can measure great in the direction of human body.

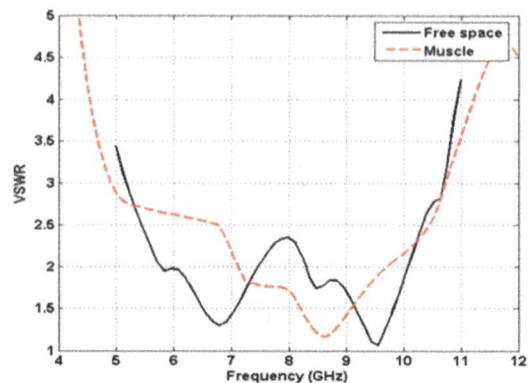

(a)

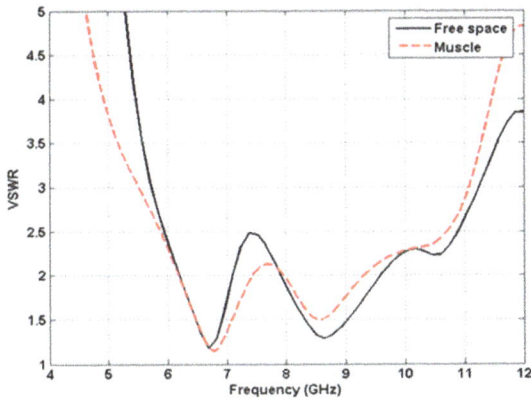

(b)

Figure 6. Simulated VSWR of the isolated antenna and on-body antenna; (a) without the reflector plate; (b) with the reflector plate.

5. Conclusion

In this paper we presented a wideband omnidirectional slot antenna. In order to minimize the impact of body on the antenna, a reflector element was improvised under the feed line. In addition, further directionality and orientation was achieved in the presence of this reflector element. We concluded that the wideband directional antenna is influenced less by proximity to human body than the omnidirectional antennas. Furthermore, changes observed for the omnidirectional antenna during proximity to human body is dependent upon its distance from the body.

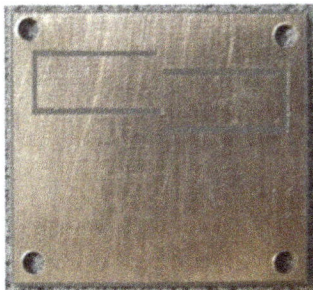

(a)

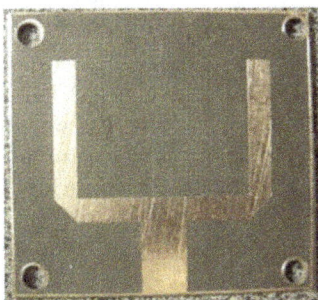

(b)

(c)

Figure 7. Images of the fabricated antenna; (a) top view, (b) bottom view, (c) 3d view.

References

[1] Klemm, M.; Kovcs, I.Z.; Pedersen, G.F.; Troster, G., "Novel small-size directional antenna for UWB WBAN/WPAN applications," Antennas and Propagation, IEEE Transactions on, vol.53, no.12, pp.3884, 3896, Dec. 2005.

[2] Peter S.Hall, and Yang Hao, "Antenna and Propagation for Body-Centric Wireless Communications" Artech House Publisher, 2006, ISBN 1-58053-493-7, pp. 93-109.

[3] Cai, A.; See, T.S.P.; Zhi Ning Chen, "Study of human head effects on UWB antenna," Antenna Technology: Small Antennas and Novel Metamaterials, 2005. IWAT 2005. IEEE International Workshop on, vol., no., pp.310, 313, 7-9 March 2005.

[4] Oppermann, I., M. Hamalainen, and J. Iinatti, (eds.), UWB: Theory and Applications, New York: John Wiley & Sons, 2004.

[5] Alomainy, A.; Hao, Y.; Hu, X.; Parini, C.G.; Hall, P.S., "UWB on-body radio propagation and system modelling for wireless body-centric networks," Communications, IEE Proceedings- , vol.153, no.1, pp.107,114, 2 Feb. 2006.

[6] K. Zhao, S. Zhang, Z. Ying, T. Bolin, and S. He, "Reduce the Hand-Effect Body Loss for LTE Mobile Antenna in CTIA Talking and Data Modes," Progress In Electromagnetics Research, PIER 137, 73-85, 2013.

[7] Balanis, C. A., Antenna Theory: Analysis and Design, 3rd ed., New York: Wiley-Interscience, 2005.

[8] Seong-Youp Suh; Stutzman, W.L.; Davis, William A., "A new ultrawideband printed monopole antenna: the planar inverted cone antenna (PICA)," Antennas and Propagation, IEEE Transactions on, vol.52, no.5, pp.1361, 1364, May 2004.

[9] Alomainy, A.; Hao, Y.; Parini, C.G.; Hall, P.S., "Comparison between two different antennas for UWB on-body propagation measurements," Antennas and Wireless Propagation Letters, IEEE , vol.4, no., pp.31,34, 2005.

[10] Samal, P.B.; Soh, P.J.; Vandenbosch, G.A.E., "UWB All-Textile Antenna With Full Ground Plane for Off-Body WBAN Communications," Antennas and Propagation, IEEE Transactions on , vol.62, no.1, pp.102,108, Jan. 2014.

12

Evaluation of Resistance Type SFCL on Reduction of Destructive Effects of Voltage Sag on Synchronous Machine Stability

Behrouz Alfi[1], Tohid Banki[2], Faramarz Faghihi[3]

[1]Department of Electrical Engineering College of Engineering Ardabil Branch, Islamic Azad University, Ardabil, Iran
[2]Department of Electrical Engineering College of Engineering Bilasouvar Branch, Islamic Azad University, Bilasouvar, Iran
[3]Department of Electrical Engineering College of Engineering Science and Research Branch, Islamic Azad University, Tehran, Iran

Email address:
beh_alfi@yahoo.com (B. Alfi), tohidbanki@gmail.com (T. Banki), faramarz_faghihi@hotmail.com (F. Faghihi)

Abstract: Electrical power quality has become an essential part of power systems and electrical machines. True identifying and regarding power quality and how to check the related destructive points are of utmost importance. Most of the time, providing the quality of power for sensitive load which requires the voltage to maintain a sinusoidal waveform at rated voltage and frequency is the main point to be considered. On the other hand, protecting power supply against resulting turbulences can be regarded. Voltage sag is one of the destructive issues in power quality problem to be taken into account. Not only can it have some undesirable effects on the sensitive load, but it also has some undesirable effects on synchronous machine as a power supply. In this study, SFCL was used in order to reduce the destructive effects of voltage sag on the performance of synchronous generators. To achieve better understanding, a simple model composed of synchronous generators, SFCL and load is used with the help of MATLAB/SIMULINK software. By creating voltage sag which results from 3-phase short-circuit voltage, current, torque, speed and angle load changes of synchronous generators were simulated and discussed. It was done in condition with/without SFCL. The results illustrate that using SFCL in synchronous generator output can improve its performance during occurrence of voltage sag.

Keywords: Superconducting Fault Current Limiters (SFCL), Synchronous Machine, Power Quality, Voltage Sag

1. Introduction

The importance of paying attention to power quality issue is increasing day by day. Power quality is usually meant to say the quality of voltage and/or the quality of current and can be defined as: the measure, analysis, and enhancement of the bus voltage to keep a sinusoidal waveform at rated voltage and frequency [1]. Nowadays, voltage sag is the main problem for huge industries and has the most pressure on power quality.

Sags are short-duration reduction in the rms voltage between 0.1 and 0.9 for the duration between 0.5 cycles and 1 minute. It is also divided into instant, momentary and temporary types to continuing time. Voltage sags are usually caused by: energization of heavy loads, starting of large induction motor and one/two/three-short circuit faults. More need for electric power results while increasing the level of

short-circuit current which makes electric equipment fail and it is because of the resulting mechanic and electric pressure.

One common way for limiting fault current with high level is using fault current limiter (FCL). FCL is fundamentally a variable resistance that is set up serially in circuit with breaker. When a fault occurs, impedance increases to an amount in which the fault current decreases to an amount with a low level as much as the breaker can undergo it. FCL efficiency in limiting short-circuit fault makes it possible for breakers to open easily and safely. The function of FCL in electric power systems is both limiting the range of short circuit currents and increasing the stability level of power system. By developing electronic power magnetic technology and superconductors, fault current limiters are developed [2-6]. Among the mentioned cases, SFCL is suggested for limiting fault current. SFCL can be a good solution for controlling fault current levels in distribution and transmission networks and it is due to less losses in normal function of superconductors [7-9]. The coordination of SFCL as a key application issue has been investigated in previous papers [10,11]. When a fault occurs, SFCL produces impedance and it is because of the loss of superconductor. It puts this impedance on the circuit which limits fault current. Therefore, SFCLs which have characteristics of superconductors can change the level of resistance from near zero (superconductor state) to high level in normal state (non-superconductor state) [12]. It is the case when current, temperature or field outreach critical amount. These kinds of limiters do not need fault recognition system; therefore, their reaction (response) is quick. Voltage sags cause apart eventual triping, large torque peak, which may cause damage to the shaft or equipment connected to the shaft. During voltage sags, the behavior of synchronous machines is explained and why there are high torque and current peak during voltage sags and approach is presented in the use of quick breaker [13]. In this study, SFCL was used in order to reduce the destructive effects of voltage sag on the performance of synchronous generators.

2. Torque and Current Formulation of Synchronous Machine

Due to the creation of peak current and huge torque, voltage sag can damage shaft or lateral equipments of synchronous machines. Stator flux, which is dependent on the time during voltage sag, explains the action of synchronous machines and the reason for emerging huge peak current and torque. Torque equation in steady state is defined as equation (1). According to this equation, torque is proportional to stator flux; if the flux is doubled, the torque will be more than twice.

$$T = P\left\{\frac{\psi_s\psi_{df}}{L_d}Sin\delta + \psi_s^2 \times \left(\frac{1}{L_d} - \frac{1}{L_q}\right)Sin2\delta\right\} \quad (1)$$

As well as, current equation in the terminal synchronous generator in temporary state is defined as equation (2).

$$i_{as} = \frac{\sqrt{2}}{2\omega}V_s\left[\frac{1}{L_d} + \frac{1}{L_q}\right]e^{-\alpha t}Sin(\theta_r(0))$$

$$i_{bs} = \frac{\sqrt{2}}{2\omega}V_s\left[\frac{1}{L_d} + \frac{1}{L_q}\right]e^{-\alpha t}Sin(\theta_r(0) - 120) \quad (2)$$

$$i_{as} = \frac{\sqrt{2}}{2\omega}V_s\left[\frac{1}{L_d} + \frac{1}{L_q}\right]e^{-\alpha t}Sin(\theta_r(0) + 120)$$

In temporary condition L'_d is replaced L_d (d axis inductance), the torque and current in temporary condition will be more because of $L'_d \ll L_d$. The effect of δ (load angle) on torque will be less because of few changes. According to equation (2), current in all three phase synchronous generator in temporary condition will increase.

3. Resistance Type SFCL

Critical current density (J_c) is one of the main parameters in superconductors. Increasing the current to more than J_c level can lead to the changing state of matter from being superconductor to normal state. Therefore, the current which passes through superconductor can resist controlling factor. One problem in conventional fault current limiter is fault recognition system. That is, it should be able to limit the fault current at the first period and before reaching the first peak. One advantage of superconductor limiters is that they do not need fault recognition system. By sudden increase in network current level which is due to fault, the current which passes through limiters traverses from critical current and so the limiter takes impedance and limits the fault current. In most case, voltage sag is temporary and can be removed immediately. Applying SFCL avoids bad effects of voltage sag. In addition, the load which makes financial losses for huge industries can be kept safe. SFCL has been proposed to limit the short circuit current and various types of SFCLs have been developed. SFCLs are of different types including saturated iron core reactor type, bridge type, flux lock type, reactor and resistance type. Equation (3) shows voltage drop in SFCLs in the superconducting state, flux flow state and normal state:

$$U_S = \begin{cases} 0 \ if \ |i_s| \leq i_c(t), T < T_c \\ R_P[i_s - sign(i_s)i_c(t)] \ if \ |i_s| > i_c(t), T < T_c \\ R_n i_s \ if \ T \geq T_c \end{cases} \quad (3)$$

In the above equation, R_p is the highest limiter resistance in normal condition. Moreover the dominant temperature is $77°K$. The model which was suggested in this study is 3-phase resistance type placed in a synchronous generator output. Fig. 1 shows the resistance type of SFCL. In this study, for simulating the condition, it was suggested that at fault moment the highest level of resistance be inserted in the line and after removing the fault, it should reach the zero level in the superconductor phase.

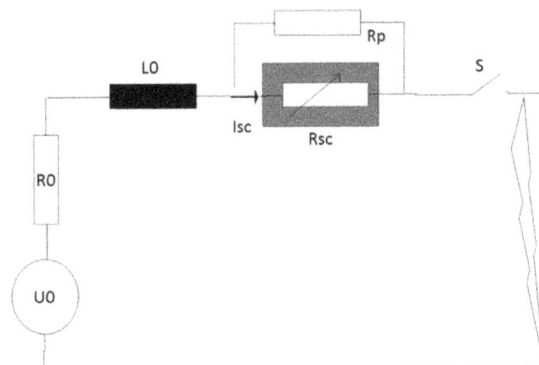

Fig. 1. *Resistance Type of SFCL [4].*

4. Result of Voltage Sag Effect on Synchronous Generator

Fig. 2 shows the single-line diagram of the discussed model. The simulation was done by MATLAB/SIMULINK software. This model includes a synchronous generator, resistance type SFCL, and load with constant power. In order to create voltage sag, 3-phase short circuit was done according to Fig. 3. Increasing peak current and torque may become 10 times more which can damage shaft or lateral equipment. Time of 1 second was selected for simulation time duration and 0.1 second was selected for voltage sag duration. Therefore, 3-phase short circuit occurred 0.7 second and the fault was removed after 0.1 second.

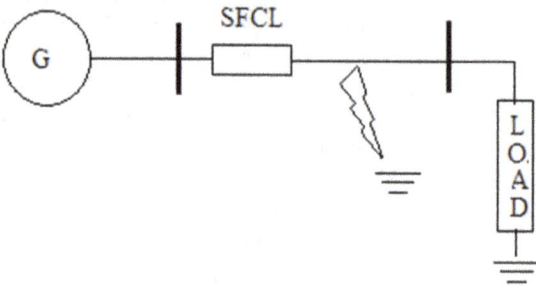

Fig. 2. *Single-Line Diagram of the Model Using SFCL.*

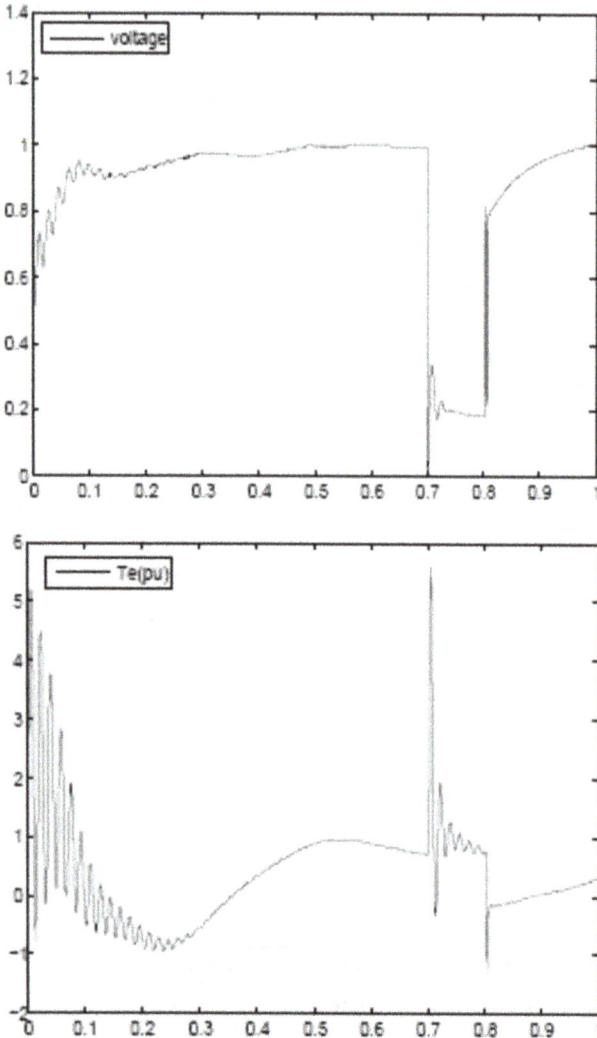

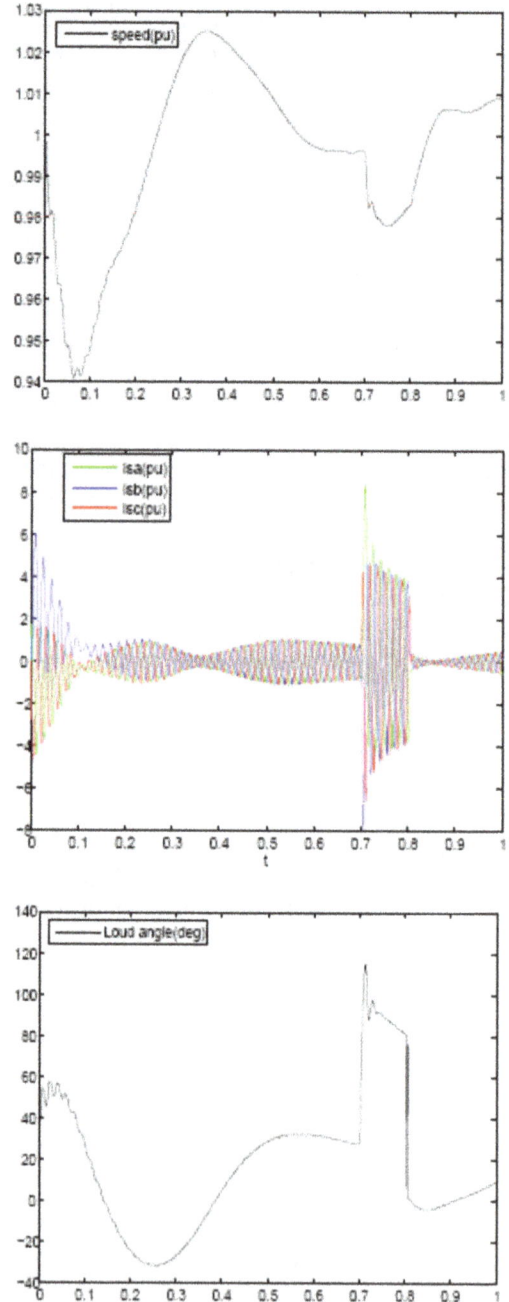

Fig. 3. *Voltage, Current, Torque, Speed and Load Angle due to Voltage Sag without SFCL.*

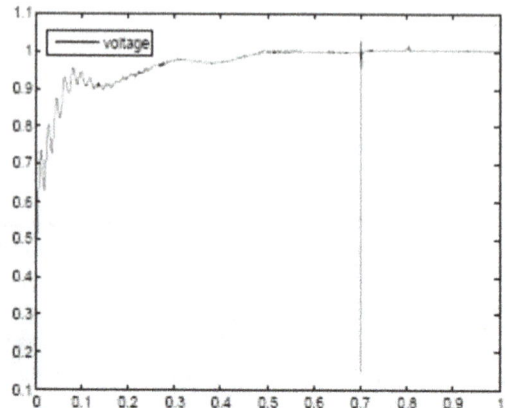

The simulation was done through two steps and voltage, current, speed, load angel and torque change results were reported in this study. At first step, voltage sag was short-duration in the rms voltage 0.2 pu, and it was without SFCL, as shown in Fig. 4. It is shown in Fig. 4 that the voltage sag has led to an increase in peak current, torque, load angle and decreasing speed. These changes were at their high level at the time of short circuit. In the second step, the fault occurred along with SFCL. In this condition, SFCL turned from superconductor phase into normal phase. Along with increasing line resistance, SFCL avoided increasing torque and current peak.

5. Conclusion

The SFCL has been used and implemented in this paper using MATLAB/SIMULINK software. SFCL can react to created turbulences immediately and avoid bad effects of voltage sag on synchronous generator such as increasing torque and current in fault period and it can stabilize generator system during a severe fault.

References

[1] Ewald F. Fuchs, Mohamad A. S. Masoum, "Power Quality in Power Systems and Electrical Machines," Elsevier Inc, ISBN 978-12-369536-9, 2008.

[2] R. F. Giese, et al, "Assessment Study of Superconducting Fault Current Limiters Operation at 77 K," *IEEE Transactions on Power Delivery*, Vol.8, No. 3, pp. 1138–1147, July 1993.

[3] R. Smith et al., "Solid State Distribution Current Limiter and Circuit Breaker Application Requirements and Control Strategies," *IEEE Transactions on Power Delivery*, Vol. 8, No. 3, pp. 1155–1164, July 1993.

[4] V. Hassenzahi, K. Johnson, T. Reis, "Electric Power Application of Superconductivity," proceeding of the IEEE, V. 92, No. 10, Oct 2004.

[5] E. Thuries, et al., "Toward the Superconducting Fault Current Limiter," IEEE Transactions on Power Delivery, Vol. 6, pp. 801–808, April 1991.

[6] M. Chen, et al., "Fabrication and Characterization of Superconducting rings for Fault Current Limiter Application," IEEE Transaction on Applied superconductivity, Vol. 282–287, pp. 2639–2642, 1997.

[7] L. Ye, L. Z. Lin, and K. P. Juengst, "Application Studies of Superconducting Fault Current Limiter in Electric Power Systems," *IEEE Transactions on Applied Super- conductivity*, Vol.12, No.1, pp. 900–903, March 2002.

[8] K. Hongesombut, "optimal location assignment and design of superconducting fault current limiter applied to loop power system", IEEE, Vol. 13, No. 2, 2003.

[9] J. S. Kim, S. H. Lim, and J. C. Kim, "Study On Protection Coordination Of Flux-lock Type SFCL with Over-current Relay," *IEEE Transactions on Applied Super- conductivity*, Vol. 20, No. 3, June 2010.

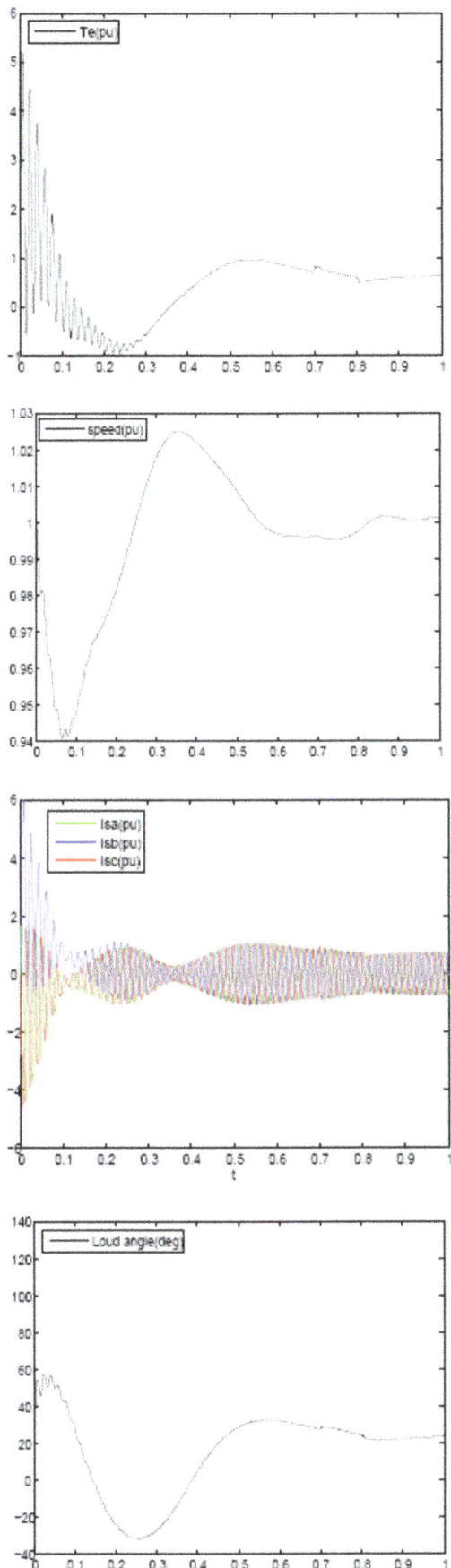

Fig. 4. *Voltage, Current, Torque, Speed and Load Angle due to Voltage Sag with SFCL.*

[10] S. Kazemia, M. Lehtoneb, "Impact of Smart Sub-transmission Level Fault Current Mitigation Solutions On Service Reliability," Electric Power System Research, Vol. 96, pp. 9-15, Mar 2013.

[11] B. Li, C. Li, F. Guo, "Coordination Superconductive Fault Current Limiters with Zero-sequence Current Protection Of Transmission Lines," IEEE Transactions on Applied Super-conductivity, Vol. 24, No. 5, pp. 5602105, Oct 2014.

[12] B. Li, C .Li, F. Guo, "Over Current Protection Coordination in a Power Distribution Network with Active Superconductive Fault Current Limiters," IEEE Transactions on Applied Super- conductivity, Vol. 24, No. 5, pp. 5602004, Oct 2014.

[13] F. Carlsson, Ch. Sadarangani, "Behavior of Synchronous Machines Subjected to Voltage Sags of Type A, B and E", EPE journal. Vol.15, 2005.

13

Study on the Effectiveness Evaluation Method Based on the Combination of AHP and Grey Interval Correlation

Xu Shi-hong[1], Zhao Wei-bin[2], Huang Guo-qing[1, 2]

[1]Department of Aeronautic Electronic Engineering, the First Aeronautical College of Air Force, Xinyang, China
[2]School of Information Engineering, Zhengzhou University, Zhengzhou, China

Email address:
zzuweibin@163.com (Zhao Wei-bin), hguoqing@163.com (Huang Guo-qing)

Abstract: Aiming at the problem of using single judging value for the expert score of the traditional effectiveness evaluation, the expert's score data are not reliable; an improved effectiveness evaluation method based on the combination of AHP and grey interval correlation analysis is proposed. Interval number is used for the expert score in the improved method, the influence of the subjective fuzziness and objective fuzziness on the expert score using the single judging value is overcame. Interval number improved the reliability of the evaluation data. Combination weighting approach considering the influence of subjective and objective factors, the weight is more reasonable. The concrete steps of the improved effectiveness evaluation method are given. The improved method is proved to be more reasonable and reliable through an example.

Keywords: AHP, Interval Number, Gray Relational Degree, Combination Weighting Approach, Effectiveness Evaluation

1. Introduction

The gray system theory proposed by Professor Deng Ju-long is a method of studying on the uncertain problem whose part of the information is known, while some of the information is unknown, "small sample" and " Poor information" [1]. Grey relational degree is often used to indicate the degree of correlation or the degree of similarity between grey system factors, and the gray relational degree plays an important role in the quantitative analysis of grey system theory. The results of the grey relational degree efficiency evaluation are based on the expert's scoring value, so the accuracy of the expert's scoring value directly affects the reliability of the results.

At present, Chen Kai, Zhao Yu-hui, Du Jun proposed that the index was given different weights according to the actual important degree of the different index, to overcome the problem of using same weight and poor objectivity of the index in the calculation of correlation degree [2]. Chang Shuang-jun and Ma Jin-ya used synthetical optimization based on gray associated method for effectiveness evaluation of weapon system to avoid the results of effectiveness evaluation which is too partial or inferior to make the evaluation results more reasonable [3]. Ning Xiao-lei et al. proposed simulation models based on the improved grey

relational analysis, which combines the geometric similarity and numerical similarity, and reduces the risk of conventional grey correlation analysis, and improves the credibility of the results of the verification [4]. ChihHsuan Wang used the grey correlation analysis method to optimize the green production project, which can use the minimum energy consumption to get lean production [5] and so on.

In the above researches the experts' score data used a single value, however, it does not meet the people's fuzzy thinking habits [6]. Because of the ambiguity of the human mind and the influence of the objective factors, the value of the index judged by many people is a certain interval [7], and the interval number is more able to reflect the reliability of expert scoring. Therefore, in order to make the evaluation result more reasonable and credible, this paper proposes a method of combining AHP with grey interval correlation.

2. Analysis on the Improved Method of Effectiveness Evaluation

The gray system is complex and information incomplete, it has a large number of factors, complex logic relationship. Therefore, it is very difficult to evaluate the effectiveness of the gray system. The analytic hierarchy process (AHP) method decomposes the complex system into several elements, such as target, criterion, index, etc., and then, according to the different attributes, these elements are arranged into a hierarchical relationship. It is the core idea of the AHP to establish an effective and reasonable hierarchical relationship.

In the process of the effectiveness evaluation, the index weight is very important, and it has a great influence on the evaluation results. Therefore, the determination of the weights of the index should be scientific and reasonable, and the reliability of the results for the effectiveness evaluation is related to it. At present, there are some subjective weighting methods, such as analytic hierarchy process, expert investigation method and so on. Objective weighting method includes the method of maximizing deviation, information entropy method, etc.. In order to make the results more reasonable and reliable, combination weighting approach can be used to not only consider the preference of the experts to the index, but also reduce the subjective randomness.

In the improved effectiveness evaluation method based on the combination of AHP and grey interval correlation analysis, reasonable hierarchical relationship is established by AHP. The expert points are unified by using the evaluation interval number so the accuracy of the evaluation data are promoted. By using the combination weight approach to determine the weight, and by taking into account the subjective and objective factors, the weight is more reasonable. Extraction of the information vector of interval numbers after weighted [8], the vector similarity algorithm has represented the gray relational degree of the comparison scheme's information vector and the reference scheme's information vector (the ideal optimal or worst scheme).

3. Concrete Steps of the Improved Effectiveness Evaluation Method

According to the above analysis, the concrete steps of the effectiveness evaluation method of the combination of AHP and grey interval correlation analysis are as follows:

Step 1: use AHP method to establish the hierarchical relationship

According to the principle of AHP method, a reasonable hierarchical relationship is set up, which includes the target layer, the criterion layer, and the index layer. The object to be evaluated is recorded as A={A1, A2, A3,... , An}; the criteria layer index is recorded as B={B1, B2,... Bf}; the index layer is denoted as C={C1, C2,... Cm}. The corresponding index interval number of the underlying index c_j in the scheme A_i is recorded as $a_{ij} = (\underline{a_{ij}}, \overline{a_{ij}})$ where $\underline{a_{ij}}$ and $\overline{a_{ij}}$ expressed the lower and upper bounds of the interval respectively.

Step 2: obtain expert evaluation interval number matrix and normalized processing

The evaluation interval number of the underlying index c_j of the expert K treatment evaluation scheme A_i is recorded as $a^k_{ij} = (\underline{a^k_{ij}}, \overline{a^k_{ij}})$ where $\underline{a^k_{ij}}$ and $\overline{a^k_{ij}}$ expressed the

lower and upper bounds of the interval respectively.

$$\underline{a_{ij}} = \frac{\sum_{m=1}^{k} \underline{a^m_{ij}}}{k} \qquad (1)$$

$$\overline{a_{ij}} = \frac{\sum_{m=1}^{k} \overline{a^m_{ij}}}{k} \qquad (2)$$

the formula (1)and(2), K is the total number of experts involved in the assessment.

The corresponding index interval number of the underlying index c_j in the scheme A_i is recorded as $a_{ij} = (\underline{a_{ij}}, \overline{a_{ij}})$ where $\underline{a_{ij}}$ and $\overline{a_{ij}}$ expressed the lower and upper bounds of the interval respectively. So the experts evaluate the interval number matrix:

$$X = \begin{bmatrix} a_{11} & a_{12} & \cdots & a_{1j} \\ a_{21} & a_{22} & \cdots & a_{2j} \\ \vdots & \vdots & \cdots & \vdots \\ a_{n1} & a_{n2} & \cdots & a_{nj} \end{bmatrix} \qquad (3)$$

Because the physical meaning of each factor is different, its mathematical dimension is not necessarily the same, thus the number of data size is different, therefore it is difficult to get the correct comparative results in comparison. So the dimension of the data in the matrix is eliminated and it is transformed into a comparable data sequence. For the benefit and the cost type index, the formula (4) and the formula (5) are used to normalize the data in order to make each index value in the range of [0, 1].

$$b_{ij} = \frac{a_{ij}}{\max_{i=1}^{n}(\overline{a_{ij}})} \qquad (4)$$

$$b_{ij} = \frac{\min_{i=1}^{n}(\underline{a_{ij}})}{a_{ij}} \qquad (5)$$

Get the score matrix of the expert after normalize

$$Y = \begin{bmatrix} b_{11} & b_{12} & \cdots & b_{1j} \\ b_{21} & b_{22} & \cdots & b_{2j} \\ \vdots & \vdots & \cdots & \vdots \\ b_{n1} & b_{n2} & \cdots & b_{nj} \end{bmatrix} \qquad (6)$$

Where $b_{ij} = (\underline{b_{ij}}, \overline{b_{ij}})$。

Step 3: determine the weights of the index
(1) Determine the subjective weight
In this paper, the subjective weight is determined by analytic hierarchy process. On the same level, the importance of each factor is relevent to the upper level of the importance

of the two comparison, getting the judgment matrix, the use of square method to calculate the eigenvalues and eigenvectors of the matrix, and the weight of the judgment matrix consistency test. When the consistency check probability CR<0.1, then the judgment matrix meets the consistency, get the subjective weight $w_s = (w_{s1} \ w_{s2} \ L \ w_{sm})$

(2) Determine the objective weight

In this paper, the objective weight is determined by entropy method. The information entropy and the weights of the index are calculated by the formula (7) and (8) respectively.

$$E_{oc} = -\frac{1}{\ln n}\sum_{m=1}^{n}b^{(m)}{}_{ij}\ln y b^{(m)}{}_{ij} \tag{7}$$

$$w_{oc} = \frac{1 - E_{oc}}{\sum_{k=1}^{n}(1 - E_{oc})} \tag{8}$$

Get the objective weight of each index $w_o = (w_{o1}, \ w_{o2}, \cdots, \ w_{om})$.

(3) combined weights

$$w_c = \frac{w_{sc}w_{oc}}{\sum_{c=1}^{m}w_{sc}w_{oc}} \tag{9}$$

Through the formula (9) to calculate the combination weight, and in the formula w_c is satisfied that $0 \le w_c \le 1, \sum_{c=1}^{m}w_c = 1$·

Step 4: weight the interval number, and find the interval number of the scheme to be compared.

The corresponding index interval number of the underlying index c_j in the scheme A_i after normalized processing is recorded as $b_{ij} = (\underline{b_{ij}}, \overline{b_{ij}})$. The weight of the index is w, and the interval number is weighted by the formula (10).

$$c_{ij} = w_{ij}b_{ij} = (w_{ij}\underline{b_{ij}}, \ w_{ij}\overline{b_{ij}}) \tag{10}$$

The interval number of the scheme A is calculated by the formula (11).

$$d = w_1(c_{11} + c_{12} + \cdots + c_{1e}) + w_2(c_{21} + c_{22} + \cdots + c_{2g}) + \cdots + w_f(c_{f1} + c_{f2} + \cdots + c_{fh}) \tag{11}$$

Step 5: extract interval number information vector

For interval numbers $d = (\underline{d}, \overline{d})$, the interval number information vector $I_d = (m_d, z_d)$ is extracted by the formula

(12) and (13).

$$m_d = \frac{d + \overline{d}}{2} \tag{12}$$

$$z_d = \frac{\overline{d} - d}{2} \tag{13}$$

Step 6: determine the reference scheme and calculate the gray relational degree

Reference vector selection $m_{cd} = \max(m^n{}_d)$, $z_{cd} = \min(z^n{}_d)$. Reference information vector is recorded as $I_{cd} = (m_{cd}, z_{cd})$.

The correlation degree between the program information vector and the reference information vector is calculated by using the formula (14) and (15) .

$$d(I_d, I_{cd}) = \sqrt{\sum_{i=1}^{2}(I_{cdi} - I_{di})^2} \tag{14}$$

$$r(I_d, I_{cd}) = \frac{\alpha}{\alpha + d(I_d, I_{cd})} \tag{15}$$

Where α is the resolution ratio, $\alpha > 0$, in the paper $\alpha = 0.1$.

Step 7: rank relational degree sequence

The relational degree between the comparison scheme's information vector and the reference information vector should be calculated respectively. And quality of the scheme is represented by the gray relational degree. According to numerical value of the gray relational degree, correlation degree sequence should be ranked. Compared with the optimal reference information vector, the greater of the gray relational degree, the expressed scheme is better. Compared with the worst reference information vector, the smaller of the gray relational degree, the expressed scheme is better. And then the efficiency of the gray system is analyzed quantitatively.

Through above steps, the complex gray system can be decomposed into some indexes. Experts use the interval number to score for each index. Overcome the problem of the reliability of the score data are not enough because of the effect of the subjective and objective fuzzy factors in the traditional gray correlation theory by using a single numerical value. Using this method, the results of effectiveness evaluation quantitative analysis are more reasonable and reliable.

4. Case Verification

In this paper, the data comes from references [9-10] to analyze the results of the trajectory planning .From the results of the trajectory planning, the two groups of L1 and L2 which are of representative significance are selected. These two groups of trajectory planning are evaluated.

L1 represents the 3 UAVs in the formation of the UAV doesn't simultaneously reach the goal for the collaborative planning of the track; L2 representing the UAVs arrives the goal at the same time. Assessment experts are five people, and the assessment schemes have two n=2.

(1) Obtain expert evaluation matrix

Using the AHP method to establish an effective and reasonable hierarchical relationship, experts score for the 12 indexes of the index layer. Score control is over the range of [0,10].Because the indexes are the benefit type index in this paper, the higher the value of the index is, the better the index will be. Thus, the expert evaluation matrix is obtained:

$$X = \begin{bmatrix} (6.7,7.0) & (4.7,5.0) & (7.2,7.5) & (6.0,6.2) & (5.1,5.3) & (7.2,7.5) \\ (6.5,6.7) & (4.8,5.2) & (7.2,7.5) & (6.3,6.6) & (6.1,6.3) & (7.0,7.2) \end{bmatrix}$$

$$\begin{matrix} (6.0,6.4) & (5.7,6.0) & (6.6,6.9) & (6.7,6.9) & (5.2,5.4) & (6.0,6.2) \\ (6.5,6.7) & (5.8,6.2) & (6.8,7.2) & (6.7,6.9) & (5.3,5.6) & (6.2,6.5) \end{matrix}$$

The value of the interval number of each index is obtained by formula (1) and (2).

(2) Get the normalized score matrix of the expert by the formula (4).

(3) Determination of the weight of the index

w1=(0.4899, 0.2239, 0.2862), w2=(0.7315, 0.1809, 0.0877),

w3=(0.3602, 0.6398), w4=(0.3266, 0.1565, 0.3671, 0.1497),

w=(0.4153, 0.3636, 0.892, 0.1318)

(4) The interval numbers of the schemes to be compared are calculated by the formula (10)and(11).

$$d_{L1} = (0.922406824486298, 0.962074864244169);$$

$$d_{L2} = (0.946046913847564, 0.989928151057143)$$

(5) In accordance with the formula (12), (13) to extract the information vector of the interval number of the scheme.

$$I_{d_{L1}} = (0.942240844365233, 0.019834019878936);$$

$$I_{d_{L2}} = (0.967987532452353, 0.021940618604790)$$

(6) Determine the reference information vector I_{cd} = （0.967987532452353, 0.019834019878936）

The gray relational degree is calculated by formula (14) and (15).

$$r_{L1} = 0.795249572940783;$$

$$r_{L2} = 0.979368632858783$$

(7) Rank relational degree sequence which is concluded that the scheme L2 is closer to the ideal optimal scheme than the scheme L1 because $r_{L1} < r_{L2}$. And the result is in line with the actual situation. The results of the difference methods are shown in table 1.

Table 1. *Comparison of results obtained from methods.*

	Traditional gray correlation degree	This paper
L1	0.7913	0.7952
L2	0.9149	0.9794

5. Conclusions

In this paper, an improved effectiveness evaluation method based on the combination of AHP and grey interval correlation analysis is proposed. The evaluation index value is replaced a single value by judging interval, which is used to solve the problem of the reliability of the data that is caused by subjective and objective fuzziness in the traditional gray correlation theory. The concrete steps of effectiveness evaluation based on the improved method are given. Example comparison shows the effectiveness evaluation method based on the combination of AHP and grey interval correlation analysis is feasible. Because of the selection of the expert data that the interval array is used, the result is more reasonable and reliable.

References

[1] Zhang Jie, Tang Hong, Su Kai. Research on the method of effectiveness evaluation [M]. Beijing, National Defense Industry Press, 2009.

[2] Chen Qin, Zhao Yu-hui, Du Jun. "Application of the Improved Grey Correlation Analysis on Evaluating Radar Low Proability of Interception Performance "[J]. Fire Control Radar Technology, Vol. 41 No. 2 Jun, 2012. pp.16 ~ 13.

[3] Chang Shuang-Jun, Ma Jin-ya. Application Study on Synthetical Optimization Based on Grey Associated Method for Evaluating Effectiveness of Weapon system [J]. Journal of Sichuan Ordnance, Vol. 32 May, 2011. Pp.16 -19.

[4] Ning Xiao-lei, Wu Ying-xia, Chen Zhan-qi. Study on Validation of Simulation Models Based on Improved Grey Relational Analysis [J]. Computer Simulation, Vol.32.Jul, 2015 pp. 259 – 263.

[5] ChihHsuan Wang. A systematic approach to select the optimal project portfolios for green manufacturing: An empirical study on TFT-LCD fabrication processes. Proceedings of the 2015 International Conference on Industrial Engineering and Operations Management Dubai, UAE, March 3 – 5, 2015.

[6] Wu Jiang, Huang Deng-shi. Review of the Research on the Method of Interval Number Ranking [J]. Systems Engineering, Vol. 22, 2004, pp.1-4.

[7] Xu Rui-li, Xu Ze-shui. Study on the practice and recognition of Mathematics in Practice and Theory [J], Vol.37 No.24 Dec, 2007. pp. 1 ~ 7.

[8] Chen Chun-fang, Zhu Chuan-xi. Method of Interval Number Ranking Based on Vector Similarity and Application [J]. *Statistics and Decision*, 2014 (3). pp.76-78.

[9] Wu Jing. Research on Trajectory Planning and Effectiveness Evaluation for Multi-UAV Cooperative [D]. Nanchang: Nanchang Hangkong University, June, 2012.

[10] Huang Guo-qing, Su Lin. "Analysis of Effectiveness Evaluation Based on Improved Gray Synthesize Relational Degree [J]". Command Control & Simulation, 2015(3). pp 90-93.

14

Challenges and Solutions for Autonomous Robotic Mobile Manipulation for Outdoor Sample Collection

Lei Cui[1], Tele Tan[1], Khac Duc Do[1], Peter Teunissen[2]

[1]Department of Mechanical Engineering, Curtin University, Perth, Australia
[2]Department of Spatial Sciences, Curtin University, Perth, Australia

Email address:

lei.cui@curtin.edu.com (Lei Cui), T.Tan@curtin.edu.com (Tele Tan), duc@curtin.edu.com (K. D. Do),
P.Teunissen@curtin.edu.au (P. Teunissen)

Abstract: In refinery, petrochemical, and chemical plants, process technicians collect uncontaminated samples to be analyzed in the quality control laboratory all time and all weather. This traditionally manual operation not only exposes the process technicians to hazardous chemicals, but also imposes an economical burden on the management. The recent development in mobile manipulation provides an opportunity to fully automate the operation of sample collection. This paper reviewed the various challenges in sample collection in terms of navigation of the mobile platform and manipulation of the robotic arm from four aspects, namely mobile robot positioning/attitude using global navigation satellite system (GNSS), vision-based navigation and visual servoing, robotic manipulation, mobile robot path planning and control. This paper further proposed solutions to these challenges and pointed the main direction of development in mobile manipulation.

Keywords: Mobile Manipulation, Autonomous, Sample Collection, GNSS, Vision, Navigation, Servoing, Path Planning, Control, Robot

1. Introduction

Sample collection is crucial to plant performance, and this practice is widely used in refinery, petrochemical and chemical plants to monitor and confirm unit operation. Process technicians are responsible for collecting uncontaminated samples that are representative of the process stream, properly labelling the samples, and taking them to the quality control laboratory to be analysed. Process technicians working a 12-hour shift may collect samples four different times during their shift.

The traditional manual sample collection causes several problems. First, process technicians expose themselves to a variety of chemical and safety hazards when collecting samples. For example, the sample may contain harmful chemical (e.g. caustic liquid used in alumina processing) or the plant vicinity may have operating machineries, etc. Second, the operation is costly. Sample collection in general occurs 24 hours a day seven days a week. A team of process technicians has to be maintained and managed for this purpose. Third, it is error-prone. Some special chemicals have to be added to certain samples to stabilize the samples. Process technicians

sometimes forget adding the correct chemicals to the right samples when they perform this tedious task.

Traditionally robotic arms are fixed in structured work cells and perform 4D tasks, namely dumb, dangerous, dull, and dirty. The maturity of mobile robot technology in recent years provides an opportunity for robotic arms to perform a wide range of tasks that require both locomotion and manipulation abilities. A universally accepted term, mobile manipulation, is used to describe the tasks performed by a mobile manipulator consisting of a robotic arm and a mobile platform.

Mobile manipulation is the ideal solution to the problems of sample collection. It eliminates the hazards exposed to personnel in the plant, and it is cost-effective considering the high labour cost in the developed countries. Further, pre-programmed procedures greatly reduce the chances of chemicals being added to wrong containers.

In the 1980s, mobile manipulation started to attract the attention of manufacturing industry, and several research prototypes were developed for a variety of purposes: delivering and handling tools and work pieces, performing simple assembly tasks, and operating in hazardous environments, [1, 2]. However, the real world application of these robots was hindered by the lack of adequate sensing technology and processing power.

In the 1990s, the application of mobile manipulation went beyond the structured industrial environments and made its way into less structured human environments. Khatib and his colleagues developed the Stanford Robotic Platform consisting of an omni-directional base and a Puma 560 robotic arm, an upper sonarring, and a lower sonar ring. This platform served as robotic assistants that were capable of obstacle avoidance in locomotion, vehicle-arm coordination in manipulation, and decentralized cooperation of multiple mobile manipulators [3-8]. Other works examined vehicle-arm coordination in terms of reactive control, motion planning, and human- robot interface [9-13].

In industry environments, efforts were put into increased reliability by error avoidance and error recovery. A combination of ultrasonic ranger sensors, laser sensors, and stereo cameras were used to ensure the reliability of navigation. Mobile manipulation was applied to move between several workstations, locate assembly parts, and perform assembly tasks autonomously[14].

Since 2000, the advance of autonomous navigation technology for example SLAM (simultaneous localization and mapping)[15] combined with the advance of sensing technology for example Lidar (light detection and ranging) brought a boom to mobile manipulation. Mobile manipulators have been used in home and healthcare [16-19], space [20, 21], and industry [22, 23].

Dynamic environments where obstacles are unpredictable and moving have been the focus of many works [23-27].The combination of task and motion planning were dealt with by probability robotics and optimization approach [28-30]. Novel human-robot interfaces such as virtual button activated by pointing an off-the-shelf green laser pointer were proposed [31].

However, the majority of the existing mobile manipulators were designed to perform tasks in indoor environments under well-controlled lightening conditions. However, mobile manipulators for sample collection will have to travel from the lab to the plant via a shared path with pedestrians and other vehicles and performing sample collection under natural lighting conditions regardless day time and night time. This full-weather all-time operation poses substantial challenges to the current technology of mobile manipulation.

This paper is to discuss these various challenges faced by mobile manipulation for sample collection in industrial environments in terms of control and path planning, localization, visual-based navigation and visual servoing, and robotic manipulation.

2. Application Requirements of Autonomous Robotic Manipulation for Outdoor Sample Collection

The application requirements for robotics can be classified into 12 areas: sustainability, configuration, adaptation,

autonomy, positioning, manipulation and grasping, robot-robot interaction, human-robot interaction, process quality, and physical properties [32].

In a typical refinery as in Fig. 1, the lab and the processing plant are connected via a path shared by pedestrians and vehicles, and a number of valves are evenly distributed in the processing plant. One sample from each of the six valves needs to be collected at an interval of six hours seven days a week.

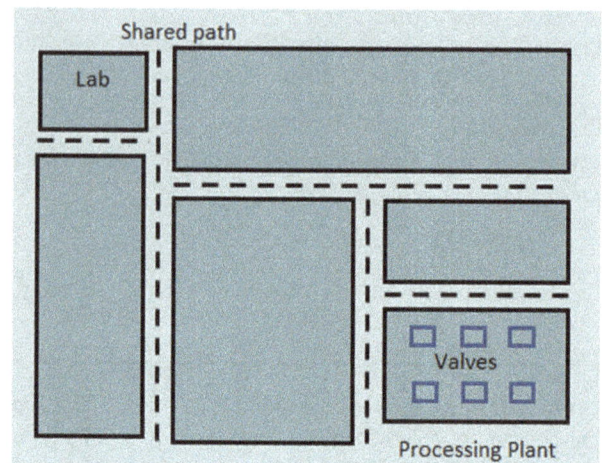

Figure 1. *A typical floor plan of a refinery.*

We propose to automate this operation of sample collection using a mobile manipulator consisting of a mobile base and an industrial robotic arm. The mobile base is equipped with GNSS (Global Navigation Satellite System) for localization and Lidar for path planning and obstacle avoidance. The robotic arm equipped with a stereo computer camera performs the manipulation task after the platform reaches the processing plant: Pick-up a jar from a basket, remove the lid of the jar, put the jar under the valve, switch on the valve, switch off the valve after liquid sample fills in the jar, screw back the lid, and put the jar on the basket.

Sample collection is a typical application of mobile manipulation in industrial environments, where robustness and error recovery under sensory variation, noise, and clutter are of paramount importance for a successful implementation.

Next we shall discuss the challenges faced by control and path planning, GNSS, Lidar and stereo computer vision, and manipulation and propose our solutions.

3. Challenges and Solutions for Autonomous Sample Collection

3.1. Mobile Robot Positioning/Attitude Using GNSS

3.1.1. Proliferation of Satellite Navigation Technologies

GNSS utilizes global navigation satellite systems for precise positioning and navigation, and it can solve real-world problems in various applications including positioning and navigation of moving platforms in land, sea, air, and space [33-35]. Next to the familiar and widely used Global

Positioning System (GPS) of the United States of America, new and modern GNSSs are currently becoming operational, like BeiDou from China, Galileo from Europe, and Glonass-K from Russia [36, 37]. Moreover, these global systems will be joined with new regional navigation satellite systems from Japan (QSZZ) and India (IRNSS), thus leading to a system of systems consisting of more than one hundred satellites. This proliferation of satellite navigation technologies will fundamentally change the positioning and navigation landscape, thus allowing for the development of exciting and challenging new applications [38, 39]. The integration of systems and the inclusion of more tracked satellites will enable robustification and improvement of the availability, reliability and accuracy of the robotic navigation solutions, in particular under constrained environments such as urban canyons [40] and open-pits [41].

3.1.2. Array-Based, Multi-GNSS Carrier-Phase Platform Navigator

In this project, these challenges will be met by developing an array-based, multi-GNSS carrier-phase platform navigator for the fetching robotic vehicle regardless of weather and time, thus facilitating continuous operation of the vehicle.

The proposed navigation system will consist of an array of GNSS antennas rigidly mounted to the robotic vehicle and a single reference station on top of a nearby building (Figure 2).

First, attitude (orientation) of the vehicle is determined using the Multi-variate Constrained Least-squares AMBiguity Decorrelation Adjustment (MC-LAMBDA) method utilizing known body frame geometry [42]. Non-linear constraints due to known antenna geometry effectively enhance carrier phase ambiguity resolution enabling instantaneous precise attitude determination. Further constraints due to the fact that the vehicle maintains levelled frame will be explored further strengthening the underlying attitude model. These constraints are further utilized using the array-aided approach for improving the positioning of the vehicle with respect to the reference station [43, 44].

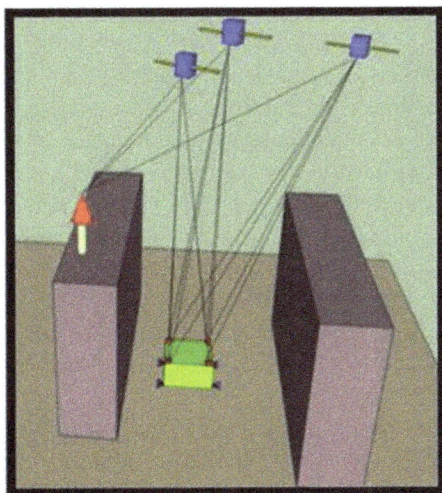

Figure 2. *Navigation System for the fetching robotic vehicle.*

As the robotic vehicle is proposed to travel along a path

surrounded by buildings and other structures (Figure 1), the satellite visibility will be hampered by masking effects of the surrounding structures. Making use of the location, the project will explore next generation multi-GNSS for navigation and positioning of the robotic vehicle. The Asia-Pacific region becomes the world testbed for new generation GNSS due to the developments of the Chinese BeiDou, and the Japanese QZSS, and the Indian IRNSS. The number of satellites is expected to increase much faster in this region than in any other area of the world (Figure 3).

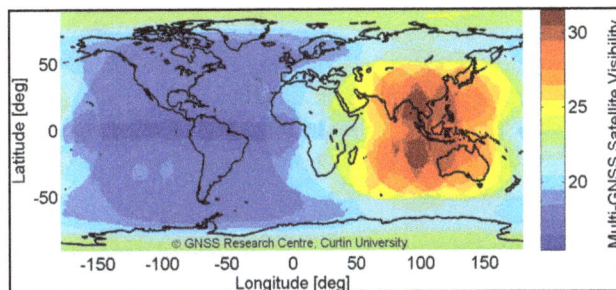

Figure 3. *Global satellite visibility for GNSS/RNSS.*

Integration of multi-GNSS data [37, 44, 45] will enhance reliability and availability of navigation solution in satellite deprived environment (Figure 2).

3.2. Vision-Based Navigation and Visual Servoing

3.2.1. Lack of Robust and Cheap Navigation Sensors

One of the key factors that have impeded the greater adoption of autonomous mobile robots is the lack of robust and cheap navigation sensors. Computer vision technology has made significant progress over the last decade that it is now becoming possible for camera to perform a range of different physical measurements (i.e. ranging, 3D reconstruction and etc.) that were traditionally done with specialised high cost sensors (i.e. laser scanner, sonar and etc.). These vision-based systems analyse the sequence of images taken from the outdoor scene from the camera/s attached to the vehicle platform and extracting the visual cues which are the used to plan the action of the vehicle [46] The actual camera system used depends on the purpose; camera can be monocular, multi-view and omnidirectional to meet various needs.

In recent years, much attention is being devoted to solve the problems of translating the visual information obtained from camera for navigating a vehicle autonomously through the environment [47]. To do this requires the determination of (1) the 3D scene geometry, and (2) the camera orientation with respect to the scene. Multi-view stereo is an established technique used to recover the 3D surface profile of the scene. It relies on capturing multiple views of the scene to provide the depth disparity needed to construct the 3D scene geometry [48].

3.2.2. Structure from Motion Techniques and Visual Simultaneous Localisation and Mapping

Advances in Structure from Motion techniques which generate 3D scene from image sequences provide the advantage of lower hardware cost suitable for navigating autonomous platforms. Visual SLAM (Simultaneous Localisation and Mapping) techniques provide the added benefit of estimating the location of the platform in addition to performing the mapping function. The robustness of VSLAM techniques have improved significantly and this technique been implemented in many navigation system both indoors and outdoors [49]. VSLAM has also been used in mobile devices in the form of applications to deliver augmented reality experiences to consumers. In the area of robotic manipulation, visual servoing techniques are commonly adopted to provide a visual feedback as the manipulation follows a sequence of predetermine path [50]. The camera can either be attached to the robotic hand (i.e. Eye-in-Hand) or fixed to the world away from the hand (Hand to Eye). This is ideal for scenarios where accurate feedback of the robotic actuator is needed for say opening a valve or the lid of the sample collection jar.

3.3. Robotic Manipulation

3.3.1. Uncertainties in Manipulation

Currently and in the foreseeable future, robotic arms will have to perform tasks [51] using uncertain, even piecemeal views of the world due to the limitation of the sensor and actuation technologies. Increasing the accuracy of robots required for fine-manipulation [52-57] in industrial environments is expensive and ultimately stifling.

Uncertainties may be caused by robot positioning errors and vision systems that can only provide piecemeal views of the world due to low resolution and/or occlusion. As a result, the uncertainties could exceed the accuracy required by sample collection tasks. Complex manipulation tasks under uncertainties have to be performed using multiple sensors such as vision and force/torque, tactile, and distance sensors to increase the robustness.

Uncertainties in manipulation have been coped with from different perspectives. Su and Lee [58] developed the propagation of uncertainty before and after a primitive action to integrate the uncertainty information into a task plan that consisted of a sequence of primitive actions. Li and Payandeh [59] design the forces exerted on the object by agents with which the object can follow a given trajectory in spite of the uncertainty on pressure distribution. Hsiao, Kaelbling and Lozano-Perez [60] provided a method for planning under uncertainty for robotic manipulation by partitioning the configuration space into a set of regions that were closed under compliant motions. Berenson, Srinivasa and Kuffner [61] present an efficient approach to generating paths for a robotic manipulator that are collision-free and guaranteed to meet task specifications despite pose uncertainty. Stulp et al [62] presented a simplified version of a model-free reinforcement learning algorithm to simultaneously learn shape parameters and goal parameters of motion primitives and use shape and goal learning to acquire motion primitives that were robust to object pose uncertainty.

In sample collection, the uncertainties mainly come from the positioning errors of the mobile base and the errors of object pose identification of the computer vision system. We propose to use guarded moves [63], in which a robot reduces the uncertainty by accomplishing relative motion and/or controlled dynamic interaction between the end-effector and the objects.

3.3.2. Automatic Generation of Guarded Moves

Automatic generation of guarded moves has been studied from the point of view of assembly tasks. Lozano-Perez, Mason and Taylor [64] presented the synthesis of compliant motion strategies from geometric description of assembly operations and explicit estimates of errors in sensing and control. Donald [65] presented a formal framework for computing motion strategies in the presence of uncertainties arising from sensing errors, control errors, and uncertainty in the geometric models of the environment and of the robot. Xiao and Zhang [66] developed a general geometric simulator allowing flexible design of task environments and modelling of nominal and uncertainty parameters to run the algorithms and simulating the kinematic robot motions guided by the replanning algorithms in the presence of uncertainties. LaValle and Hutchinson [67] developed a general framework for determining sensor-based robot plans by blending ideas from stochastic optimal control and dynamic game theory with traditional preimage backchaining concepts.

For the sample collection in a refinery, the key steps in robotic manipulation include putting a jar under a tap, switching on a valve, and switching it off after liquid sample filling the jar. These operations need accurate positioning. Hence it is indispensable for the robotic arm to cope with the uncertainties arising from the positioning/attitude errors of the mobile base and the errors of object identification of the computer vision system. Guarded moves provide very precise information about the relative poses of robotic arm and environment, thus are essential to mobile manipulation in a refinery.

3.4. Mobile Robot Path Planning and Control

3.4.1. Path Following of Holonomic and Noholonomic Mobile Platforms

As discussed in Section 1, the autonomous mobile manipulator consists of a manipulator mounted on a mobile platform. It combines the dextrous manipulation capability offered by fixed-base manipulators and the mobility offered by mobile platforms. However, the mobile manipulator brings about a number of challenging problems in path-planning and control. The following fundamental issues need to be addressed for carrying out tasks of the proposed mobile manipulators:

(i) How can we plan the effective motion trajectory of a mobile manipulator under both holonomic and nonholonomic constraints?

(ii) How do we design the hybrid motion/force control and hybrid position/ force for the mobile manipulator since it needs to interact with the environments including collision avoidance, carrying out operations at valve locations, recharge batteries, and so on?

A wheeled mobile manipulator is fundamentally an

underactuated system subject to nonholonomic constraints. A combination of a wheeled robot and a multi-link manipulator also creates kinematic redundancy. Moreover, the wheeled robot and the manipulator dynamically interact with each other. In addition, the environment contains both stationary and moving obstacles. A good survey of the recent development in terms of nonholonomic motion planning is given by Li and Canny [68]. There are many studies on motion planning of mobile robots using various approaches, e.g., potential field [69], graph search algorithms [70], the A* algorithm [71], Bellman-Ford algorithms [72], the wavefront algorithm [73], and visibility graph approaches [74]. For the motion generation plan planning for mobile manipulators, since mobility is the main concern, the approaches are similar to motion planning for mobile robots. However, the problem is to choose/derive an appropriate path-planning method so that it can be incorporated in a design of the motion control system.

The mobile manipulator needs to park at valve, sample returning, and battery recharging locations, and to move along the planned-path. Thus, we need to solve the control problems consisting of fixed-point stabilization, and [75-79]; trajectory-tracking, which deals with the design of controllers that force a mobile robot to reach and follow a time parameterized reference trajectory (i.e., a geometric path with an associated timing law) [78-85]; or path-following, in which the robot is required to converge to and follow a reference path that is specified without a temporal law (i.e., dealing with the design of controllers driving the robot's trajectories to a maneuver up to time re-parameterization) [75, 86-98]. The path-following task is more suitable for the proposed mobile manipulator since it can use the path-derivative as an additional control input (giving more robustness) and the sample collection time is not so strict.

Existing solutions to path-following control of mobile robots can be roughly classified into four main methods. In the first method referred to as the Serret-Frenet one, the Serret-Frenet frame is used to define the path-following errors, i.e., the cross-track and heading angle errors, then the control inputs (velocities when only kinematics is considered or torques applied to the driving wheels of the robot when both kinematics and dynamics are considered) are designed to stabilize these errors at the origin (e.g., [75, 86, 87]). Due to singularity in the cross-track error dynamics, this approach requires the robot's position to be within a tube, of which the center-line is the path and the radius is less than inverse of the path's curvature., i.e., only local results are obtained for curved paths. To resolve the singularity, a combination of the trajectory-tracking and path-following using the Serret-Frenet approach in the sense that the lateral path-following error is not always set to zero (to avoid singularity in the cross-track error dynamics) and that the path-parameter is used as an additional input to control the lateral path-following error. Thus, global control results are usually obtained. Exemplary works includes [88-90].

The second method defines the path-following objective as the one of forcing a robot to follow a virtual robot moving on the reference path (e.g., [91-93] and [94], Chapter 14, Section 14.1.4). Polar coordinates are used to interpret the path-following errors, i.e., the distance and angle between the real and virtual robots. Roughly speaking, the approach is to steer the robot such that it heads toward the virtual robot and eliminates the distance between itself and the virtual robot. This approach requires the robot not to be too close to the path to avoid singularity due to the polar coordinate representation.

The third method is referred to as the transverse feedback linearization (TFL) method [96-98] (see also [99] for an extension to N-trailer robots, and [100] for a consideration of PVTOL aircraft). This method involves with conversion of the path-following problem to an input-output feedback linearization problem (cascaded with a zero dynamics problem) with respect to an appropriate output, which usually defines the reference path. In this context, the TFL method is related to the differential flatness approach [101]. This method usually achieves local results.

The fourth method referred to as the level curve one has been recently introduced in [95] (see also [102] for an extension to a three dimensional path-following problem) for design of a path-following controller for unicycle-type mobile robot. This method is based on the observation: if the position of the robot satisfies the equation of the reference path, then the robot must be on the path. Thus, similar to the TFL method neither distance from the robot to the reference path nor virtual moving robot is needed. Although it is related to the TFL method, there is a vital difference between these two methods. The suitable output is differentiated till the control inputs appear in the TFL method while the level curve method directly control this output. This difference can be understood as the difference between the feedback linearization control design [103] (Chapter 13) and the backstepping method [104] (Chapter 2). The level curve method is also used to design a global path-following controller for underactuated ships in [105]. An initial work in this approach for global path-following control of mobile robot is given in [106].

3.4.2. Integrated Approach to Path Planning

Path-planning: The floor and structure maps from the operational office, and various curve fitting algorithms are used to generate a preplanned-path. Valves, sample returning, and batteries recharging locations are marked on this preplanned-path. The preplanned-path is served as a preliminary path for the mobile manipulator to follow and to park at the above locations. The preplanned-path is then on-line deformed if necessary for unforeseen stationary or moving obstacles by an algorithm embedded in the mobile robot motion control system.

Mobile robot motion control: The positioning (GNSS, computer vision, and local sensing devices: ultrasonic, infrared, etc.) and manipulator-health information is used for the mobile robot motion control design. The artificial potential field is incorporated into the level curve motion control approach for both mobile robot motion control and obstacle avoidance. The initial work in [106, 107] is to be further explored in conjunction with [79] for a design of a mobile robot motion control system that can perform both stabilization (parking) and path-following objectives.

Robot manipulator control: Information from the manipulator-health, and location activity are used to activate this control. When the mobile manipulator is parked at a desired location, a signal string is sent to the robot manipulator to activate a preprogramed software embedded in the manipulator for valve or sample returning or battery recharging operations.

4. Conclusion

This paper reviewed the challenges faced by robotic manipulation for sample collection and proposed solutions to these challenges. Mobile robot would explore next generation multi-GNSS for navigation and positioning / attitude of the robotic vehicle. Vision-based navigation and servoing would use the advances in structure from motion (SfM) techniques which generate 3D scene from image sequences. Robotic Robotic manipulation would make use of guarded moves to reduce uncertainties in unstructured environments. Mobile robot path planning and control would integrate the artificial potential field into the level curve motion control approach for both mobile robot motion control and obstacle avoidance. These proposed solutions pointed a road to a successful implementation of mobile manipulation for sample collection.

References

[1] J. Schuler, *Integration von Förder- und Handhabungseinrichtungen*: Springer Berlin Heidelberg, 2013.

[2] C. R. Weisbin, J. Barhen, G. de Saussure, W. R. Hamel, C. Jorgensen, J. L. Lucius, E. M. Oblow, and T. Swift, *HERMIES-I: a mobile robot for navigation and manipulation experiments*, 1985.

[3] O. Khatib, "Mobile manipulation: The robotic assistant," *Robotics and Autonomous Systems*, vol. 26, no. 2–3, pp. 175-183, 2/28/, 1999.

[4] O. Khatib, K. Yokoi, K. Chang, D. Ruspini, R. Holmberg, and A. Casal, "Coordination and decentralized cooperation of multiple mobile manipulators," *Journal of Robotic Systems*, vol. 13, no. 11, pp. 755-764, 1996.

[5] O. Khatib, K. Yokoi, K. Chang, D. Ruspini, R. Holmberg, A. Casal, and A. Baader, "Force Strategies for Cooperative Tasks in Multiple Mobile Manipulation Systems," *Robotics Research*, G. Giralt and G. Hirzinger, eds., pp. 333-342: Springer London, 1996.

[6] O. Brock, and O. Khatib, "Elastic Strips: Real-Time Path Modification for Mobile Manipulation," *Robotics Research*, Y. Shirai and S. Hirose, eds., pp. 5-13: Springer London, 1998.

[7] O. Khatib, K. Yokoi, K. Chang, D. Ruspini, R. Holmberg, and A. Casal, "Vehicle/arm coordination and multiple mobile manipulator decentralized cooperation." pp. 546-553 vol.2.

[8] O. Khatib, K. Yokoi, O. Brock, K. Chang, and A. Casal, "Robots in Human Environments: Basic Autonomous Capabilities," *The International Journal of Robotics Research*, vol. 18, no. 7, pp. 684-696, July 1, 1999, 1999.

[9] Y. Yamamoto, and X. Yun, "Coordinating locomotion and manipulation of a mobile manipulator." pp. 2643-2648 vol.3.

[10] J. M. Cameron, D. C. MacKenzie, K. R. Ward, R. C. Arkin, and W. J. Book, "Reactive control for mobile manipulation." pp. 228-235 vol.3.

[11] U. D. Hanebeck, C. Fischer, and G. Schmidt, "ROMAN: a mobile robotic assistant for indoor service applications." pp. 518-525 vol.2.

[12] W. Feiten, B. Magnussen, G. Hager, and K. Toyama, "Modeling and Control for Mobile Manipulation in Everyday Environments," *Robotics Research*, Y. Shirai and S. Hirose, eds., pp. 14-22: Springer London, 1998.

[13] U. M. Nassal, and R. Junge, "Fuzzy control for mobile manipulation." pp. 2264-2269 vol.3.

[14] T. C. Lueth, U. M. Nassal, and U. Rembold, "Reliability and integrated capabilities of locomotion and manipulation for autonomous robot assembly," *Robotics and Autonomous Systems*, vol. 14, no. 2–3, pp. 185-198, 5//, 1995.

[15] H. A. Wurdemann, E. Georgiou, L. Cui, and J. S. Dai, "SLAM Using 3D Reconstruction via a Visual RGB and RGB-D Sensory Input," in ASME 2011 International Design Engineering Technical Conferences and Computers and Information in Engineering Conference, 2011, pp. 615-622.

[16] C. Borst, T. Wimböck, F. Schmidt, M. Fuchs, B. Brunner, F. Zacharias, P. R. Giordano, R. Konietschke, W. Sepp, and S. Fuchs, "Rollin'Justin-Mobile platform with variable base." pp. 1597-1598.

[17] P.-G. Plöger, and B. Nebel, "The DESIRE Service Robotics Initiative," *KI*, vol. 22, no. 4, pp. 29-32, 2008.

[18] J. Bohren, R. B. Rusu, E. G. Jones, E. Marder-Eppstein, C. Pantofaru, M. Wise, L. Mosenlechner, W. Meeussen, and S. Holzer, "Towards autonomous robotic butlers: Lessons learned with the PR2." pp. 5568-5575.

[19] B. Graf, U. Reiser, Ha, x, M. gele, K. Mauz, and P. Klein, "Robotic home assistant Care-O-bot® 3 - product vision and innovation platform." pp. 139-144.

[20] R. O. Ambrose, R. T. Savely, S. M. Goza, P. Strawser, M. A. Diftler, I. Spain, and N. Radford, "Mobile manipulation using NASA's Robonaut." pp. 2104-2109 Vol.2.

[21] P. Schenker, T. Huntsberger, P. Pirjanian, E. Baumgartner, and E. Tunstel, "Planetary Rover Developments Supporting Mars Exploration, Sample Return and Future Human-Robotic Colonization," *Autonomous Robots*, vol. 14, no. 2-3, pp. 103-126, 2003/03/01, 2003.

[22] R. Bischoff, U. Huggenberger, and E. Prassler, "KUKA youBot - a mobile manipulator for research and education." pp. 1-4.

[23] M. Hvilshj, S. Bgh, O. Madsen, and M. Kristiansen, "The mobile robot "Little Helper": Concepts, ideas and working principles." pp. 1-4.

[24] Y. Yang, and O. Brock, "Elastic roadmaps—motion generation for autonomous mobile manipulation," *Autonomous Robots*, vol. 28, no. 1, pp. 113-130, 2010/01/01, 2010.

[25] O. Brock, O. Khatib, and S. Viji, "Task-consistent obstacle avoidance and motion behavior for mobile manipulation." pp. 388-393 vol.1.

[26] P. Ogren, M. Egerstedt, and X. Hu, "Reactive mobile manipulation using dynamic trajectory tracking." pp. 3473-3478 vol.4.

[27] S. Chitta, E. G. Jones, M. Ciocarlie, and K. Hsiao, "Perception, planning, and execution for mobile manipulation in unstructured environments," *IEEE Robotics and Automation Magazine, Special Issue on Mobile Manipulation,* vol. 19, no. 2, pp. 58-71, 2012.

[28] D. Berenson, J. Kuffner, and H. Choset, "An optimization approach to planning for mobile manipulation." pp. 1187-1192.

[29] M. Egerstedt, and H. Xiaoming, "Coordinated trajectory following for mobile manipulation." pp. 3479-3484 vol.4.

[30] C. Dornhege, M. Gissler, M. Teschner, and B. Nebel, "Integrating symbolic and geometric planning for mobile manipulation." pp. 1-6.

[31] H. Nguyen, A. Jain, C. Anderson, and C. C. Kemp, "A clickable world: Behavior selection through pointing and context for mobile manipulation." pp. 787-793.

[32] R. Bischoff, and T. Guhl, "The strategic research agenda for robotics in europe [industrial activities]," *IEEE Robotics & Automation Magazine,* vol. 1, no. 17, pp. 15-16, 2010.

[33] P. J. Teunissen, and A. Kleusberg, *GPS for Geodesy*: Springer Science & Business Media, 2012.

[34] B. Hoffman-Wellenhof, H. Lichtenegger, and E. Wasle, "GNSS—Global Navigation Satellite Systems," *GPS, GLONASS, Galileo and more. Wien: Springer-Verlag,* 2008.

[35] P. J. Teunissen, G. Giorgi, and P. J. Buist, "Testing of a new single-frequency GNSS carrier phase attitude determination method: land, ship and aircraft experiments," *GPS solutions,* vol. 15, no. 1, pp. 15-28, 2011.

[36] C. Rizos, M. Higgins, and G. Johnston, "Impact of next generation GNSS on Australasian geodetic infrastructure," 2010.

[37] R. Odolinski, P. J. Teunissen, and D. Odijk, "Combined BDS, Galileo, QZSS and GPS single-frequency RTK," *GPS Solutions,* vol. 19, no. 1, pp. 151-163, 2015.

[38] V. Ashkenazi, D. Park, and M. Dumville, "Robot positioning and the global navigation satellite system," *Industrial Robot: An International Journal,* vol. 27, no. 6, pp. 419-426, 2000.

[39] W. Stempfhuber, and M. Buchholz, "A precise, low-cost RTK GNSS system for UAV applications," *International Archives of Photogrammetry, Remote Sensing and Spatial Information Science,* vol. 38, pp. 1-C22, 2011.

[40] D. N. Aloi, and O. V. Korniyenko, "Comparative performance analysis of a Kalman filter and a modified double exponential filter for GPS-only position estimation of automotive platforms in an urban-canyon environment," *Vehicular Technology, IEEE Transactions on,* vol. 56, no. 5, pp. 2880-2892, 2007.

[41] L. Johnson, and F. Van Diggelen, "Advantages of a combined GPS+ GLONASS precision sensor for machine control applications in open pit mining." pp. 549-554.

[42] P. Teunissen, "A general multivariate formulation of the multi-antenna GNSS attitude determination problem," *Artificial Satellites,* vol. 42, no. 2, pp. 97-111, 2007.

[43] P. J. Teunissen, "A-PPP: array-aided precise point positioning with global navigation satellite systems," *Signal Processing, IEEE Transactions on,* vol. 60, no. 6, pp. 2870-2881, 2012.

[44] N. Nadarajah, J.-A. Paffenholz, and P. J. Teunissen, "Integrated GNSS attitude determination and positioning for direct geo-referencing," *Sensors,* vol. 14, no. 7, pp. 12715-12734, 2014.

[45] P. Teunissen, R. Odolinski, and D. Odijk, "Instantaneous BeiDou+ GPS RTK positioning with high cut-off elevation angles," *Journal of geodesy,* vol. 88, no. 4, pp. 335-350, 2014.

[46] G. N. DeSouza, and A. C. Kak, "Vision for mobile robot navigation: A survey," *Pattern Analysis and Machine Intelligence, IEEE Transactions on,* vol. 24, no. 2, pp. 237-267, 2002.

[47] R. Hartley, and A. Zisserman, *Multiple view geometry in computer vision*: Cambridge university press, 2003.

[48] H.-H. Vu, P. Labatut, J.-P. Pons, and R. Keriven, "High accuracy and visibility-consistent dense multiview stereo," *Pattern Analysis and Machine Intelligence, IEEE Transactions on,* vol. 34, no. 5, pp. 889-901, 2012.

[49] A. Kim, and R. M. Eustice, "Real-time visual SLAM for autonomous underwater hull inspection using visual saliency," *Robotics, IEEE Transactions on,* vol. 29, no. 3, pp. 719-733, 2013.

[50] C. Teuliere, and E. Marchand, "A dense and direct approach to visual servoing using depth maps," *Robotics, IEEE Transactions on,* vol. 30, no. 5, pp. 1242-1249, 2014.

[51] J. S. Dai, D. Wang, and L. Cui, "Orientation and Workspace Analysis of the Multifingered Metamorphic Hand-Metahand," *IEEE Transactions on Robotics,* vol. 25, no. 4, pp. 942-947, 2009.

[52] L. Cui, and J. S. Dai, "A Darboux-Frame-Based Formulation of Spin-Rolling Motion of Rigid Objects with Point Contact," *IEEE Transactions on Robotics,* vol. 26, no. 2, pp. 383-388, 2010.

[53] L. Cui, and J. S. Dai, "Posture, Workspace, and Manipulability of the Metamorphic Multifingered Hand With an Articulated Palm," *ASME Journal of Mechanisms and Robotics,* vol. 3, no. 2, pp. 021001_1-7, 2011.

[54] L. Cui, and J. S. Dai, "Reciprocity-Based Singular Value Decomposition for Inverse Kinematic Analysis of the Metamorphic Multifingered Hand," *ASME Journal of Mechanisms and Robotics,* vol. 4, no. 3, pp. 034502_1_6, 2012.

[55] L. Cui, U. Cupcic, and J. S. Dai, "An Optimization Approach to Teleoperation of the Thumb of a Humanoid Robot Hand: Kinematic Mapping and Calibration," *Journal of Mechanical Design,* vol. 136, no. 9, pp. 091005_1_7, 2014.

[56] L. Cui, and J. S. Dai, "A Polynomial Formulation of Inverse Kinematics of Rolling Contact," *Journal of Mechanisms and Robotics,* vol. 7, no. 4, pp. 041003_1-9, 2015.

[57] L. Cui, and J. S. Dai, "From sliding–rolling loci to instantaneous kinematics: An adjoint approach," *Mechanism and Machine Theory,* vol. 85, no. 0, pp. 161-171, 3//, 2015.

[58] S.-F. Su, and C. S. G. Lee, "Manipulation and propagation of uncertainty and verification of applicability of actions in assembly tasks," *IEEE Transactions on Systems, Man, and Cybernetics,* vol. 22, no. 6, pp. 1376-1389, 1992.

[59] L. Qingguo, and S. Payandeh, "Planning for dynamic multiagent planar manipulation with uncertainty: a game theoretic approach," *IEEE Transactions on Systems, Man and Cybernetics, Part A: Systems and Humans,* vol. 33, no. 5, pp. 620-626, 2003.

[60] K. Hsiao, L. P. Kaelbling, and T. Lozano-Perez, "Grasping POMDPs." pp. 4685-4692.

[61] D. Berenson, S. S. Srinivasa, and J. J. Kuffner, "Addressing pose uncertainty in manipulation planning using Task Space Regions." pp. 1419-1425.

[62] F. Stulp, E. Theodorou, M. Kalakrishnan, P. Pastor, L. Righetti, and S. Schaal, "Learning motion primitive goals for robust manipulation." pp. 325-331.

[63] J. A. Bagnell, F. Cavalcanti, L. Cui, T. Galluzzo, M. Hebert, M. Kazemi, M. Klingensmith, J. Libby, L. Tian Yu, N. Pollard, M. Pivtoraiko, J. S. Valois, and R. Zhu, "An integrated system for autonomous robotics manipulation." pp. 2955-2962.

[64] T. Lozano-Pérez, M. T. Mason, and R. H. Taylor, "Automatic Synthesis of Fine-Motion Strategies for Robots," *The International Journal of Robotics Research,* vol. 3, no. 1, pp. 3-24, March 1, 1984, 1984.

[65] B. R. Donald, "A geometric approach to error detection and recovery for robot motion planning with uncertainty," *Artificial Intelligence,* vol. 37, no. 1–3, pp. 223-271, 12//, 1988.

[66] X. Jing, and Z. Lixin, "A geometric simulator SimRep for testing the replanning approach toward assembly motions in the presence of uncertainties." pp. 171-177.

[67] S. M. LaValle, and S. A. Hutchinson, "An objective-based stochastic framework for manipulation planning." pp. 1772-1779 vol.3.

[68] Z. X. Li, and J. Canny, "Motion of Two Rigid Bodies with Rolling Constraint," *IEEE Transactions on Robotics and Automation,* vol. 6, no. 1, pp. 62-72, 1990.

[69] O. Khatib, "Real-Time Obstacle Avoidance for Manipulators and Mobile Robots," *The International Journal of Robotics Research,* vol. 5, no. 1, pp. 90-98, March 1, 1986, 1986.

[70] T. Lozano-Pérez, and M. A. Wesley, "An algorithm for planning collision-free paths among polyhedral obstacles," *Communications of the ACM,* vol. 22, no. 10, pp. 560-570, 1979.

[71] J. J. Kuffner, and S. M. LaValle, "RRT-connect: An efficient approach to single-query path planning." pp. 995-1001.

[72] D. Kortenkamp, R. P. Bonasso, and R. Murphy, *Artificial intelligence and mobile robots: case studies of successful robot systems*: MIT Press, 1998.

[73] L. Dorst, and K. Trovato, "Optimal path planning by cost wave propagation in metric configuration space." pp. 186-197.

[74] J.-C. Latombe, *Robot motion planning*: Springer Science & Business Media, 2012.

[75] C. C. de Wit, B. Siciliano, and G. Bastin, *Theory of robot control*: Springer Science & Business Media, 2012.

[76] A. Astolfi, "Discontinuous control of the Brockett integrator." pp. 4334-4339.

[77] Z. Cao, L. Yin, Y. Fu, and J. S. Dai, "Adaptive dynamic surface control for vision-based stabilization of an uncertain electrically driven nonholonomic mobile robot," *Robotica,* pp. 1-19.

[78] K. D. Do, Z.-P. Jiang, and J. Pan, "Simultaneous tracking and stabilization of mobile robots: an adaptive approach," *IEEE Transactions on Automatic Control,* vol. 49, no. 7, pp. 1147-1152, 2004.

[79] K. D. Do, Z.-P. Jiang, and J. Pan, "A global output-feedback controller for simultaneous tracking and stabilization of unicycle-type mobile robots," *IEEE Transactions on Robotics and Automation,* vol. 20, no. 3, pp. 589-594, 2004.

[80] T.-C. Lee, K.-T. Song, C.-H. Lee, and C.-C. Teng, "Tracking control of unicycle-modeled mobile robots using a saturation feedback controller," *Control Systems Technology, IEEE Transactions on,* vol. 9, no. 2, pp. 305-318, 2001.

[81] W. E. Dixon, D. M. Dawson, E. Zergeroglu, and A. Behal, *Nonlinear control of wheeled mobile robots*: Springer-Verlag New York, Inc., 2001.

[82] W. E. Dixon, M. S. de Queiroz, D. M. Dawson, and T. J. Flynn, "Adaptive tracking and regulation of a wheeled mobile robot with controller/update law modularity," *Control Systems Technology, IEEE Transactions on,* vol. 12, no. 1, pp. 138-147, 2004.

[83] S. Blažič, "A novel trajectory-tracking control law for wheeled mobile robots," *Robotics and Autonomous Systems,* vol. 59, no. 11, pp. 1001-1007, 2011.

[84] J. Huang, C. Wen, W. Wang, and Z.-P. Jiang, "Adaptive output feedback tracking control of a nonholonomic mobile robot," *Automatica,* vol. 50, no. 3, pp. 821-831, 2014.

[85] J. M. Toibero, F. Roberti, R. Carelli, and P. Fiorini, "Switching control approach for stable navigation of mobile robots in unknown environments," *Robotics and Computer-Integrated Manufacturing,* vol. 27, no. 3, pp. 558-568, 2011.

[86] C. Samson, "Velocity and torque feedback control of a nonholonomic cart," *Advanced robot control,* pp. 125-151: Springer, 1991.

[87] R. Fierro, and F. L. Lewis, "Control of a nonholonomic mobile robot: backstepping kinematics into dynamics." pp. 3805-3810.

[88] D. Soetanto, L. Lapierre, and A. Pascoal, "Adaptive, non-singular path-following control of dynamic wheeled robots." pp. 1765-1770.

[89] K. D. Do, "Bounded controllers for global path tracking control of unicycle-type mobile robots," *Robotics and Autonomous Systems,* vol. 61, no. 8, pp. 775-784, 2013.

[90] K. Do, and J. Pan, "Global output-feedback path tracking of unicycle-type mobile robots," *Robotics and Computer-Integrated Manufacturing,* vol. 22, no. 2, pp. 166-179, 2006.

[91] M. Aicardi, G. Casalino, A. Bicchi, and A. Balestrino, "Closed loop steering of unicycle like vehicles via Lyapunov techniques," *Robotics & Automation Magazine, IEEE,* vol. 2, no. 1, pp. 27-35, 1995.

[92] M. Egerstedt, X. Hu, and A. Stotsky, "Control of mobile platforms using a virtual vehicle approach," *Automatic Control, IEEE Transactions on,* vol. 46, no. 11, pp. 1777-1782, 2001.

[93] S. Sun, and P. Cui, "Path tracking and a practical point stabilization of mobile robot," *Robotics and Computer-Integrated Manufacturing,* vol. 20, no. 1, pp. 29-34, 2004.

[94] K. D. Do, and J. Pan, *Control of ships and underwater vehicles: design for underactuated and nonlinear marine systems*: Springer Science & Business Media, 2009.

[95] A. Morro, A. Sgorbissa, and R. Zaccaria, "Path following for unicycle robots with an arbitrary path curvature," *Robotics, IEEE Transactions on,* vol. 27, no. 5, pp. 1016-1023, 2011.

[96] C. Nielsen, and M. Maggiore, "Maneuver regulation via transverse feedback linearization: Theory and examples." pp. 59-66.

[97] C. Nielsen, and M. Maggiore, "Output stabilization and maneuver regulation: A geometric approach," *Systems & control letters,* vol. 55, no. 5, pp. 418-427, 2006.

[98] A. Doosthoseini, and C. Nielsen, "Coordinated path following for a multi-agent system of unicycles." pp. 2894-2899.

[99] M. M. Michałek, "A highly scalable path-following controller for N-trailers with off-axle hitching," *Control Engineering Practice,* vol. 29, pp. 61-73, 2014.

[100] L. Consolini, M. Maggiore, C. Nielsen, and M. Tosques, "Path following for the PVTOL aircraft," *Automatica,* vol. 46, no. 8, pp. 1284-1296, 2010.

[101] M. Fliess, J. Lévine, P. Martin, and P. Rouchon, "Flatness and defect of non-linear systems: introductory theory and examples," *International journal of control,* vol. 61, no. 6, pp. 1327-1361, 1995.

[102] A. Sgorbissa, and R. Zaccaria, "3D path following with no bounds on the path curvature through surface intersection." pp. 4029-4035.

[103] H. K. Khalil, and J. Grizzle, *Nonlinear systems*: Prentice hall New Jersey, 1996.

[104] M. Krstic, P. V. Kokotovic, and I. Kanellakopoulos, *Nonlinear and adaptive control design*: John Wiley & Sons, Inc., 1995.

[105] K. Do, "Global Path-Following Control of Stochastic Underactuated Ships: A Level Curve Approach," *Journal of Dynamic Systems, Measurement, and Control,* vol. 137, no. 7, pp. 071010, 2015.

[106] K. Do, "Global output-feedback path-following control of unicycle-type mobile robots: A level curve approach," *Robotics and Autonomous Systems,* vol. 74, pp. 229-242, 2015.

[107] K. D. Do, "Formation tracking control of unicycle-type mobile robots with limited sensing ranges," *Control Systems Technology, IEEE Transactions on,* vol. 16, no. 3, pp. 527-538, 2008.

15

Low Power Parallel Prefix Adder Design Using Two Phase Adiabatic Logic

Alireza Hassanzadeh[*], Ahmad Shabani

ECE Dept., Shahid Beheshti University, Tehran, Iran

Email address:

a_hassanzadeh@sbu.ac.ir (A. Hassanzadeh), a.shabani@mail.sbu.ac.ir (A. Shabani)

Abstract: In this paper low power implementation of parallel prefix adders using two phase adiabatic logic has been investigated. A new structure has been proposed for the main blocks of parallel prefix adder. Three parallel prefix adders including Kogge-Stone, Brent-Kung and Ripple Carry have been considered. The effects of power clock frequency and loading capacitance on the new blocks have also been considered. Simulation results using 180nm technology parameters and trapezoidal waveform show an average of 34% power reduction in the main building blocks of the adder at 200MHz clock frequency. This power reduces to 54% for sine wave power clock waveform. This research suggests adiabatic implementation of parallel prefix adders for low power microprocessor and signal processing applications.

Keywords: Low Power, Adiabatic Logic, Parallel Prefix, 2PASCL

1. Introduction

In recent years wide use of portable devices, such as cell phones, tablets and GPSs, has demanded for low power electronic circuit. In contrast, increasing the number of transistors in chip area and higher clock frequencies have resulted in higher power consumption of electronic devices. Therefore, design of low power electronic circuits using new nanometer technologies is an important factor in dealing with this phenomena.

Adders are widely used in microprocessors and digital signal processing systems. Since adders are used in many computational algorithms such as multiplication, division, floating point calculations and address generators, the efficiency of the processor can be dependent on the efficiency of the adder module. Many adders have been discussed in the literature that have different speed, power and chip area. Parallel prefix adders are among the high speed methods that are used for large number of bits and high speed applications [1]. Unfortunately, parallel prefix adders consume large chip area and power. Therefore, design of low power parallel prefix adders is important. Many techniques have been introduced to decrease power consumption of parallel prefix adders using lower power supply voltage, node capacitance and switching activity factor. These techniques are not very effective in large scale applications.

Therefore, adiabatic techniques have been considered for these applications [2]. Adiabatic family circuits have offered lower power consumption comparing to standard CMOS logic [3]. Adiabatic circuits are divided into fully adiabatic and quasi-adiabatic circuits. In [4] SCRL (Split-Level Charge Recovery Logic), a charge recovery logic circuit has been introduced that uses a split level method to recover charge in the logic circuit. This method is similar to standard CMOS, but replacing VDD and GND with clock pulses. In [5] 2N-2N2D method has been introduced that uses diodes to implement quasi-adiabatic circuit. Using diodes causes irreversible charge transfer and reduces output voltage. In [2, 6, 7], complimentary pass transistor logic (CPAL) reduces power consumption using pass transistor. In [6] asynchronous adiabatic logic has been implemented using Dual Rail Domino Logic (DRDAAL). This method brings the advantage of asynchronous circuits to adiabatic logic. DRDAAL uses Dual Rail Domino for higher speed and lower Power Delay Product (PDP) comparing to static CMOS and quasi adiabatic logic.

Authors in [8] and [9] introduce another family of adiabatic circuits that have better energy recovery comparing to other methods and uses single phase sine wave for the clock. The current spikes are reduced by a factor of 4 that makes the circuits attractive for power attach resistant and cryptography applications.

This paper discusses power reduction in parallel prefix adders using 2 phase clock pulse (2PASCL) adiabatic method. Instead of using DC power supply a clocked power supply is used and by omitting the series diodes, output voltage has been increased and the voltage drop has been canceled [10,11]. New structures of the Brent-kung, Kogge-Stone blocks have been redesigned using 2PASCL and the simulation results have been compare with static CMOS technology. Conclusion remarks are at the end.

2. Adiabatic Logic

Adiabatic technique has been used for saving power during charge and discharge method. The Adiabatic is a Greek term meaning a change with no loss or heat generation. The term is widely used in thermodynamic systems. In electronics it means charge conservation in each transfer [6]. In adiabatic systems the energy stored in the circuit can return to power supply in certain phases of the circuit operation. If all nodes in a circuit are charged with constant current the circuit will dissipate the least amount of energy. To implement this, the power supply of the circuit waveform is changed from DC to AC waveform [10]. Figure 1 shows charge and discharge method of an adiabatic switching. The rate of switching reduces in adiabatic logic compared to static CMOS, because the power supply waveform has changed.

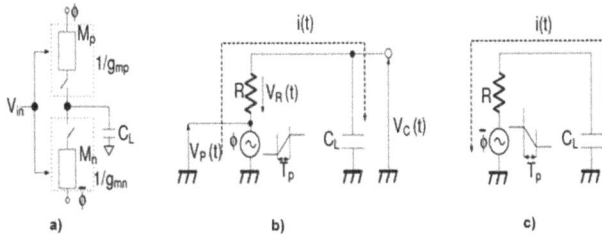

Figure 1. *a) AC powered adiabatic gate, b) charge model c) discharge model.*

The supply voltage is applied during Φ and $\overline{\Phi}$ phases. If Im is the average of charging current of CL, the total power dissipation in charging phase can be expressed in equation (1).

$$Wt = I_m^{\ 2} R T_p = \left(\frac{R C_L}{T_p} \right) C_L V_{dd}^{\ 2} \qquad (1)$$

Where R is the transistor effective resistance and wiring resistance, CL is load capacitance, T_p is time for the supply to reach from 0 to V_{dd} during Φ. Considering (1) the energy dissipated in the system is inversely proportional to the charging time, RC_l / T_p. Therefore, in theory, if the charging time approaches infinity, the energy dissipated approaches zero.

2.1. Clocked Power Adiabatic Circuits

Short circuit power dissipation is a large part of total

power dissipation in static CMOS circuits. This type of dissipation happens during rise and fall time of the input clock. In adiabatic logic by modifications of the clock waveform this type of power is reduced. In adiabatic logic with clocked power, the clock also serves as the power supply of the circuit. One of the common waveforms used as clocked power is trapezoidal waveform. This waveform has four phases of charge, evaluation, discharge and idle. Using trapezoidal waveforms stabilizes the output logic at VDD and GND comparing to other waveforms. However, different phases of clock make the circuit slow. Besides, additional circuits are required for linear ramp generation [12]. Sinusoidal waveform is also used for the clocked power adiabatic circuits. Using sine wave saves more energy comparing to trapezoidal and linear waveforms. Most sinewave designs use two phase clock that makes larger circuit implementations easier [13].

Depending on charge recovery method, adiabatic circuits may or may not have diode in the circuit structure. Using diode makes circuit implementation easier, but reduces output voltage swing and also increases power dissipation across diode. A logic gate based on Adiabatic Dynamic CMOS Logic (ADCL) that uses diodes is shown in figure 2.

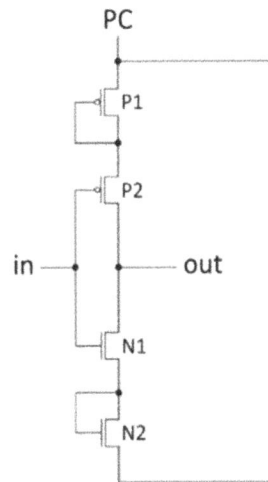

Figure 2. *ADCL Inverter gate [14].*

P1 and N2 diodes have been added comparing to static CMOS and a sine wave clock has been used. Diodes separate charging and discharge paths and will have power dissipation. The speed of the ADCL reduces with increasing number of gates and in complex logic circuits [14].

2.2. Two Phase clock Adiabatic Split Charge Logic (2PASCL)

In 2PASCL logic family the charging current passes through a transistor instead of diodes in contrast to diode based adiabatic circuits. This removes voltage drop across diodes and increases output amplitude [11]. Figure 3 shows an inverter using 2PASCL that uses two complementary Φ and $\overline{\Phi}$ for the power supply and ground. Didoes D1 and D2 make charge recovery from output nodes and discharging of internal nodes easier. Φ and $\overline{\Phi}$ are complement of each other and when Φ changes form 0.9V to 1.8V, $\overline{\Phi}$ changes from 0.9V to 0V.

As shown in figure 4, there will be no transition at the output when output is high and the pull up network is on, or when the output is low and the pull down network is on. However, if the output is high and the pull down is on, the output capacitor is discharged through transistor and D2. This happens when Φ is rising and $\overline{\Phi}$ is falling (evaluation phase). In contrast, if Φ is falling and $\overline{\Phi}$ is rising (hold phase) and the output is high and the pull up is on, output is discharged through D1 [11].

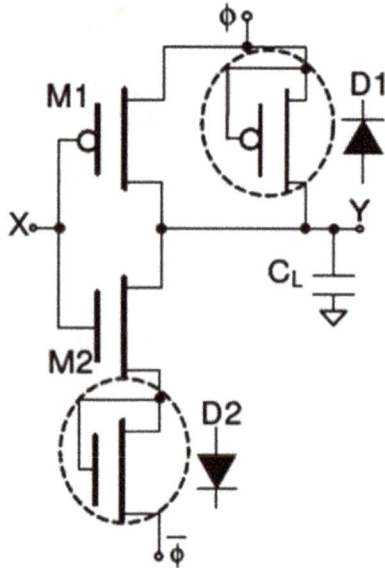

Figure 3. 2PASCL inverter circuit.

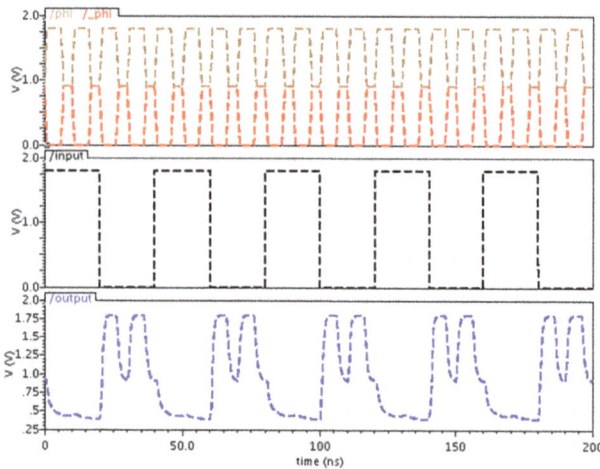

Figure 4. Input output waveforms of the 2PASCL.

3. Parallel Prefix Adders

Parallel prefix adders are fast and are used in large number of bits and high performance adders. The three main blocks of parallel adders are: pre-computing, carry generate, post-computing [1]. The block diagram of parallel prefix adder is shown in figure 5.

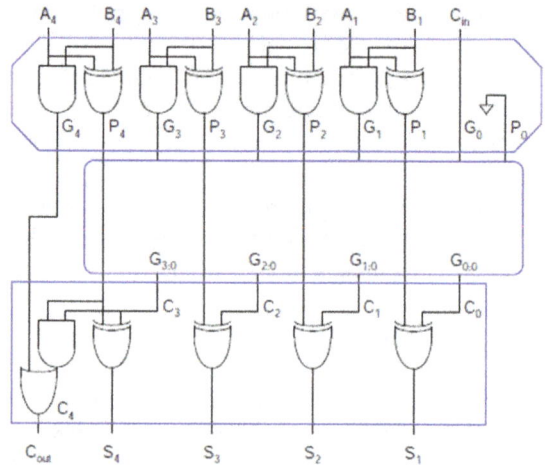

Figure 5. Parallel prefix adder block diagram.

In the first stage, Pi (Propagate) and Gi (Generate) signals are produced. These signals are fed to carry generate block. Different structures of parallel prefix adders are different with respect to carry generation method. The elements of the carry generate adders are shown in figure 6 consisting of black and gray and white operators. Buffers are used to cancel loading effects in these blocks. The last stage consists of two bit XOR gates to produce output bits. In the next section two common parallel prefix methods for Brent-Kung [15] and Kogge-Stone [16] are introduced.

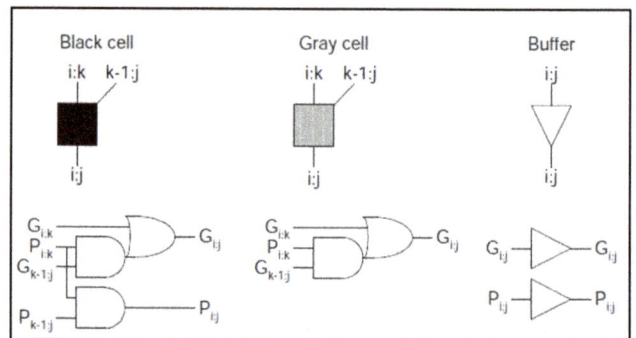

Figure 6. Main building blocks of parallel prefix adders [17].

3.1. Brent-Kung Parallel Prefix Adder

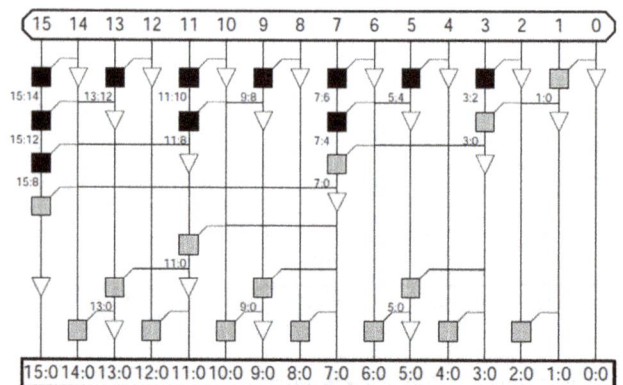

Figure 7. Brent-Kung parallel prefix adder [17].

The Brent-Kung parallel prefix adder has more logic level and smallest chip area comparing to Kogge-Stone and Lander-Fisher. The number of blocks in this adder is obtained using $2(n-1)$ -$Log2^n$ and n is the number of bits. The fan out is two and has $2*\log_2^n - 1$ levels. The block diagram for carry generate section of this adder is shown in figure 7 for a 16-bit adder [1].

3.2. Kogge-Stone Parallel Prefix Adder

The Kogge-Stone is a fast parallel prefix adder and is widely used in VLSI applications. The adder is used in large number of bits, since this adder has the least delay comparing to other methods. However, the adder consumes relatively large chip area. The carry generate block of such adder is shown in figure 8. This adder has Log_2^N level and the fan out of two in each stage. Large wiring is required between the logic levels. Since this adders has more blocks comparing to Brent-Kung method the power dissipation of the adder is high. If the layout is carefully designed, it may not increase the chip are drastically [1].

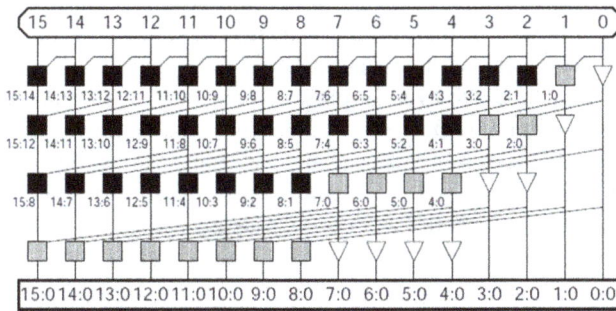

Figure 8. *Kogge-Stone Parallel Prefix adder [17].*

4. Parallel Prefix Adders Using 2PASCL Adiabatic Logic

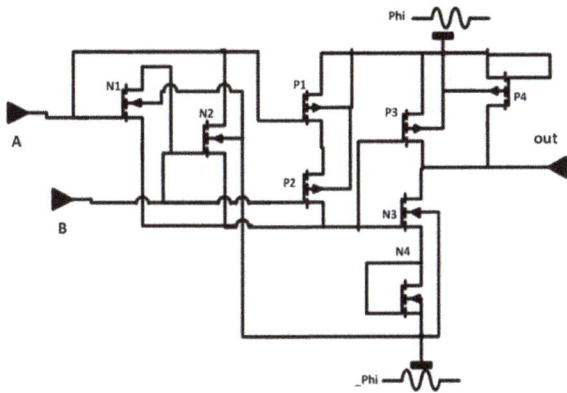

Figure 9. *XOR gate using 2PASCL adiabatic method.*

In this section the two parallel prefix adders Brent-Kung and Kogge-Stone, and Ripple Carry adder are built using new adiabatic structures. The XOR is widely used in the adder structures and its transistor level schematic is shown in figure

9. To investigate the operation of the XOR gate in the 2PASCL adiabatic logic, two trapezoidal and sine waveforms have been applied to the power supply of the gate. The power signals are complementary and Φ and $\overline{\Phi}$ have been used. The simulation results for the XOR gate using 180nm technology parameter, 1.8V power supply voltage, 200fF load capacitor, at 200MHz clock frequency are shown in figure 10 and 11. As is shown in figure 10, the output signal is discharged to threshold voltage of D2.

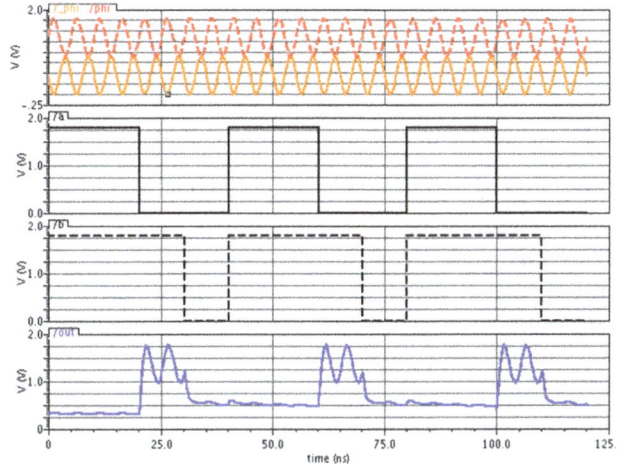

Figure 10. *Sine power clock waveform for the XOR gate.*

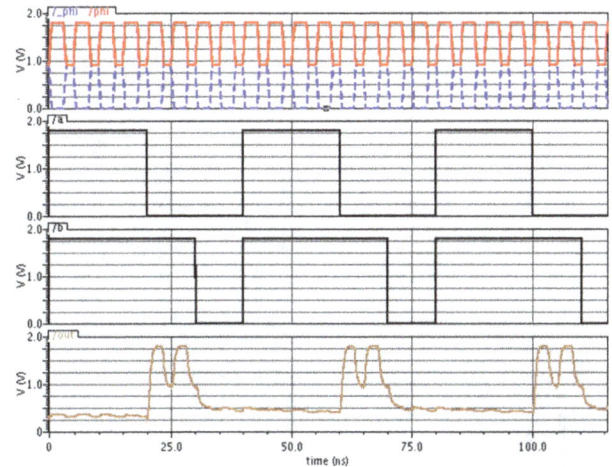

Figure 11. *Trapezoidal power clock waveform for the XOR gate.*

The same power supply waveforms are also applied for all the blocks in the parallel prefix adder. The simulation result for the critical path delay, power dissipation with sine and trapezoidal waveforms are shown in tables 1 and 2.

Table 1. *Power dissipation and delay for the trapezoidal waveform of the power supply.*

		Pd	Delay	# of Tr.
XOR gate	2PASCL	11.41µW	1.125ns	8
	CMOS	18.83µW	301.3ps	6
Gray operator	2PASCL	22.28µW	1.05ns	10
	CMOS	32.44µW	339ps	8
Black operator	2PASCL	24.21µW	1.05ns	18
	CMOS	35.16µW	339ps	14

Table 2. Power dissipation and delay for adder blocks for sine wave power clock.

	XOR gate	Gray operator	Black operator
Pd	10.23µW	10.05µW	18.41µW
Delay	3.87ns	2.65ns	2.65ns

Table 3. Power dissipation for sine waveform of the power supply.

	Ripple Carry		Brent-Kung		Kogge-Stone	
	2PASCL	CMOS	2PASCL	CMOS	2PASCL	CMOS
Pd	250µW	270µW	306µW	388.4µW	328.6µW	484.6µW
#Transistor	140	104	200	138	206	180

Using the new blocks, 4-bit adders for Brent-Kung, Kogge-Stone and Ripple Carry adders have been simulated using 2PASCL adiabatic logic. Simulation results are shown in table 3 and are compared to static CMOS logic. The effects of power supply clock frequency and the load capacitor on the power consumption of the adders have also been investigated as shown in figure 12. A comparison plot has also been shown in figure 13.

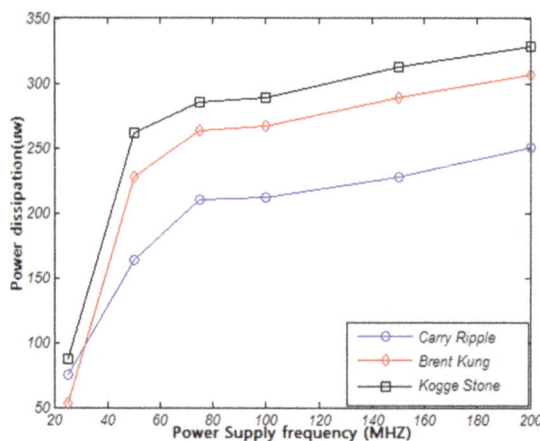

Figure 12. Power dissipation of adders versus clock frequency.

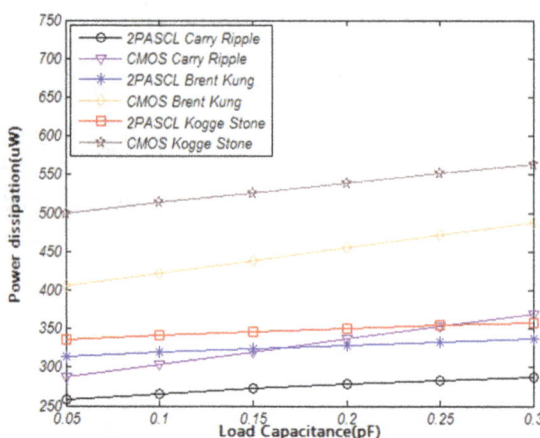

Figure 13. Power dissipation of 3 type of adders using 2PASCL and static CMOS with various load capacitors.

5. Discussion

Using simulation results listed in table 1, the power

dissipation of XOR, gray and black operators for the trapezoidal waveform has been reduced by 31.32%, 39.4% and 31.14% respectively comparing to the static CMOS logic. However this method increases critical path delay and increases number of transistors. Using table 2 results, it can be concluded that sine wave clock waveform reduces power dissipation comparing to trapezoidal waveform [10, 11]. This result also holds true for the maximum rise time and fall time of trapezoidal waveform. The disadvantage of using sine wave is increasing delay comparing to trapezoidal waveform. Figure 12 shows that increasing clock frequency increases power dissipation in all 3 adders and if the frequency increases above 100MHz, the rate of change in power dissipation will be the same for all adders. Considering figure 13 results it can be concluded that increasing load capacitance has little effect on power dissipation using 2PASCL and the slope of power dissipation approaches zero with large load capacitance. In contrast in static CMOS logic the power dissipation directly increases with load capacitance [10, 11].

Parallel prefix adders are faster for large number of bits comparing to Ripple Carry adders. However, parallel prefix adders consume more power. As shown in the figure 13 simulations results, the power dissipation of parallel prefix adders decreases and approaches Ripple Carry adders using 2PASCL method and even surpasses as the load capacitance increases. Static power dissipation also exists in the 2PASCL adiabatic logic, but its effects are less important in long channel devices and increases with short channel technologies. Considering static power dissipation in 2PASCL adiabatic circuit can be investigated as future work.

6. Conclusion

In this paper 2PASCL adiabatic circuit has been proposed for parallel prefix adders to reduce power consumption. Diode based adiabatic circuits have smaller output voltage and power dissipation due to diode voltage drop. Using the proposed method in this paper the diode is by passed by a transistor and the above mentioned disadvantages are canceled out. A new structure has been introduced for parallel prefix adder building blocks using 2PASCL adiabatic logic. Using 180nm technology parameters, simulation results show that the power dissipation of parallel prefix adders using the 2PASCL adiabatic circuit has been reduced by 34% in average using trapezoidal waveform and 54% using sine power clock waveforms. Furthermore, the capacitive load has minimum effect on the power dissipation using this method. Therefore this method can be used for parallel prefix adders in low power microprocessor and signal processing applications.

References

[1] P. Chaitanya kumari and R. Nagendra, "Design of 32 bit Parallel Prefix Adders," Journal of Electronics and Communication Engineering (IOSR-JECE), Vol. 6, pp. 01-06, 2013.

[2]   A. K. Kumar, D. Somasundareswari, V. Duraisamy, and M. Pradeepkumar, "Low power multiplier design using complementary pass-transistor asynchronous adiabatic logic," International Journal on Computer Science and Engineering, vol. 2, pp. 2291-2297, 2010.

[3]   D. J. Willingham, "Asynchrobatic logic for low-power VLSI design," University of Westminster, 2010.

[4]   S. G. Younis, "Asymptotically zero energy computing using split-level charge recovery logic," Massachusetts Institute of Technology, 1994.

[5]   A. Kramer, J. S. Denker, S. C. Avery, A. G. Dickinson, and T. R. Wik, "Adiabatic computing with the 2N-2N2D logic family," in VLSI Circuits, Digest of Technical Papers, Symposium on, pp. 25-26, 1994.

[6]   A. K. Kumar, D. Somasundareswari, V. Duraisamy, and M. G. Nair, "Asynchronous adiabatic design of full adder using dual-rail domino logic," in Computational Intelligence & Computing Research (ICCIC), IEEE International Conference on, 2012, pp. 1-4.

[7]   H. Jianping, X. Tiefeng, and L. Hong, "A lower-power register file based on complementary pass-transistor adiabatic logic," IEICE transactions on information and systems, vol. 88, pp. 1479-1485, 2005.

[8]   M. Cutitaru and L. Belfore, "A partially-adiabatic energy-efficient logic family as a power analysis attack countermeasure," in Signals, Systems and Computers, 2013 Asilomar Conference on, 2013, pp. 1125-1129.

[9]   M. Cutitaru and L. A. Belfore II, "New Single-Phase Adiabatic Logic Family," in Proceedings of the International Conference on Computer Design (CDES), pp. 9-14, 2012.

[10]   N. Anuar, Y. Takahashi, and T. Sekine, "4-bit ripple carry adder of two-phase clocked adiabatic static CMOS logic: a comparison with static CMOS," Proc. IEEE ECCTD, pp. 65-68, 2009.

[11]   N. Anuar, Y. Takahashi, and T. Sekine, "Adiabatic logic versus CMOS for low power applications," in ITC-CSCC: International Technical Conference on Circuits Systems, Computers and Communications, pp. 302-305, 2009.

[12]   A. Kramer, J. S. Denker, B. Flower, and J. Moroney, "2nd order adiabatic computation with 2N-2P and 2N-2N2P logic circuits," in Proceedings of the 1995 international symposium on Low power design, pp. 191-196., 1995.

[13]   S. Kim and M. C. Papaefthymiou, "True single-phase adiabatic circuitry," Very Large Scale Integration (VLSI) Systems, IEEE Transactions on, vol. 9, pp. 52-63, 2001.

[14]   K. Takahashi and M. Mizunuma, "Adiabatic dynamic CMOS logic circuit," Electronics and Communications in Japan (Part II: Electronics), vol. 83, pp. 50-58, 2000.

[15]   R. P. Brent and H.-T. Kung, "A regular layout for parallel adders," 1979.

[16]   R. E. Ladner and M. J. Fischer, "Parallel prefix computation," Journal of the ACM (JACM), vol. 27, pp. 831-838, 1980.

[17]   W. N. HE, CMOS VLSI Design: A Circuits and Systems Perspective, 3/E: Pearson Education India, 2006.

16

New Delta Sigma modulator structure using second order filter in one stage technique

M. Dashtbayazi[1], M. Sabaghi[2, *], M. Rezaei[3], S. Marjani[1]

[1]Department of Electrical Engineering, Ferdowsi University of Mashhad, Mashhad, Iran
[2]Laser and Optics Research School, Nuclear Science and Technology Research Institute (NSTRI), Tehran, Iran
[3]Department of Electrical Engineering and Computer Engineering, Laval University, Quebec City, QC, Canada

Email address:

msabaghi@aeoi.org.ir (M. Sabaghi)

Abstract: In this paper, a new structure of delta sigma modulator has been proposed in which a second order filter in one stage technique has been taken into the account and a second order noise shaping method and a third order one have been designed by the proposed idea. In the proposed structure an OPAMP has been saved and since in a conventional delta sigma modulator OPAMPs consumes most of the power and takes very large area, the proposed idea causes less power consumption and also the chip area is minimized.

Keywords: Delta Sigma Modulator, Noise Shaping, Second Order Filter in One Stage Technique

1. Introduction

Delta sigma modulators are categorized in communicational modulators. It has been widely used in Transceivers [1]. It has also many applications in signal processing [2]-[3]. In fact these modulators convert analog input signal U(z) to a digital representation Y(z). The difference between output digital signal and input analog signal of modulators is so called quantization error E(z). Quantization error is known as noise which limits analog to digital modulators accuracy. In this type of modulators, the quantization noise is fed back to the input of the modulator and being attenuated while passing through loop filters. In addition, any other noise sources which are located in the forward path of the modulator could be attenuated by this technique of fed back the noise signals to the input. Attenuating quantization noise causes an increase in accuracy of the modulator. Concept of a conventional delta sigma modulator is shown in figure 1.

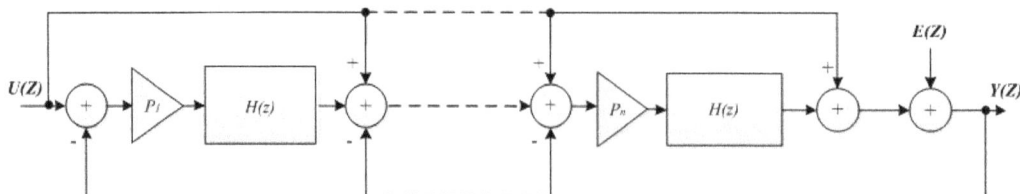

Fig. 1. The conventional delta sigma modulator.

The signal transfer function STF(z) and noise transfer function NTF(z) of the delta sigma modulator is as the following:

$$Y(Z) = (1 - Z^{-1})^n E(Z) + U(Z) \qquad (1)$$

$$\begin{cases} S_{TF}(Z) = \frac{Y(Z)}{U(Z)} = 1 \\ N_{TF}(Z) = \frac{Y(Z)}{E(Z)} = (1 - Z^{-1})^n \end{cases} \qquad (2)$$

In the conventional delta sigma modulators, increasing forward path filters more than 3 usually leads to instability [4]. There has been introduced many techniques which provide the same noise transfer function and signal transfer function as the conventional has for the high order noise shaping by mathematical techniques [5][6][7].

In this paper, a new mathematical technique has been introduced in which provides a second order noise shaping and third order noise shaping by a second order filter in one stage technique.

The paper has been organized as the following. Sections II and III describe the proposed second order noise shaping method and results. It has been followed by the proposed third order noise shaping method and its results in sections IV and V. Finally, the paper presents the conclusion.

2. Proposed Second Order Noise Shaping Method

A second order noise shaping is possible even in one filter. The proposed method presents a system in which one filter block can provide a second order noise shaping. The proposed technique is depicted in figure 2. In the proposed method one OPAMP is enough to provide a second order noise shaping. In fact, instead of using two OPAMPs only one OPAMP will be used. Since power consumption of a delta sigma modulator is highly related to power consumption of its OPAMPs, saving an OPAMP means saving half of the power approximately. Furthermore, removing an OPAMP means smaller chip area.

In the figure 2 the signal X(z) is being calculated by:

$$X(Z) = \frac{Z^{-1}}{1-Z^{-1}}\big((1-Z^{-1})X(Z) + (2-Z^{-1})(U(Z)-Y(Z))\big) \quad (3)$$

Obviously, this equation could be simplified in the way that collecting X(z) in left side and the other parameters in the right side of the equation.

$$X(Z)(1 - \frac{Z^{-1}(1-Z^{-1})}{1-Z^{-1}}) = \frac{Z^{-1}}{1-Z^{-1}}((2-Z^{-1})(U(Z)-Y(Z)) \quad (4)$$

$$X(Z)(\frac{(1-Z^{-1})-Z^{-1}(1-Z^{-1})}{1-Z^{-1}}) = \frac{Z^{-1}}{1-Z^{-1}}((2-Z^{-1})(U(Z)-Y(Z)) \quad (5)$$

$$X(Z)(\frac{(1-Z^{-1})^2}{1-Z^{-1}}) = \frac{Z^{-1}}{1-Z^{-1}}((2-Z^{-1})(U(Z)-Y(Z)) \quad (6)$$

$$X(Z)(1-Z^{-1}) = \frac{Z^{-1}}{1-Z^{-1}}((2-Z^{-1})(U(Z)-Y(Z)) \quad (7)$$

Then, $(1-Z^1)$ could be moved from left side to the right side of the equation.

$$X(Z) = \frac{Z^{-1}}{(1-Z^{-1})^2}((2-Z^{-1})(U(Z)-Y(Z)) \quad (8)$$

Also $(2-Z^1)$ in the right side of the equation could be rewritten as $(2-Z^1+Z^1-Z^1)$, so the equation (8) turns to:

$$X(Z) = \frac{Z^{-1}}{(1-Z^{-1})^2}(((2-2Z^{-1})+Z^{-1})(U(Z)-Y(Z)) \quad (9)$$

$$X(Z) = \frac{Z^{-1}}{(1-Z^{-1})^2}\big(2(1-Z^{-1})(U(Z)-Y(Z))\big) + \frac{Z^{-1}}{(1-Z^{-1})^2}Z^{-1}\big(U(Z)-Y(Z)\big) \quad (10)$$

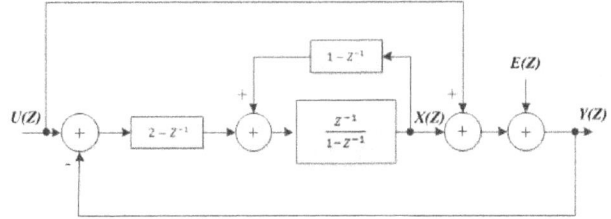

Fig. 2. *The proposed method.*

$$X(Z) = \frac{2Z^{-1}}{1-Z^{-1}}\big(U(Z)-Y(Z)\big) + \frac{Z^{-2}}{(1-Z^{-1})^2}\big(U(Z)-Y(Z)\big) \quad (11)$$

Furthermore, digital output signal Y(z) is being calculated by:

$$Y(Z) = U(Z) + E(Z) + \frac{2Z^{-1}}{1-Z^{-1}}\big(U(Z)-Y(Z)\big) + \frac{Z^{-2}}{(1-Z^{-1})^2}\big(U(Z)-Y(Z)\big) \quad (12)$$

All the Y(z) parameters could be collected in the left side of the equation and other parameters in the right side.

$$Y(Z)\left(1 + \frac{2Z^{-1}}{1-Z^{-1}} + \frac{Z^{-2}}{(1-Z^{-1})^2}\right) =$$
$$(1 + \frac{2Z^{-1}}{1-Z^{-1}} + \frac{Z^{-2}}{(1-Z^{-1})^2})U(Z) + E(Z) \quad (13)$$

Then, the equation (13) could be simplified as:

$$Y(Z)\left(\frac{(1-Z^{-1})^2 + 2Z^{-1}(1-Z^{-1}) + Z^{-2}}{(1-Z^{-1})^2}\right) =$$
$$\left(\frac{(1-Z^{-1})^2+2Z^{-1}(1-Z^{-1})+Z^{-2}}{(1-Z^{-1})^2}\right)U(Z) + E(Z) \quad (14)$$

$$Y(Z)\left(\frac{1 - 2Z^{-1} + Z^{-2} + 2Z^{-1} - 2Z^{-2} + Z^{-2}}{(1-Z^{-1})^2}\right) =$$
$$\left(\frac{1-2Z^{-1}+Z^{-2}+2Z^{-1}-2Z^{-2}+Z^{-2}}{(1-Z^{-1})^2}\right)U(Z) + E(Z) \quad (15)$$

$$Y(Z)\left(\frac{1}{(1-Z^{-1})^2}\right) = \left(\frac{1}{(1-Z^{-1})^2}\right)U(Z) + E(Z) \quad (16)$$

Finally the output signal is simplified as:

$$Y(Z) = U(Z) + (1-Z^{-1})^2 E(Z) \quad (17)$$

Obviously, equation (17) shows a second order noise shaping.

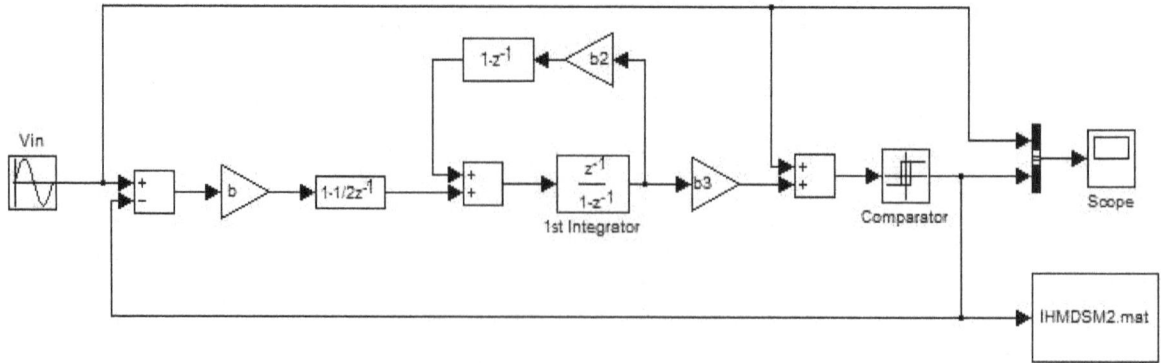

Fig. 3. *Ideal implementation of the proposed method.*

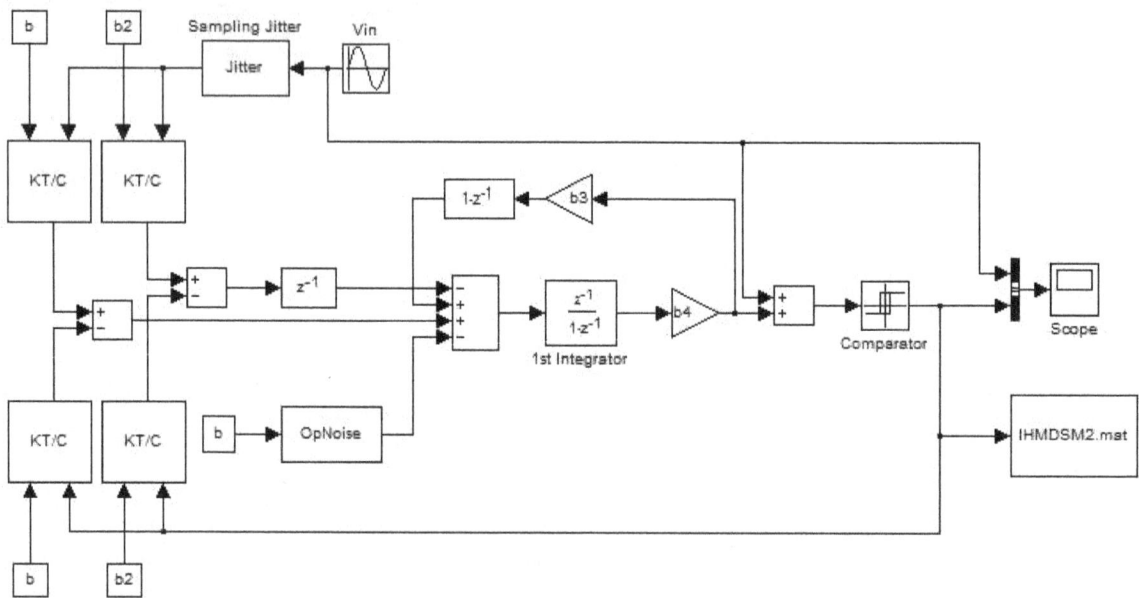

Fig. 4. *Actual implementation of the proposed method.*

3. Simulation Results of the Proposed Second Order Delta Sigma Modulator

MATLAB implementation of the proposed method is shown in figure 3. Obviously only one integrator has been located in the forward path and a second order filter technique has been taken into the account to perform a second order noise shaping.

Considering actual effects in implementation of this method leads to the MATLAB circuit diagram is shown in figure 4. In this figure, sampling jitter, thermal noise of switched capacitors and OPAMP noise have been taken into the account as presented in [5].

To simulate MATLAB models presented in Figure 3 and 4 all the parameters are considered as the parameters presented in table 1.

The Power Spectral Density (PSD) of the proposed second order noise shaping method is shown in figure 5. Also, the comparison between the actual implementation and the ideal implementation method has been shown in the figure 5.

Table 1. *Simulating parameters are used to simulate the proposed idea (second order delta sigma modulator).*

Parameter	Value
Signal bandwidth	$f_{BW}=100Hz$
Sampling Frequency	$F_S= 50kHz$
Oversampling ratio	$OSR=250$
Number of Samples	$N=65536$
Ideal Delta-Sigma Coefficient	$b=2, b2=b3=1$
Actual Delta-Sigma Coefficient	$b=0.5, b2=0.25, b3=b4=1$

Another simulation result in which Signal to Noise and Distortion Ratio (SNDR) has been plotted versus input analog signal amplitude has been depicted in figure 6 and figure 7. In figure 6, variation of SNDR versus input amplitude has been studied by different sampling jitters. And in figure 7, variation of this parameter has been studied by different OPAMP saturation levels.

Obviously, maximum SNDR has been reached in -8db for the input analog signal amplitude. Also, sampling jitter has not a considerable effect on the SNDR in the proposed method.

Comparison between the result obtained from the proposed method and ideas presented in [5], [6], [7] and [8] has been presented in table 2.

been saved and power consumption and chip area will be minimized.

Table 2. *Comparison between related works and the proposed idea.*

Parameters	[5]	[6]	[7]	[8]	This Work
f_{BW}(Hz)	100	256	10k	500	100
Order	2	2	2	2	2
SNDR(dB)	90.2	72	87.8	76	91.6
ENOB[1](bit)	14.7	11.7	14.3	12.3	14.9

1 Effective number of bits [4]

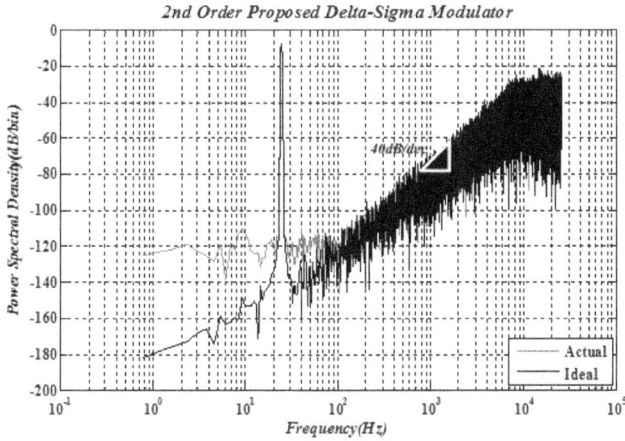

Fig. 5. *The PSD diagram of the proposed second order noise shaping method.*

4. Proposed Third Order Noise Shaping Method

It is also possible to extend this idea for third order noise shaping. The proposed technique for third order noise shaping is shown in figure 8. In this idea also an OPAMP has

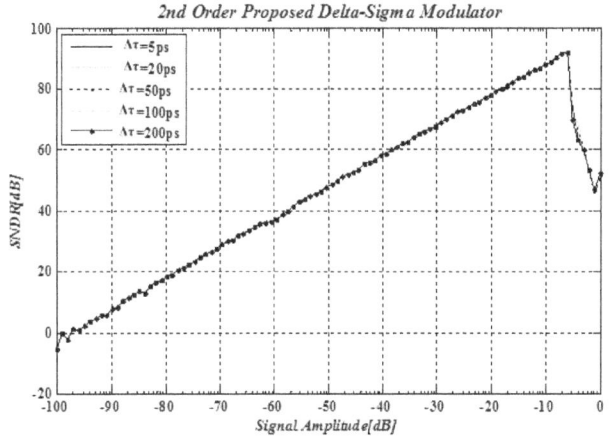

Fig. 6. *SNDR versus analog input amplitude per different sampling jitters.*

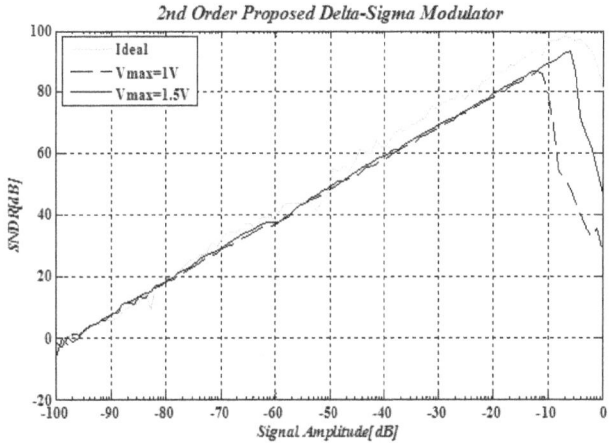

Fig. 7. *SNDR versus analog input amplitude per different OPAMP saturation levels.*

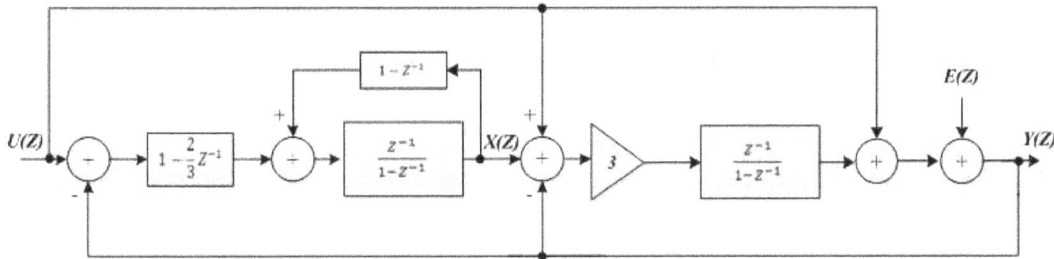

Fig. 8. *The proposed third order noise shaping method.*

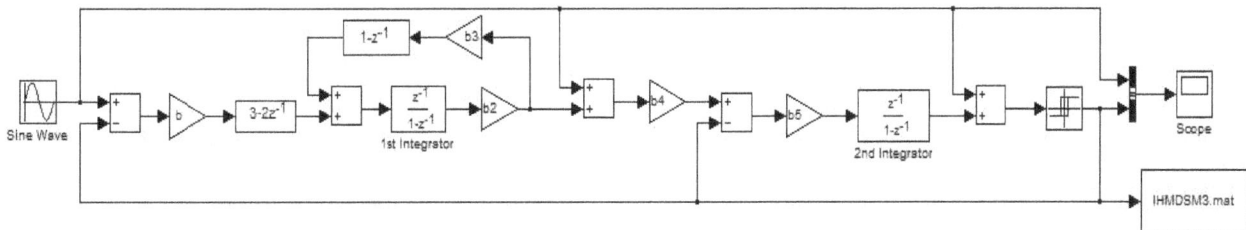

Fig. 9. *MATLAB implementation of the proposed third order noise shaping method.*

In the figure 8 the signal X(z) is being calculated by:

$$X(Z) = \frac{Z^{-1}}{1-Z^{-1}}((1-Z^{-1})X(Z) + \left(1 - \frac{2}{3}Z^{-1}\right)(U(Z) - Y(Z))) \quad (18)$$

Collecting all X(z) in the left side of equation gives:

$$X(Z)(1 - \frac{Z^{-1}(1-Z^{-1})}{1-Z^{-1}}) = \frac{Z^{-1}}{1-Z^{-1}}(\left(1 - \frac{2}{3}Z^{-1}\right)(U(Z) - Y(Z)) \quad (19)$$

$$X(Z)(\frac{(1-Z^{-1})-Z^{-1}(1-Z^{-1})}{1-Z^{-1}}) = \frac{Z^{-1}}{1-Z^{-1}}((1 - \frac{2}{3}Z^{-1})(U(Z) - Y(Z)) \quad (20)$$

$$X(Z)(\frac{(1-Z^{-1})^2}{1-Z^{-1}}) = \frac{Z^{-1}}{1-Z^{-1}}((1 - \frac{2}{3}Z^{-1})(U(Z) - Y(Z)) \quad (21)$$

Simplifying equation (21) gives:

$$X(Z)(1-Z^{-1}) = \frac{Z^{-1}}{1-Z^{-1}}((1 - \frac{2}{3}Z^{-1})(U(Z) - Y(Z)) \quad (22)$$

Moving (1-Z⁻¹) from left side of the equation 22 leads to:

$$X(Z) = \frac{Z^{-1}}{(1-Z^{-1})^2}((1 - \frac{2}{3}Z^{-1})(U(Z) - Y(Z)) \quad (23)$$

Also the output signal Y(z) could be calculated as:

$$Y(Z) = U(Z) + E(Z) + \frac{3Z^{-1}}{1-Z^{-1}}\left((U(Z) - Y(Z)) + X(Z)\right) \quad (24)$$

Replacing X(z) from (24) by equation (23) gives:

$$Y(Z) = U(Z) + E(Z) + \frac{3Z^{-1}}{1-Z^{-1}}\left((U(Z) - Y(Z)) + \frac{Z^{-1}}{(1-Z^{-1})^2}((1 - \frac{2}{3}Z^{-1})(U(Z) - Y(Z))\right) \quad (25)$$

Replacing (1-2/3Z⁻¹) by (1-Z⁻¹+1/3Z⁻¹) gives:

$$Y(Z) = U(Z) + E(Z) + \frac{3Z^{-1}}{1-Z^{-1}}\left((U(Z) - Y(Z)) + \frac{Z^{-1}}{(1-Z^{-1})^2}((1 - Z^{-1}) + \frac{1}{3}Z^{-1})(U(Z) - Y(Z)) \quad (26)$$

$$Y(Z) = U(Z) + E(Z) + \frac{3Z^{-1}}{1-Z^{-1}}\left(U(Z) - Y(Z)\right) +$$

$$\frac{3Z^{-2}}{(1-Z^{-1})^2}\left(U(Z) - Y(Z)\right) + \frac{Z^{-3}}{(1-Z^{-1})^3}(U(Z) - Y(Z))) \quad (27)$$

Collecting all Y(z) to the left side and the others to the right side gives:

$$Y(Z)\left(1 + \frac{3Z^{-1}}{1-Z^{-1}} + \frac{3Z^{-2}}{(1-Z^{-1})^2} + \frac{Z^{-3}}{(1-Z^{-1})^3}\right) =$$

$$(1 + \frac{3Z^{-1}}{1-Z^{-1}} + \frac{3Z^{-2}}{(1-Z^{-1})^2} + \frac{Z^{-3}}{(1-Z^{-1})^3})U(Z) + E(Z) \quad (28)$$

Simplifying equation (28) leads to:

$$Y(Z)\left(\frac{1-3Z^{-1}+3Z^{-2}-Z^{-3}+3Z^{-1}-6Z^{-2}+3Z^{-3}+3Z^{-2}-3Z^{-3}+Z^{-3}}{(1-Z^{-1})^3}\right) =$$
$$\left(\frac{1-3Z^{-1}+3Z^{-2}-Z^{-3}+3Z^{-1}-6Z^{-2}+3Z^{-3}+3Z^{-2}-3Z^{-3}+Z^{-3}}{(1-Z^{-1})^3}\right)U(Z) + E(Z) \quad (29)$$

$$Y(Z)\left(\frac{1}{(1-Z^{-1})^3}\right) = \left(\frac{1}{(1-Z^{-1})^3}\right)U(Z) + E(Z) \quad (30)$$

Finally moving (1-Z⁻¹)³ from left side of the equation to the right side gives:

$$Y(Z) = U(Z) + (1 - Z^{-1})^3 E(Z) \quad (31)$$

The equation (31) is output signal of a third order noise shaping delta sigma modulator.

5. Simulation Results the Proposed Third Order Delta Sigma Modulator

Ideal MATLAB model of the proposed third order noise shaping method has been shown in the figure 9. If noise sources such as thermal noise from switched capacitors, sampling jitter and OPAMP noise being considered, a model such as what is shown in figure 10 will be reached.

To simulate MATLAB models presented in Figure 9 and 10 all the parameters are considered as the parameters presented in table 3.

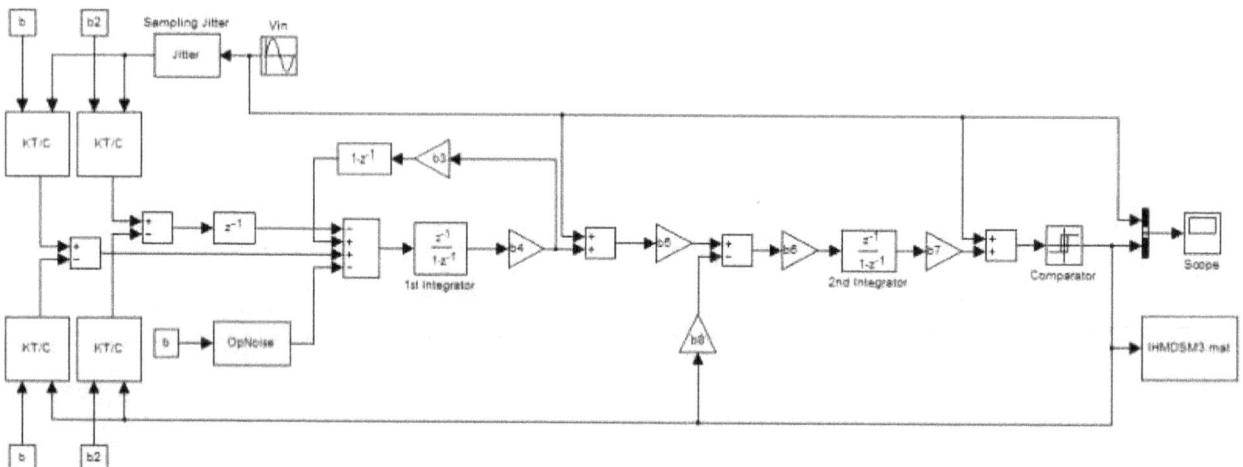

Fig. 10. The actual MATLAB model of the proposed third order noise shaping method.

Table 3. *Simulating parameters are used to simulate the proposed idea (third order delta sigma modulator).*

Parameter	Value
Signal bandwidth	$f_{BW}=300$
Sampling Frequency	$F_S= 50kHz$
Oversampling ratio	$OSR=83.3$
Number of Samples	$N=65536$
Ideal Delta-Sigma Coefficient	$b=b4=0.33, b2=b3=1, b5=3$
Actual Delta-Sigma Coefficient	$b=3/7, b2=2/7, b3=b4=b5=b7=1, b6=0.5, b8=1.5$

The PSD diagram of the proposed third order noise shaping method and comparison between the actual implementation and the ideal implementation method have been shown in figure 11.

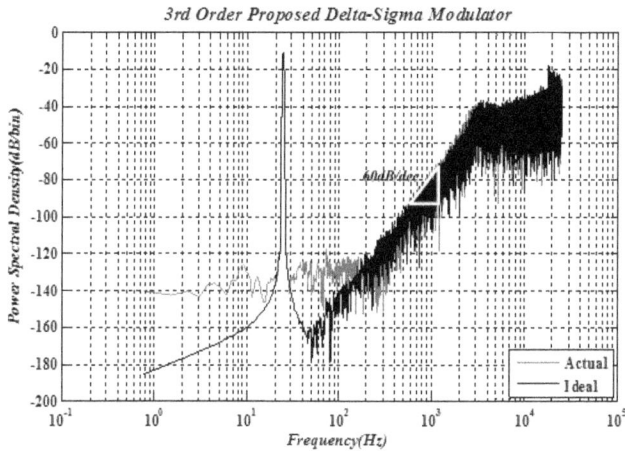

Fig. 11. *The PSD diagram of the proposed third order noise shaping method.*

Same as simulation results for the proposed second order noise shaping, simulation results in which SNDR versus analog input signal amplitude has been depicted in figure12 and figure 13. In figure 12, variation of SNDR is studied by different input sampling jitters. And figure 13 shows this parameter versus different OPAMP saturation levels.

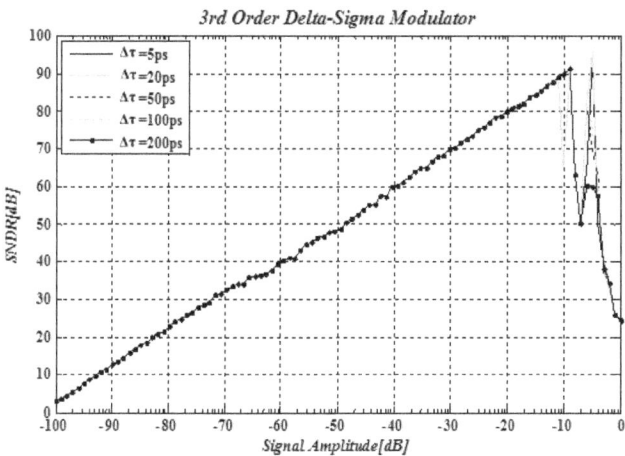

Fig. 12. *SNDR versus analog input amplitude per different sampling jitters.*

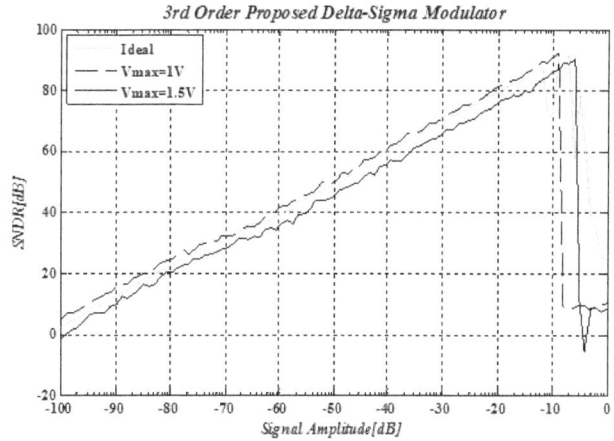

Fig. 13. *SNDR versus analog input amplitude per different sampling jitters.*

Obviously, same as the results obtained from the proposed second order noise shaping idea, maximum SNDR has been reached in -8db for the input analog signal amplitude.

Comparison between the result obtained from the proposed method and ideas presented in [9], [10] and [11] has been presented in table 4.

Table 4. *Comparison between related works and the proposed idea.*

Parameters	[9]	[9]	[10]	[11]	This Work
$f_{BW}(Hz)$	120	20k	20k	100k	300
Order	3	3	3	3	3
SNDR(dB)	65	81	61.4	84	91.5
ENOB(bit)	10.5	13.2	9.9	13.7	14.9

6. Conclusion

In this paper a method has been proposed in which an OPAMP has been saved compared with conventional delta sigma modulator and still the functionality of the delta sigma modulator is correct. Removing one OPAMP leaded to smaller chip area and less power consumption. The proposed idea has been studied in second order noise shaping and third order noise shaping delta sigma modulator. Also the comparison between the proposed idea and other techniques shows better SNDR and effective number of bits.

References

[1] L. Bos, G. Vandersteen, P. Rombouts, A. Geis, A. Morgado, Y. Rolain, G.V. Plas and J. Ryckaert, "Multirate cascaded discrete-time low-pass ΔΣ modulator for GSM/Bluetooth/UMTS," *IEEE J. Solid-State Circuits*, vol. 45, pp. 1198-1208, June 2010.

[2] L. Yao, M. S. J. Steyaert, and W. Sansen, "A 1-V 140-μW 88-dB audio sigma-delta modulator in 90-nm CMOS," *IEEE J. Solid-State Circuits*, vol. 39, pp. 1809-1818, Nov. 2004.

[3] E. Fogleman, J. Welz, and I. Galton, "An audio ADC delta-sigma modulator with 100-dB peak SINAD and 102-dB DR using a second order mismatch-shaping DAC," *IEEE J. Solid-State Circuits*, vol. 36, pp. 339-348, Mar. 2001.

[4] D. Johns and K. Martin, *Analog Integrated Circuit Design.* New York: Wiley, 1997.

[5] P. Malcovati, *et al.,* "Behavioral modeling of switched-capacitor sigma-delta modulators," in *IEEE Trans. Circuits Syst. I,* vol. 50, pp. 352 - 364, Mar 2003.

[6] F. Cannillo, *et al.*, " 1.4 V 13 W83dBDRCT- modulator with dual-slope quantizer and PWM DAC for bio potential signal acquisition," in *Proc. ESSCIRC,* pp. 267-270, Sept. 2011.

[7] A. Nilchi and D. A. Johns, "A low-power Delta-Sigma modulator is using a charge-pump integrator," in *IEEE Trans. Circuits Syst. I,* vol.60, pp.1310-1321, May 2013.

[8] A.F. Yeknami, *et al.*, "Low-Power DT $\Delta\Sigma$ Modulators Using SC Passive Filters in 65 nm CMOS," in *IEEE Trans. Circuits Syst. I,* vol.61, pp.358-370, Feb. 2014.

[9] Y. Chae and G. Han, "Low voltage, low power, inverter-based switched-capacitor Delta-Sigma modulator," in *IEEE J. Solid-State Circuits*, vol. 44, pp. 458-472, Feb. 2009.

[10] F. Michel and M. Steyaert, " A 250 mV 7.5 W 61 dB SNDR SC modulator using near threshold voltage biased inverter amplifiers in 130 nm CMOS " in *IEEE J. Solid-State Circuits*, vol. 47, pp.709-721, May 2012.

[11] A.P. Perez, E. Bonizzoni and F. Maloberti, "A84dB SNDR 100 kHz Bandwidth Low Power Single Op-Amp Third-Order $\Delta\Sigma$ Modulator Consuming 140µW," in *IEEEISSCC Dig. Tech. Papers*, pp.478-480, Feb. 2011.

17

Theoretical Investigation of Transmission and Dispersion Properties of One Dimensional Photonic Crystal

Ouarda Barkat

Department of Electronics, University of Constantine 1, Constantine, Algeria

Email address:

barkatwarda@yahoo.fr

Abstract: In this work, we demonstrate via numerical simulation the general design for one dimensional photonic crystal (1D-PC). In the design procedure, the transfer matrix method and Bloch theorem are used to determine the transmission coefficient and dispersion relation of (1D- PC) structure for both TE (transverse electric) and TM (transverse magnetic) modes. Results obtained showing the effect of the filling factor as well as the incident angle on the photonic band gap width. The analysis is carried out using MATLAB software tool. The accuracy of the analysis is tested by comparing the computed results with measurements published data.

Keywords: Photonic Band Gap, Transfer Matrix Method, Bloch Theorem, Transmission, Dispersion

1. Introduction

Photonic crystals (PCs) structures have attracted increasing interest in recent years [1-5]. Various materials and techniques have been employed in order to obtain one, two, and three dimensional photonic crystals [6-9]. These structures designed to control the propagation of electromagnetic waves in the same way as the periodic potential in semiconductor crystals [10-12]. Particularly, one dimensional photonic crystals have been known for several decades as Bragg mirror. Interference of the Bragg scattering is considered as a cause of the Bragg gap or band gap. The periodicity creates the band gaps depend on some parameters, as the dielectric contrast between the employed materials, and the filling factor of the elementary cell [13].

One dimensional photonic crystal (1D- PC) structures have a number of useful properties, which are employed as low-loss optical waveguides, dielectric reflecting mirrors, optical switches, optical limiters, optical filters etc. It has been demonstrated experimentally and theoretically that one-dimensional PBG have complete omnidirectional PBGs. Various materials have been employed in order to obtain one photonic crystals (1D) as: Si/SiO$_2$, SiO$_2$/TiO$_2$, Na$_3$AlF$_6$/ZnSe, Na$_3$AlF$_6$/Ge [5, 14-18]. Silicon has been the choice for microelectronics technology because of various reasons such as its cost, compatible with mass production and availability. For this reason, we are study the transmission coefficient and dispersion curves of (1D- PC) structures composed of Si/SiO$_2$.

Nowadays, the numerical modeling of photonics crystals is based on the calculation of the transmission, the reflection coefficients properties [19-23]. These methods including the plane wave expansion (PWE) method, the generalized Rayleigh identity method, the finite-difference time-domain (FDTD) method, and the transfer matrix method (TMM). Each method has its own limitations for finding the band structure. The transfer matrix method is most popular because of its simplicity in algorithm and capability to model complex structures. It is recently introduced by Pendry and MacKinnon, to calculate the EM transmission through the PBG materials [24].

In this paper, we combine the transfer matrix method (TMM) to the Block theorem in order to find the characteristics of transmission spectra and diagrams of dispersion of one dimensional photonic crystal (1D- PC). Several simulation cases by Matlab will be given to show the performance of this approach. The results obtained from this approach are in very good agreement with reported by the practical study of H. Tian et al. [14].

2. Theory

2.1. Study of Transmissions in the (1D- PC) Structure

Let us consider first the (1D-PC) structure consisting of alternating multilayer shown in Fig. 1, there are N layer or (N/2) period made up of dielectric materials. Every layer has to be d_l thicknesses, and index n_l. In order to find the formulation of the structure, we supposed that the incident electromagnetic wave from air to Si and SiO_2 medium.

Let the layers be in the x-y plane, the z direction being normal to interface of layers.

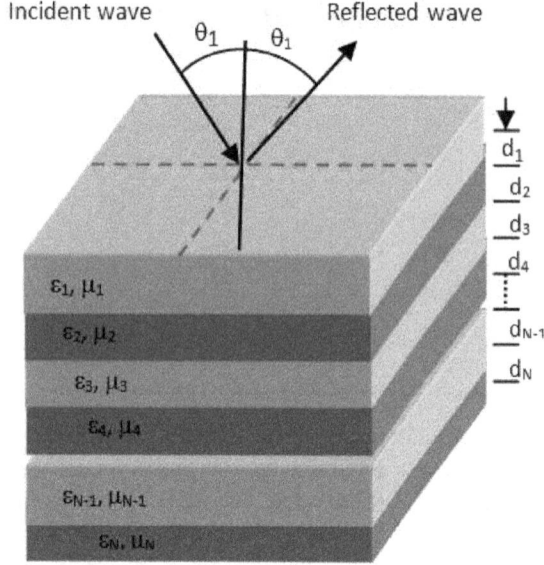

Fig. 1. Structure of one-dimensional photonic crystal.

The refractive index profile of considered structure can be given as

$$\varepsilon_l = \begin{cases} \varepsilon_1 \ 0 < z < d_1 \\ \varepsilon_2 \ d_1 < z < d_2 \end{cases} \quad (1)$$

and

$$\varepsilon_l(z) = \varepsilon_l(z + d) \quad (2)$$

where

l: is number of layer

$d = d_1 + d_2$: is period

Based on the Maxwell equations and the boundary conditions, the TMM has been widely used to calculate the amplitude and phase spectra of the light wave propagating in a (1D- PC) structure. We will suppose that a space time dependence of all the components of the kind $e^{i(\vec{k}.\vec{r}-\omega t)}$. The transverse components of the E and H fields from Maxwell's equations in the lth layer, for TM polarization, are given by:

$$H_{ly} = A_l e^{i(\omega t - k_l(Z_l.\cos\theta_l + X_l.\sin\theta_l))} + B_l e^{i(\omega t + k_l(Z_l.\cos\theta_l + X_l.\sin\theta_l))}) \quad (3)$$

$$E_{lx} = \eta_l \cos\theta_l \left(A_l e^{i(\omega t - k_l(Z_l.\cos\theta_l + X_l.\sin\theta_l))} - B_l e^{i(\omega t + k_l(Z_l.\cos\theta_l + X_l.\sin\theta_l))} \right) \quad (4)$$

$$E_{lz} = -\eta_l \cos\theta_l \left(A_l e^{i(\omega t - k_l(Z_l.\cos\theta_l + X_l.\sin\theta_l))} + B_l e^{i(\omega t + k_l(Z_l.\cos\theta_l + X_l.\sin\theta_l))} \right) \quad (5)$$

Where A_l and B_l are the amplitudes of the forward and backward travelling waves in the lth layer.

The transverse components of the E and H fields from Maxwell's equations in the lth layer, for TE polarization, are given by:

$$E_{ly} = A_l e^{i(\omega t - k_l(Z_l.\cos\theta_l + X_l.\sin\theta_l))} + B_l e^{i(\omega t + k_l(Z_l.\cos\theta_l + X_l.\sin\theta_l))}) \quad (6)$$

$$H_{lx} = -\frac{\eta_l}{\cos\theta_l} \left(A_l e^{i(\omega t - k_l(Z_l.\cos\theta_l + X_l.\sin\theta_l))} - B_l e^{i(\omega t + k_l(Z_l.\cos\theta_l + X_l.\sin\theta_l))} \right) \quad (7)$$

$$H_{lz} = \frac{\eta_l}{\cos\theta_l} \left(A_l e^{i(\omega t - k_l(Z_l.\cos\theta_l + X_l.\sin\theta_l))} + B_l e^{i(\omega t + k_1(Z_l.\cos\theta_l + X_l.\sin\theta_l))} \right) \quad (8)$$

Where the wave numbers and intrinsic impedances are:

$$k_l = \omega\sqrt{\varepsilon_0\mu_0\varepsilon_l\mu_l} \quad (9)$$

$$\eta_l = \frac{k_l}{\omega\varepsilon_l\varepsilon_0} = \sqrt{\frac{\mu_0\mu_l}{\varepsilon_0\varepsilon_l}} \quad (10)$$

By using the boundary conditions and the condition of continuity of E and H fields at the interfaces of z = 0 and $z = d_1, d_2, d_3 \dots \dots d_N$, we can found the relationship between the fields (1D- PC) structure consisting of l layer, this relation is given by:

$$\begin{bmatrix} E_1 \\ H_1 \end{bmatrix} = M_1 M_2 \dots \dots M_k \dots M_{l-1} M_l \begin{bmatrix} E_l \\ H_l \end{bmatrix} \quad (11)$$

The matrix M_{l-1} of the lth layer can be written in the form

$$M_{(l-1)} = \begin{bmatrix} \cos(\delta_{(l-1)}) & i\gamma_{(l-1)}\sin(\delta_{(l-1)}) \\ i\gamma_{(l-1)}^{-1}\sin(\delta_{(l-1)}) & \cos(\delta_{(l-1)}) \end{bmatrix} \quad (12)$$

$\delta_{(l-1)}$ and $\gamma_{(l-1)}$ being the matrix parameters and depending on the incident angle of light, the optical constants and the layer thickness, are expressed as:

$$\delta_{(l-1)} = k_{(l-1)}.d_{(l-1)}.\cos\theta_{(l-1)} \quad (13)$$

$$\gamma_{(l-1)} = \begin{cases} \dfrac{\eta_{(l-1)}}{\cos\theta_{(l-1)}} \ TE \ mode \\ \eta_{(l-1)}\cos\theta_{(l-1)} \ TM \ mode \end{cases} \quad (14)$$

We note that $\theta_{(l-1)}$ is related to the angle of incidence θ_0 by the Snell's Descart's low, that is

$$n_{(l-1)}sin\theta_{(l-1)} = n_0 sin\theta_0 \qquad (15)$$

By considering the transmission matrix of each layer, we can obtain the transmission matrix of whole structure. For 1 number of multilayers; the corresponding transfer matrix can be defined as a product of matrices, is obtained to be

$$\prod_{k=1}^{l} M_k = \begin{bmatrix} m_{11} & m_{12} \\ m_{21} & m_{22} \end{bmatrix} \qquad (16)$$

Where

m_{11}, m_{12}, m_{21} and m_{22} are the complex numbers

The transmittance t and reflectance r are defined as the ratios of the fluxes of the transmitted and reflected waves, respectively, to the flux of the incident wave. After some derivations, the total transmission and reflection coefficients are given by

$$r = \frac{(m_{11}+p_s^{-1}m_{12})p_0^{-1}-(m_{21}+p_s^{-1}m_{22})}{(m_{11}+p_s^{-1}m_{12})p_0^{-1}+(m_{21}+p_s^{-1}m_{22})} \qquad (17)$$

$$t = \frac{2.p_0^{-1}}{(m_{11}+p_s^{-1}m_{12})p_0^{-1}+(m_{21}+p_s^{-1}m_{22})} \qquad (18)$$

Here p_0 and p_s are the first and last medium of the structure which given as

$$p_s^{-1} = \begin{cases} \dfrac{\eta_s \cos\theta_s}{Z_0} & TE\ mode \\[2mm] \dfrac{\cos\theta_s}{\eta_s Z_0} & TM\ mode \end{cases} \qquad (19)$$

$$p_0^{-1} = \begin{cases} \dfrac{\eta_0 \cos\theta_0}{Z_0} & TE\ mode \\[2mm] \dfrac{\cos\theta_0}{\eta_0 Z_0} & TM\ mode \end{cases} \qquad (20)$$

Where

$$Z_0 = \sqrt{\frac{\mu_0}{\varepsilon_0}}$$

Hence the reflectance R and transmittance T spectrums of can be obtained by using the expressions:

$$T = |t|^2 \qquad (21)$$

$$R = |r|^2 \qquad (22)$$

2.2. The Dispersion Relation

In general, wave propagation in periodic media can be described in terms of Bloch waves. For a determination of the dispersion surfaces of a periodic crystal, it is necessary only to integrate the wave-field through a periodic media [6].

According to Bloch theorem, fields in a periodic structure satisfy the following equations:

$$E(z + d) = e^{-ikd}E(z) \qquad (23)$$

The parameter k is called the Bloch wave number or dispersion relation. In order to determinate k, we can use relation between the electric field amplitudes of two layers. From Equation (11), we obtain:

$$\begin{bmatrix} E_1 \\ H_1 \end{bmatrix} = M_1 M_2 \begin{bmatrix} E_2 \\ H_2 \end{bmatrix} \qquad (24)$$

We can put the product matrix as:

$$M_1.M_2 = \begin{bmatrix} M_{11} & M_{12} \\ M_{21} & M_{22} \end{bmatrix} \qquad (25)$$

$Tr[M_1.M_2]$ is the trace of the transfer matrix characterizing the wave scattering in a periodic structure, is given by;

$$Tr[M_1.M_2] = M_{11} + M_{22} = 2\cos(kd) \qquad (26)$$

Where

$$M_{11} = \cos(\delta_1) * \cos(\delta_2) - (\gamma_1/\gamma_2)\sin(\delta_1)\sin(\delta_2) \qquad (27)$$

$$M_{22} = \cos(\delta_1) * \cos(\delta_2) - (\gamma_2/\gamma_1)\sin(\delta_1)\sin(\delta_2) \qquad (28)$$

Substituting (27) and (28) into (26), we obtain the following equation:

$$\cos(kd) = \cos(\delta_1) * \cos(\delta_2) - (\tfrac{\gamma_2^2+\gamma_1^2}{2\gamma_1\gamma_2})\sin(\delta_1)\sin(\delta_2) \qquad (29)$$

The quantity $\cos(kd)$ determines the band structures of the (1D- PC) structure. In the region where $|\cos(kd)| < 1$, k takes a real value and this leads to propagating Bloch waves (pass band). In the region where $|\cos(kd)| > 1$, the value of k become complex which consists of an imaginary and a real part corresponding to the evanescent and propagating Bloch waves. The band edges are the regions where $|\cos(kd)| = 1$.

3. Numerical Calculation and Discussion

3.1. Transmission Properties Under Different Incident Angles

In this subsection, we consider only normal incidence of the electromagnetic wave on the (1D- PC) structure. To check the correctness of our computer program, numerical results are compared with those obtained from the real values in the practical uses [14]. The structure is restructured as $(H_n L_n)^m$ where n = 1 . . . m, m is chosen as 15. We have kept constant the dielectric permittivity's of the layers, are fixed to be $n_H = 3.7$ and $n_L = 1.5$ (Si and the second layer is SiO_2). There are three different values of the thickness of the considered layers that we will use in our studies, are: ($d_H = 112.1$ nm, $d_L = 276.4$ nm), ($d_H = 112.1$ nm, $d_L = 281.4$ nm) and ($d_H = 117.1$ nm, $d_L = 281.4$ nm).

The transmission spectra in figures 2-4 is computed and plotted with wavelength centered at 2.5 µm taking into account the different values of the filling ratio (F=d_H/d). It may be seen from these results, that the structure exhibits various band gaps (or stop band) where the photonic states are forbidden in the structure, can be seen in the transmission spectrum. In this study we are considered only the larger band gap width, is defined as the frequency range when $T \leq 0.01\%$. When the thickness of the layer is $d_H = 112.1$ nm, and $d_L = 276.4$ nm, the band gap rang is 1294nm to 2321nm. When $d_H = 112.1$ nm and $d_L = 281.4$ nm, the band gap rang is 1394nm to 2437nm. When $d_H = 112.1$ nm and $d_L = 281.4$ nm, the band gap rang is 1463nm to 2699nm. The photonic band gap is also same for both the TE and TM modes. So that, we can observe that the figure 4 has wider band gap than the figures 2-3. Also we can say that the change in the filling ratio allows us to obtain a new structure which can have adjustable gap width. It is found that the agreement between our results and those obtained via practical uses [14] is very good since the discrepancies between the two sets of results are below 0.8%.

In this subsection, we consider only normal incidence of the electromagnetic wave on the (1D- PC) structure. To check the correctness of our computer program, numerical results are compared with those obtained from the real values in the practical uses [14]. The structure is restructured as $(H_n L_n)^m$ where $n = 1 \ldots m$, m is chosen as 15. We have kept constant the dielectric permittivity's of the layers, are fixed to be $n_H = 3.7$ and $n_L = 1.5$ (Si and the second layer is SiO_2). There are three different values of the thickness of the considered layers that we will use in our studies, are: ($d_H = 112.1$ nm, $d_L = 276.4$ nm), ($d_H = 112.1$ nm, $d_L = 281.4$ nm) and ($d_H = 117.1$ nm, $d_L = 281.4$ nm).

The transmission spectra in figures 2-4 is computed and plotted with wavelength centered at 2.5 μm taking into account the different values of the filling ratio (F=d_H/d). It may be seen from these results, that the structure exhibits various band gaps (or stop band) where the photonic states are forbidden in the structure, can be seen in the transmission spectrum. In this study we are considered only the larger band gap width, is defined as the frequency range when $T \leq 0.01\%$. When the thickness of the layer is $d_H = 112.1$ nm, and $d_L = 276.4$ nm, the band gap rang is 1294nm to 2321nm. When $d_H = 112.1$ nm and $d_L = 281.4$ nm, the band gap rang is 1394nm to 2437nm. When $d_H = 112.1$ nm and $d_L = 281.4$ nm, the band gap rang is 1463nm to 2699nm. The photonic band gap is also same for both the TE and TM modes. So that, we can observe that the figure 4 has wider band gap than the figures 2-3. Also we can say that the change in the filling ratio allows us to obtain a new structure which can have adjustable gap width. It is found that the agreement between our results and those obtained via practical uses [14] is very good since the discrepancies between the two sets of results are below 0.8%.

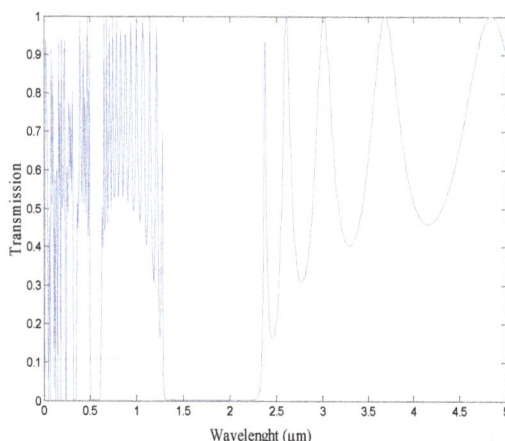

Fig. 2. Transmission spectra of 1D, $n_H = 3.7$, $n_L = 1.5$, $d_H = 112.1$ nm, $d_L = 276.4$ nm and F=0.288.

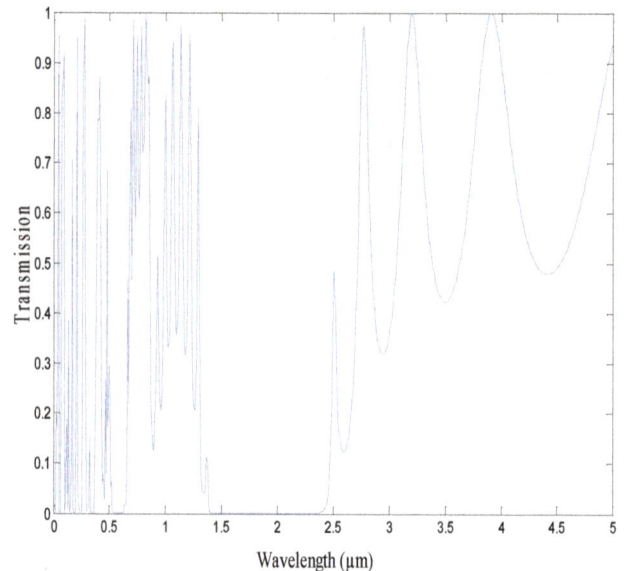

Fig. 3. Transmission spectra of 1D, $n_H = 3.7$, $n_L = 1.5$, $d_H = 112.1$ nm, $d_L = 281.4$ nm, and F=0.284.

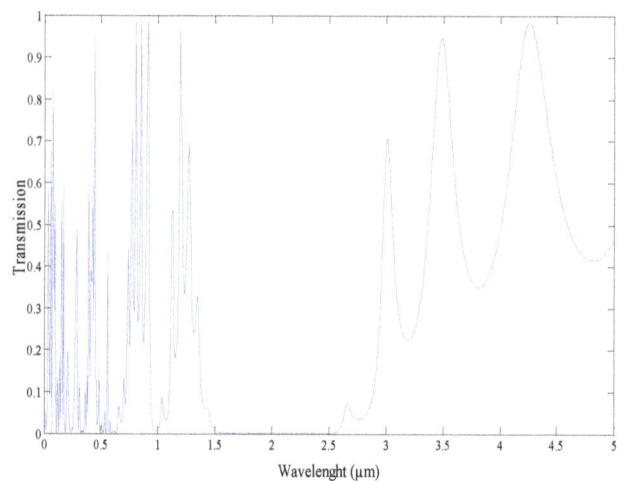

Fig. 4. Transmission spectra of 1D, $n_H = 3.7$, $n_L = 1.5$, $d_H = d_H = 117.1$ nm, $d_L = 276.4$ nm, and F=0.297.

The angle of the incidence is an important factor in the enlargement of gap width that is omitted in the previous studies [1, 4, 5, 14]. In the following, we present the results obtained of band gap width as functions of incident angle for two kinds (TE and TM modes). From the Figures 5-7, it is clearly seen that the band gap width for the TM mode is less than the TE mode for all the three cases. The band gap width of the structure compared for the incident angle with that in case of normal incidence is better, and the band gap width is more suitable when the angle is important for mode TE. The gap width of TM mode decreases with the increase of the angle for all the three cases. In addition, we can see that the band gap width of both modes (TE or TM) increases with the increase of filling ratio.

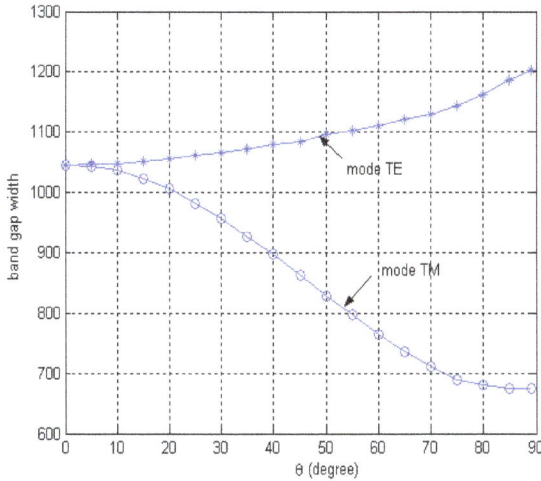

Fig. 5. Band gap width as functions of incident angle, $n_H = 3.7$, $n_L = 1.5$, $d_H = 112.1$ nm and $d_L = 276.4$ nm.

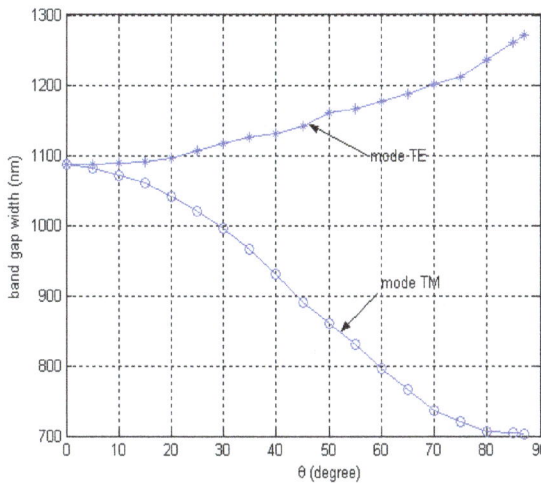

Fig. 6. Band gap width as functions of incident angle, $n_H = 3.7$, $n_L = 1.5$, $d_H = 112.1$ nm and $d_L = 281.4$ nm.

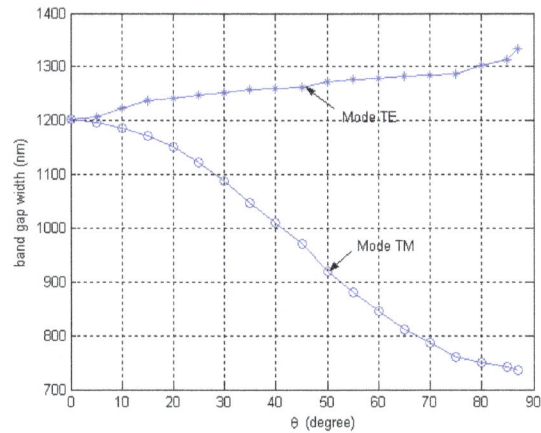

Fig. 7. Band gap width as functions of incident angle, $n_H = 3.7$, $n_L = 1.5$, $d_H = 117.1$ nm and $d_L = 276.4$ nm.

3.2. Band Gap Structure of the (1D- PC) Structure

To predict the propagation of an electromagnetic wave in a (1D- PC) structure, it is necessary to know its dispersion

relation or $\omega(k)$, where ω is the frequency of a wave with wavenumber k. In order to determine the variation of dispersion diagram, we computed the normalized frequency versus the wave vector for transverse electric (TE) and transverse magnetic (TM) modes. It has been calculated for the first Brillouin zone by employing the transfer matrix method and Bloch's theorem.

In figures 8-10, results are presented for the dispersion diagram of the structure analyzed in fig 1 for normal incidence for TE (or TM) mode. The band diagram of figure 8 is extends between the normalized frequencies (0.4344 (λ/d) to 0.7713 (λ/d)). In figure 9, we have (0.4294 (λ/d) to 0.7631 (λ/d)). In figure 10, we have (0.4245 (λ/d) to 0.7532 (λ/d)).

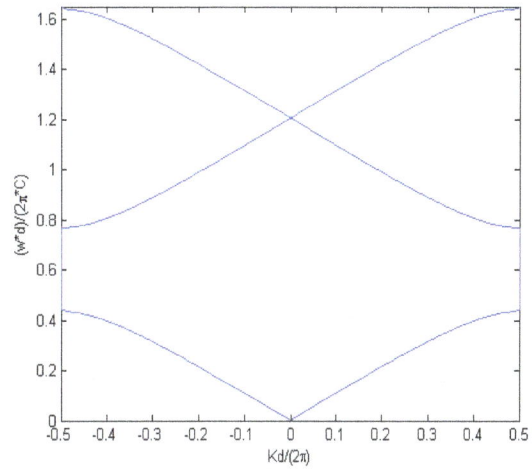

Fig. 8. Photonic band structure of (1D- PC), $n_H = 3.7$, $n_L = 1.5$, $d_H = 112.1$ nm and $d_L = 276.4$ nm, $\theta=0°$.

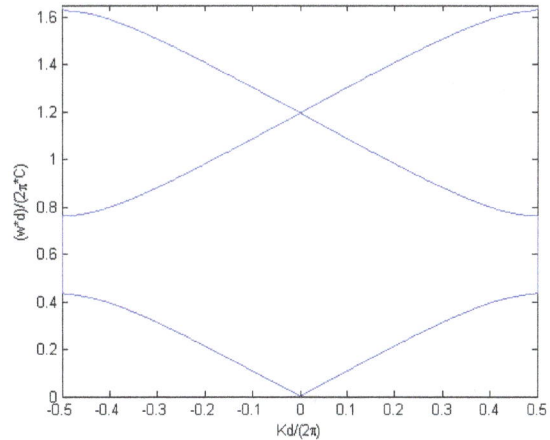

Fig. 9. Photonic band structure of (1D- PC), $n_H = 3.7$, $n_L = 1.5$, $d_H = 117.1$ nm and $d_L = 281.4$ nm, $\theta=0°$.

As seen in these figures, a gap is present between the allowed states which can be easily identified, this corresponds to the band gap of the structure, and comparison shows that these gaps occur over the same frequency range as the transmission stop band. The existence of photonic band gap appeared when the value of Bloch wave vector becomes complex. However, the real values of Bloch wavevector k are corresponding to the pass band and imaginary values are corresponding to the forbidden band gap.

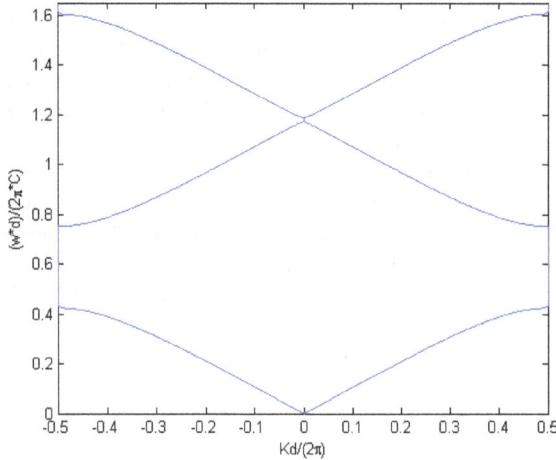

Fig. 10. *Photonic band structure of (1D- PC) (1D- PC), $n_H = 3.7$, $n_L = 1.5$, $d_H = 117.1$ nm and $d_L = 276.4$ nm, $\theta=0°$.*

3.3. Photonic Gap Map of the (1D- PC) Structure

In this section, we have demonstrated that it is possible to modify the photonic gap (PBG) of structure by varying the incident angle and filling factor f. The gap map shown in Figures 11–13 represent results in terms of frequencies ($\frac{fd}{c}$) (where d is the period constant, C the speed of light and f the frequency) as function of the filling factor ($f = d_H/d$). The photonic gap map for the TE or TM mode (normal incidence), presented in figure 11, the red region indicates the variation of band gap (PBG), the empty space regions represent the ranges of transmission. From simulation analysis, we observe that the increasing of f has the effect of increasing photonic gap. Also, we found that the first band gap is appear when the value of f is greater than 0.001.

In the figures 12-13, the red region indicates the variation of TE PBG, and the blue region indicates the variation of TM PBG for oblique incidence ($\theta=20°$). It may be seen from the plot that the forbidden bandwidth varies with the ratio filling factor (d_H/d) of the two modes. As the angle of the incidence increases the forbidden bandwidth was also found to be increased and shifted to higher frequency regions

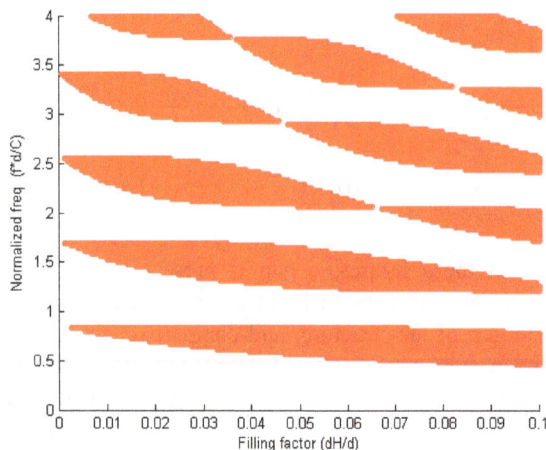

Fig. 11. *Photonic gap map (normalized frequency wa/2πc versus the filling factor(d_H/d) of (1D- PC), $n_H = 3.7$, $n_L = 1.5$, d = 388.5nm, $\theta=0°$.*

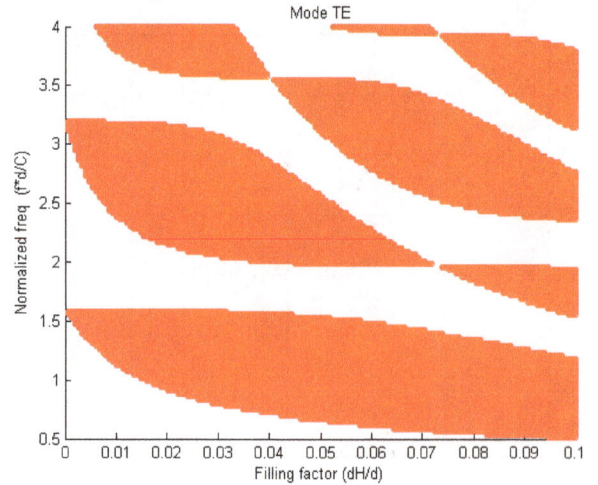

Fig. 12. *Photonic gap map (normalized frequency wa/2πc versus the filling factor(d_H/d) of (1D- PC), $n_H = 3.7$, $n_L = 1.5$, d = 388.5nm, Mode TE, $\theta=20°$.*

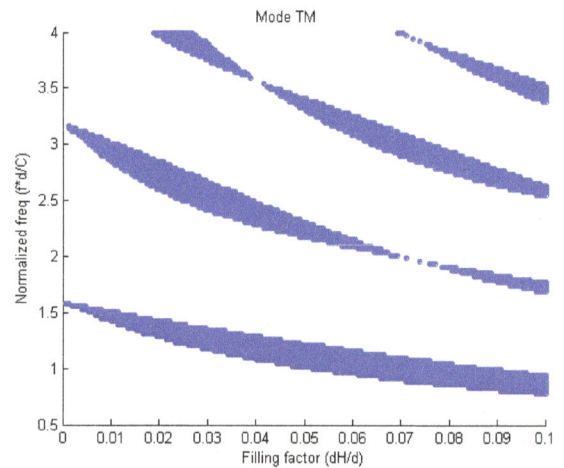

Fig. 13. *Photonic gap map (normalized frequency wa/2πc versus the filling factor(d_H/d) of (1D- PC), $n_H = 3.7$, $n_L = 1.5$, d = 388.5nm, Mode TM, $\theta=20°$.*

4. Conclusion

In summary, we have used the transfer matrix method and Bloch theorem to study one-dimensional photonic crystals. We have discussed the calculation of transmissivity, reflectivity, and dispersion relation for both TE and TM modes at different incidence angles. Our results show that the PBG sensitivity depend on many factors, such as the filling factor and polarization modes. The angle of incidence of the wave is also another factor which affects the width of band gaps. Further, theory approach TMM and Bloch theorem can allow for better characterization of the one-dimensional photonic crystals with defect.

References

[1] H. Oraizi A. Abdolali, "Several theorems for reflection and transmission coefficients of plane wave incidence on planar multilayer metamaterial structures", IET Microw. Antennas Propag., Vol. 4, Iss. 11, pp. 1870–1879, 2010.

[2] I. V. Shadrivov, A. A. Sukhorukov, and Y. S. Kivshar, "Complete band gaps in one-dimensional left-handed periodic structures," *Physical Review Letters*, Vol. 95, pp. 1-4, 2005.

[3] Z. Wang, D. Liu "A few points on omnidirectional band gaps in one- dimensional photonic crystals," Applied Physics B, Appl. Phys. B 86, pp 473–476, 2007.

[4] A. Gharaati and Z. Zare, "Photonic band structures and enhancement of omnidirectional reflection bands by using a ternary 1D photonic crystal including left-handed materials", Progress In Electromagnetics Research M, Vol. 20, pp 81-94, 2011.

[5] S. K. Srivastava, and S. P. Ojha, "Enhancement of omnidirectional reflection bands in one-dimensional photonic crystals with left-handed materials," *Progress In Electromagnetics Research*, Vol. 68, pp 91-111, 2007.

[6] Sakoda K: Optical properties of Photonic Crystals. New York: Springer Berlin Heidlberg; 2005.

[7] M. Skorobogatiy, and J. Yang, *Fundamentals of Photonic Crystal Guiding*, Cambridge University Press, New York, 2009.

[8] I. A. Sukhovanov, and I. V. Guryev, *Photonic Crystals Physics and Practical Modeling*, Springer Series in Optical Science, New York, 2009.

[9] R. Gonalo, P. D. Maagt, M. Sorrolo, "Enhanced Patch antenna perfomnce by suppressing surface waves using PBG substrates,," *IEEE* transactions on microwave theory and techniques. Vol. 47, No. 11, pp. 2131-2138, November 1999.

[10] E. Yablonovitch, T.J. Gmitter, "Photonic band structure: The face centered cubic case employing nonspherical atoms," J. Physical review letters, Vol.67, N° 17, pp. 2295– 2298, 1991.

[11] K. M. Ho, C. T. Chan, and C. M. Soukoulis, "Existence of a Photonic Gap in Periodic Dielectric Structures,"Phys Rev Lett, 1990 Dec 17, vol 65, N°25, pp 3152-3155.

[12] Costas M. Soukoulis, "Photonic Band Gap Materials," Proceedings of the NATO Advanced Study Institute on Photonic Band Gap Materials, Elounda, Crete, Greece , June 18-30, 1995.

[13] C. v. Mee, P. Contu, P. Pintus, "One-dimensional photonic crystal design", Journal of Quantitative Spectroscopy & Radiative Transfer, Vol. 111, pp. 214–225, 2010.

[14] H. Tian, Y. Ji, C. Li, H. Liu, "Transmission properties of one-dimensional graded photonic crystals and enlargement of omnidirectional negligible transmission gap", Elsevier, Optics Communications Vol 275, N°1, pp 83–89, July 2007.

[15] B. Gallas S. Fission, E. Charron, A. Brunet-Bruneau, R. Vuye and J. Revory, " Making an omnidirectional reflector", Appl. Opt. 40, 5056-5063, 2001.

[16] K. M.Chen, A. W. Sparks, H. C. Luan, D. R. Lim, K. wada, and L. C. Kimerling, "SiO2/TiO2 omni-directional reflector and microcavity resonator via the sol-gel method," *Appl. Phys. Lett.*, Vol. 75, pp. 3805-3807, 1999.

[17] D. N. Chigrin, A. V. Lavrinenko, D. A. Yarotsky, and S. V. Gaponenko, "Observation of total omni-directional reflection from a one-dimensional dielectric lattice," *Appl. Phys. A: Mater. Sci. Process.*, Vol. 68, pp. 25–28, 1999.

[18] D. N. Chigrin, A. V. Lavrinenko, D. A. Yarotsky, and S. V. Gaponenko, "All dielectric one dimensional periodic structures for total omnidirectional reflection and partial spontaneous emission control, " *J. Lightwave Tech.*, Vol. 17, pp.2018–2024, 1999.

[19] F. Scotognella, "Four-material one dimensional photonic crystals," Elsevier, Optical Materials, Vol. 34, N° 9, July 2012, pp 1610–1613.

[20] K. M. Leung, and Y. F. Liu, "Photon band structures: The plane wave method", Physical Review B, 41, pp. 10188-10190, 1990.

[21] S. Şimşek, "A novel method for designing one dimensional photonic crystals with given bandgap characteristics," Elsevier, Int. J. Electron. Commun. (AEÜ), Vol. 67, pp827– 832, 2013.

[22] L. C. Botten, T. P. White, A. A. Asatryan, T. N. Langtry, C. M. de Sterke and R. C. McPhedran, "Bloch mode scattering matrix methods for modeling extended photonic crystal structures. Part I: Theory," Phys. Rev. E, Vol. 70, 056606, 2004.

[23] L. C. Botten, N. A. Nicorovici, R. C. McPhedran, A. A. Asatryan, and C. M. de Sterke, "Photonic band structure calculations using scattering matrices", *Phys. Rev. E*, Vol. 64, 046603, pp. 1-20, 2001.

[24] J. B. Pendry, A. Mackinnon, "Calculation of photon dispersion, "Phys. Rev. Lett., 69, (3), pp. 2772–2775, 1992.

18

Design and Simulation of TCR-TSC in Electric Arc Furnace for Power Quality Improvement at Steel Making Plant (No-1 Iron and Steel Mill, Pyin Oo Lwin, Myanmar)

Thet Mon Aye, Soe Win Naing

Dept. of Electrical Power Engineering, Mandalay Technological University, Mandalay, Myanmar

Email address:

thetmon.pp@gmail.com (T. M. Aye), soewinnaing2011@gmail.com (S. W. Naing)

Abstract: Electric Arc Furnaces (EAFs) are unbalanced nonlinear and time varying loads, which can cause many problems in the power system quality. As the use of arc furnace loads increases in industry, the important of the power quality problems also increase. So, in order to optimize the usage of electric power in EAFs, it is necessary to minimize the effect of arc furnace loads on power quality in power systems as much as possible. Therefore, in this paper, design and simulation of an electric plant supplying an arc furnace is considered. Then by considering the high changes of reactive power and voltage flicker of nonlinear furnace load, a thyristor controlled reactor compensation with thyristor switched capacitor (TCR-TSC) are designed and simulated. Finally, simulation results verify the accuracy of the load modeling and show the effectiveness of the proposed TCR-TSC model for reactive power compensating of the EAF. The installation site for this proposed system is No (1) Iron and Steel Mill (Pyin- Oo- Lwin). And data is taken from this Steel Mill. Simulation results will be provided by using MATLAB/ Simulink.

Keywords: Electric Arc Furnaces, Power Quality, Voltage Flicker, MATLAB/Simulink, TCR-TSC Compensation

1. Introduction

The use of electric arc furnaces (EAFs) for steel making has grown dramatically in the last decade. Of the steel made today 36% is produced by the electric arc furnace route and this share will increase to 50% by 2030. The electric arc furnaces are used for melting and refining metals, mainly iron in the steel production. AC and DC arc furnaces represent one of the most intensive disturbing loads in the sub-transmission or transmission electric power systems; they are characterized by rapid changes in absorbed powers that occur especially in the initial stage of melting, during which the critical condition of a broken arc may become a short circuit or an open circuit. In the particular case of the DC arc furnaces, the presence of the AC/DC static converters and the random motion of the electric arc, whose nonlinear and time-varying nature is well known, are responsible for dangerous perturbations such as waveform distortions and voltage fluctuations [2].

Nowadays, arc furnaces are designed for very large power input ratings and due to the nature of both, the electrical arc and the melt down process, these devices can cause large power quality problems on the electrical network, mainly harmonics, inter-harmonics, flicker and voltage imbalances. The EAFs are among the largest electrical loads in power systems. Regarding the fast and heavy deviations of electric power in these loads, the bus voltage of these furnaces is unbalanced and has large oscillations. Moreover, the EAFs cause deteriorating of the power quality, making voltage flicker, unbalancing in voltage and current, and occurring odd and even harmonics in power systems. In order to study in this field and improve the above mentioned factors, exact and complete design of the power system with arc furnaces should be performed [1].

2. Power Quality and Electrical Arc Furnaces

Power quality can be interpreted by the existence of two components:

- Voltage quality. It expresses the voltage deviation from the ideal one and can be interpreted as the product quality delivered by the utilities.

• Current quality. It expresses the current deviation from the ideal one and can be interpreted as the product quality received by the customers.

The main Power quality disturbances are:

• Harmonics
• Under-voltages or Over-voltages
• Flicker
• Transients
• Voltage sags
• Interruptions.

An electric arc furnace (EAF) transfers electrical energy to thermal energy in the form of an electric arc to melt the raw materials held by the furnace [3]. Typical steelmaking cycles are:

• Arc ignition period (start of power supply)
• Boring period
• Molten metal formation period
• Main melting period
• Meltdown period
• Meltdown heating period respectively.

3. Model of Power System with AC Electric Arc Furnace

The electric diagram of a source supplying an EAF is illustrated in Figure 1.

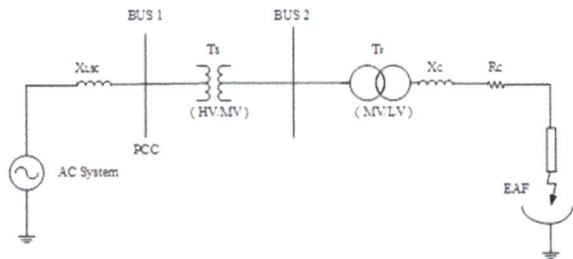

Figure 1. Diagram of an EAF connected to the rest of power system.

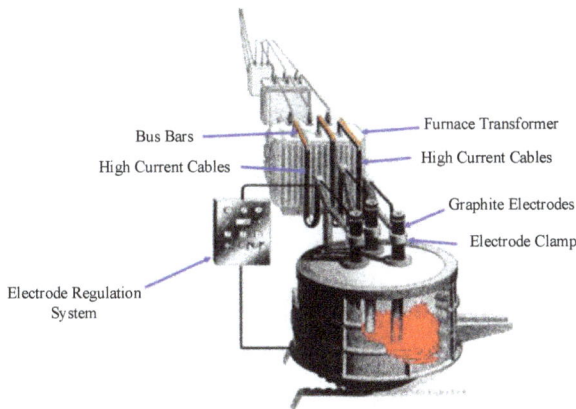

Figure 2. Electrical Power Supply Network for EAF in No(1) Iron and Steel Mill.

In this figure, bus 1 is the point of common coupling (PCC) which is the supplying bus of the EAF transformer. In order to

change the input active power of the EAF, transformer T_F, (MV/LV) is used. This transformer is equipped with a tap changer located at the secondary winding to have the ability of changing the voltage of the furnace. The arc furnace is also connected to the PCC through the transformer T_S, (HV/MV). X_C and R_C are the reactance and resistance of the connecting cable line to the furnace electrodes, respectively. Also, X_{Lsc} is the short circuit reactance at bus PCC.

In this work, the steel mill No.1, (Pyin Oo Lwin, Mandalay, Myanmar) is composed of the various departments, namely, Electric Arc Furnace (EAF), Ladle Refining Furnace (LRF), Continuous Casting Machine (CCM), Water Treatment Plant (WTP), Fumes Treatment Plant (FTP), Air Separation Plant (ASP), Oxygen Plant, Wire Production Plant, Precipitated Calcium Carbonate (PCC) $CaCO3$, Lime, Air Compressor, Shredder and Bolt Nut. Among these departments, EAF is studied as research area.

Figure 3. Practical Site of No(1) Iron and Steel Mill.

4. Design of TCR-TSC for Electric Arc Furnace

In order to compensate reactive power and voltage flicker improvement in power system including the EAF and on the basis of Figure 1, in this section, an optimal design of TCR compensator along with a parallel capacitor will be presented. The one-line diagram of this structure is illustrated in Figure 4.

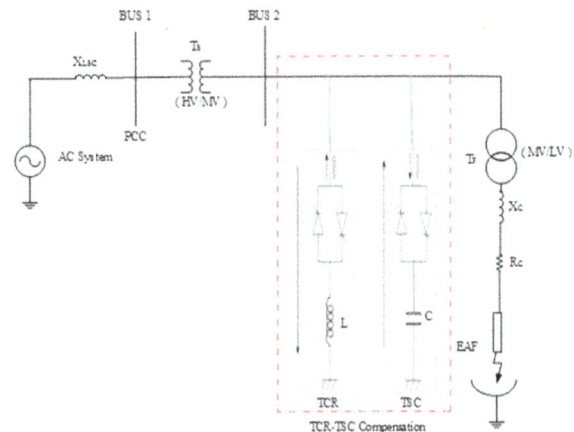

Figure 4. Configuration of a TCR-TSC Connected to an EAF.

As it can be seen in Figure 4, the effective inductance of the TCR compensator can be varied by changing the firing angle of the thyristors. If the thyristors are fired exactly at the voltage peak of the supply, then they will conduct perfectly. The TCR current is essentially reactive with almost 90 degrees lag phase.

The operation of EAF Plant, steelmaking cycle is represented in Table 1.

Table 1. *Typical Steelmaking Cycle of Proposed System.*

Stage of Melting	Time (min)	Voltage (kV)	Reactive Power (MVAR)	Real Power (MW)
Bucket 1	0	11	0	0
	2.91	11	0	0
Initial Melting	2.92	9.84	16.49	21.99
	3.68	9.84	16.49	21.99
Bore Down	3.69	10.87	13.81	18.41
	4.32	10.87	13.81	18.41
Melting	4.33	9.71	22.635	30.18
	8.51	9.71	22.635	30.18
Melting	8.52	8.81	20.87	27.83
	9.09	8.81	20.87	27.83
Melting	9.10	9.88	18.795	25.06
	10.52	9.88	18.795	25.06
Melting	10.53	9.17	25.29	33.72
	16.27	9.17	25.29	33.72
End Melting	16.28	10.10	24.06	32.09
	17.93	10.10	24.06	32.09
Finish Bucket 1	17.94	10.97	0	0
	19.22	10.97	0	0
Initial Melting/Boring	19.23	10.68	15.42	20.57
	19.94	10.68	15.42	20.57
Melting	19.95	9.73	25.18	33.57
	20.53	9.73	25.18	33.57
Melting	20.54	9.00	21.27	28.36
	24.03	9.00	21.27	28.36
Melting	24.04	8.56	24.74	32.99
	29.69	8.56	24.74	32.99
Finish Bucket 2	29.70	10.97	0	0
	31.89	10.97	0	0
Initial Melting/Boring	31.90	7.81	16.03	21.38
	32.51	7.81	16.03	21.38
Bore Down	32.52	10.17	24.07	32.09
	33.40	10.17	24.07	32.09
Melting	33.41	8.87	22.38	29.84
	36.13	8.87	22.38	29.84
Melting	36.14	8.99	25.39	33.86
	41.97	8.99	25.39	33.86
Melting	41.98	10.99	0	0
	42.23	10.99	0	0
Melting	42.24	10.26	24.50	32.67
	48.80	10.26	24.50	32.67
End Melting	48.81	9.96	24.315	32.42
	50.64	9.96	24.315	32.42
End Melting/ Superheating	50.65	9.81	23.46	31.28
	51.84	9.81	23.46	31.28
Finish Bucket 3	51.85	10.98	0	0
	53.56	10.98	0	0
	53.57	11	0	0
End Melting	57.17	11	0	0
	57.18	11	0	0

The compensated values for the capacitance and the compensated TCR inductance are calculated based on this setting.

$$X_{SVC} = \frac{V_{bus}^2}{Q_{SVC}} \tag{1}$$

$$\therefore X_{SVC} = 4.7845\Omega$$

$$X_{SVC} = \frac{X_{TSC}X_{TCR}}{X_{TSC} + X_{TCR}} \tag{2}$$

$$X_{TSC} = 2X_{TCR} \tag{3}$$

$$X_{SVC} = 0.6667X_{TCR}$$
$$\therefore X_{TCR} = 7.1764\Omega$$

$$X_{TCR} = 2\pi f L \tag{4}$$

$$X_{TSC} = \frac{1}{2\pi f C} \tag{5}$$

Table 2. *Compensated Values of Inductor and Capacitor.*

No	Load (EAF)	Compensating Inductance mH	Compensating Capacitance μF
1	2.0 ton	23.759	213.224
2	3.0 ton	23.579	214.848
3	4.0 ton	23.351	216.954
4	5.0 ton	22.843	221.780
5	7.0 ton	22.753	222.650
6	0.5 ton	41.832	121.104
7	1.0 ton	37.464	135.223
8	1.5 ton	36.039	140.573
9	1.8 ton	35.030	144.686
10	1.9 ton	30.796	164.500

$$V_{bus} = 11kV \text{ (From Table 1)}$$

$$Q_{SVC} = 25.39MVAR \text{ (From Table 2)}$$

The maximum reactive capacity of the TCR-TSC is set at $Q_{SVC} = 25.39MVAR$

From Equation 4 and Equation 5, considering a fundamental frequency is of 50Hz. The capacitance and inductances are $C = 222.65\mu F$ and $L = 22.7532mH$.

According to the inductive and capacitive reactance, each SVC has its own firing angle reactive power characteristic, $Q_{SVC}(\alpha)$ which is a function of the inductive and capacitive reactance.

It is necessary to obtain the effective reactance X_{SVC} as a function of the firing angle α, using the fundamental frequency TCR equivalent reactance X_{TCR}.

$$X_{TCR} = \frac{\pi X_L}{\sigma - \sin\sigma} \tag{6}$$

$$\sigma = 2(\pi - \alpha) \tag{7}$$

Where X_L is the reactance of the linear inductor and σ and α are the thyristors' conduction angle and firing angle repectively.

At $\alpha = 90°$ the TCR conducts fully and its equivalent reactance X_{TCR} becomes X_L.

At $\alpha = 180°$ the TCR is blocked and its equivalent reactance becomes extremely large. It is also expressed in Table 3.

Table 3. *Variation of Value of* B_L *with Firing Angle.*

Firing Angle α (Degree)	Conduction Angle σ (Degree)	Susceptance $B_L(\alpha)$ (S)	Perunit of $B_L(\alpha)$
90°	180°	0.028	1
100°	160°	0.022	0.8
110°	140°	0.016	0.6
120°	120°	0.011	0.4
130°	100°	6.7×10^{-3}	0.2
140°	80°	3.63×10^{-3}	0.1
150°	60°	1.59×10^{-3}	0.06
160°	40°	4.88×10^{-3}	0.02
170°	20°	6.21×10^{-3}	0.002
180°	0	0	0

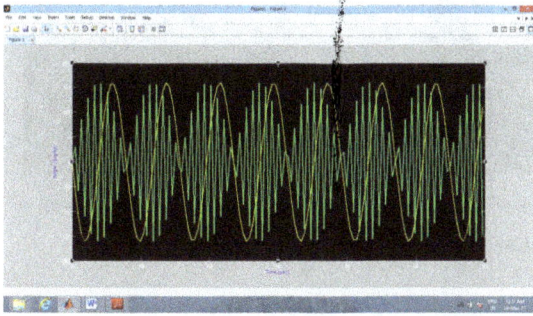

Figure 5. *Sinusoidal Waveform of Conduction Angle and Firing Angle.*

The total effective reactance of the SVC, including the TCR and capacitive reactance, is determined by the parallel combination of both components and which as a function of the conduction angle σ becomes

$$X_L = \frac{X_{TCR}(\sigma - \sin\sigma)}{\pi} \quad (8)$$

And finally as a function of the firing angle α becomes

$$X_{SVC} = \frac{X_{TSC}\pi X_L}{X_{TSC}[2\pi - 2\alpha + \sin 2\alpha] + \pi X_L} \quad (9)$$

As expected, the effective reactance of the SVC is a function of the firing angle α. Equation 9 may be used to determine the firing angle α, using the fundamental relationship.

$$Q_{SVC} = V_{bus}^2 \left[\frac{X_{TSC}[2\pi - 2\alpha + \sin 2\alpha] + \pi X_L}{X_{TSC}\pi X_L}\right] \quad (10)$$

At initial melting conduction, the reactive power of the load is set 16.49MVAR and bus voltage is 9.84kV respectively. The

calculation of firing angle value is 156° and the next stages are also calculated as before.

By using the iteration method, the next step to go on calculate until the voltage level is stable.

$$V_{bus} = V_{ref} - V_{drop} \quad (11)$$

By iteratively, the voltage level is stable with 10.2kV at the step (18). In this way, another condition will also calculate till stable.

The new total melting time has been reduced from 51.85 min down to 47.16 min which is 9.05% reduction. The power on time has been reduced from 48.36 min down to 43.68 min which is 9.7% reduction.

The new melting times have been calculated assuming that furnace is able to transfer increased power into melting process by the similar efficiency than in the system without TCR-TSC compensator. The comparison of times without and with TCR-TSC is presented in Table 4 and Table 5.

Table 4. *Total Melting Times.*

Time (min) Without TCR-TSC	Time (min) With TCR-TSC
51.85	47.16
Reduction in time	9.05%

Table 5. *Power On Times.*

Time (min) Without TCR-TSC		Time (min) With TCR-TSC
Bucket 1	17.94	15.94
Bucket 2	10.47	9.05
Bucket 3	19.95	18.69
Total	48.36	43.68
Reduction in time		9.7%

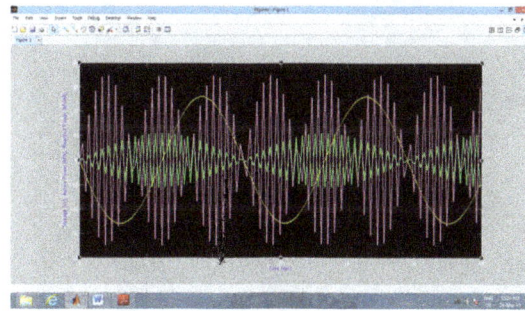

Figure 6. *Sinusoidal Waveform of Voltage and Power for EAF Load System without TCR-TSC.*

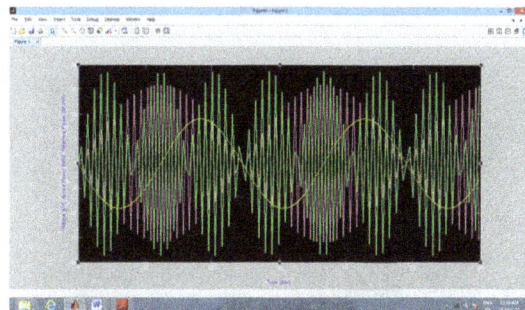

Figure 7. *Sinusoidal Waveform of Voltage and Power for EAF Load System with TCR-TSC.*

5. Simulation Result for Voltage and Current

The power quality means voltage quality for this work and to enhance the power quality i.e., to keep the voltage stability is the aim. Voltage stability problems usually encounter in heavily loaded systems.

EAF is operating to melt down the scrap in a moment. During melting cases, generally, the system voltage decreased from the constant level. If the running condition stopped, the voltage will suddenly rise.

The simulation diagram and simulation result without and TCR-TSC are described in the following Figure 8, Figure 9 and Figure 10.

The inserted values of inductance and resistance of EAF model are $0.148mH$ and 0.053Ω. Figure 8 shows Simulink block diagram of EAF without TCR-TSC. The result without TCR-TSC was found in Figure 9 for voltage and in Figure 10 for current. The TCR-TSC compensator operates on single phase basis and thus it is balancing the voltage.

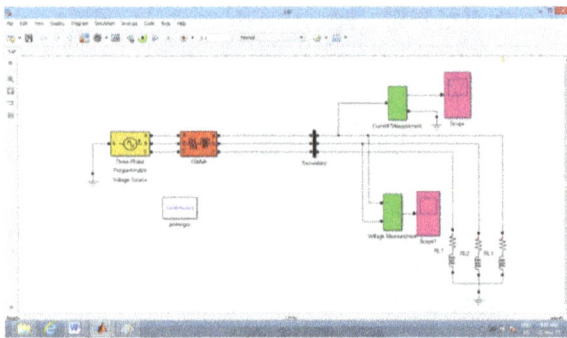

Figure 8. Simulink Block Diagram of EAF without TCR-TSC.

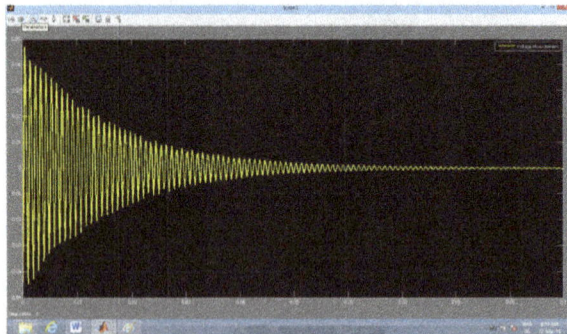

Figure 9. Simulation Result of Voltage without TCR-TSC.

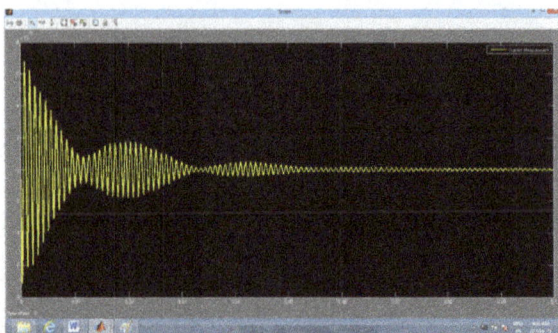

Figure 10. Simulation Result of Current without TCR-TSC.

To obtain the desire results, a TCR-TSC is added to the EAF. The Simulink block diagram of EAF with TCR-TSC is shown in Figure 11. The system voltage of bore down process with 25.29 MVAR, 33.72 MW and 9.17 kV is stable as can be seen in Figure 12. The current flow is expressed in Figure 13.

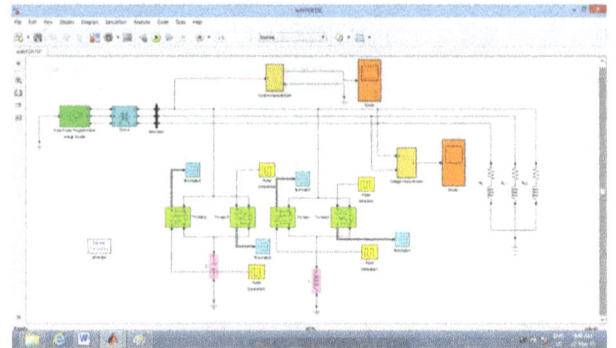

Figure 11. Simulink Block Diagram of EAF with TCR-TSC.

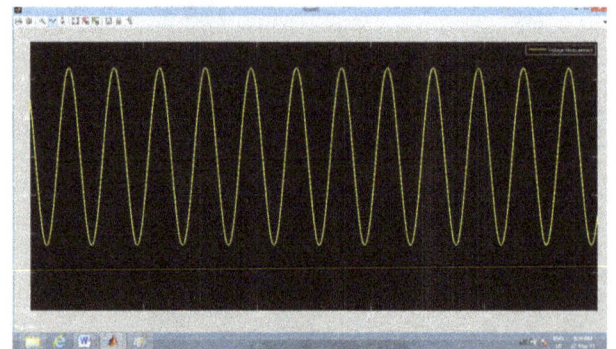

Figure 12. Simulation Result of Voltage with TCR-TSC.

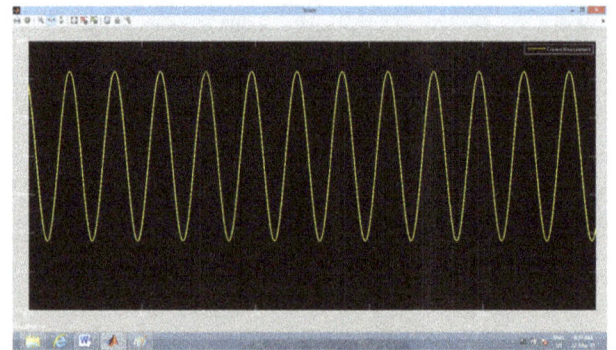

Figure 13. Simulation Result of Current with TCR-TSC.

The TCR-TSC will stabilize the voltage. This is advantageous not only for the furnace itself but also for the control and protection system of the steel making plant because modern electronics and process instruments are very sensitive to voltage fluctuations. The TCR-TSC can reduce the flicker down to the value requested by the power utility.

6. Conclusion

This paper provides a detailed description of a modern, TCR-TSC type Static Var Compensator installed in steel making plant. The Electric Arc Furnace (EAF) for steel production is analyzed. The reactive power compensation requirements are quantitatively analyzed. The TCR-TSC is a controller for voltage regulation and for maintaining constant voltage at a bus. This can result saving in operational costs, reduced line losses, reduction operating time, rise the steel tons, etc.

Acknowledgements

The author would like to express grateful thanks to her supervisor Dr. Soe Win Naing, Associate Professor, Department of Electrical Power Engineering, Mandalay Technological University, for his help, guidance and advice. The author wishes to thank to all of her teachers from Mandalay Technological University. The author's special thanks are sent to her father, mother and sister for their support and encouragement.

References

[1]   S. Y. Lee, C. J .Wu and W. N. Change, A compact control algorithm for compensator, Electric Power System Research, 58, 2001, pp. 63-70.

[2]   G. E1-Saady, Adaptive static VAR controller for simultaneous elimination of voltage flicker and phase current imbalances due to arc furnace loads, Electric Power Systems Research, 58, 2001, pp.133-140.

[3]   T. J. E. Miller, Reactive Power Compensation and the Electric Arc Furnace, 2002.

[4]   Robertas Staniulis., Reactive Power Valuation, Department of Industrial Electrical Engineering and Automation Lund University, Lund, 2001.

[5]   Md M. Biswas, Kamol K. Das, Voltage level improving by using static VAR compensator, Global Journal of Research in Engineering Vol.11, Issue 5, Version 1.0, 2011.

[6]   Wangha L., Taewon K., Control of the Thyristor-Controlled Reactor for reactive power compensation and load balancing, IEEE Conference on Industrial Electronics and Applications (ICIEA), 2007, pp. 201-206.

[7]   Hingorani, N.G., Flexible AC Transmission, IEEE Spectrum, 1993, pp.40-44.

[8]   M. N. Murthy, Director, Reactive Power Fundamentals, PSTI, Bangalore, January 2003.

[9]   Golkar M.A., Meschi S., MATLAB modeling of arc furnace for flicker study, IEEE Conference on Industrial Technology (ICIT), 2008, p.1-6.

[10]   Hingornai, N.G. and Gyugyi, L, Understanding FACTS, IEEE Press, Newyork, 1999.

19

A Self-Adaptive DC/DC Buck Converter Control Modulation Design

Pengcheng Xu, Zhigang Han

Department of Electronic Science and Technology, Tongji University, Shanghai, China

Email address:

13xpc@tongji.edu.cn (Pengcheng Xu), hanzhg@tongji.edu.cn (Zhigang Han)

Abstract: DC/DC converter is widely used in many electronic application power supplies. Usually, in the previously DC/DC converter control modulation, the duty cycle can be changed according the feedback signal in pulse width modulation (PWM) or the frequency be changed with a constant ON time or OFF time in pulse frequency modulation (PFM). A self-adaptive DC/DC converter control modulation is proposed in this paper. Based on the outputs of two uniform operational transconductance amplifiers which are influenced by the feedback voltage, both of the pulse ON time and pulse OFF time will be changed simultaneously. A self-adaptive frequency can be achieved in this control modulation. It can get a same output voltage ripple with a lower control frequency.

Keywords: DC/DC, ON Time, OFF Time, Frequency

1. Introduction

In a previously typical DC/DC converter with pulse frequency regulation (PFM) [1], usually, the OFF time of the pulse can be changed with a constant ON time or the ON time of the pulse can be changed with a constant OFF time. In the other widely used control modulation, pulse width regulation (PWM) [2], the duty cycle will be changed with a constant frequency. PWM control modulation has lower conversion efficiency in light load, while the PFM control modulation has lower conversion efficiency in high load. For a wide range high efficiency DC/DC converter, from light load to high load, dual-mode converter is been proposed previously. But there is no doubt that it increases the design complication [3-6].

A self-adaptive DC/DC converter regulation is proposed with a simple control mode. The feedback voltage from output will be used to adjust the ON time and OFF time of the pulse simultaneously. The simulation result shows that this control regulation can get a same voltage ripple with a lower control frequency under the same conditional. And also the self-adaptive frequency achieves the wide range high efficiency. Because lower switching frequency reduces switching losses.

As depicted in Fig.01, it is schematic of synchronous buck converter with self-adaptive control modulation. More and more DC/DC converter uses synchronous rectifier for its irreplaceable advantages, like fast transient response and high power density [7]. According the varying feedback voltage from output, both drive voltage pulse frequency and width will be changed. In other words, both ON time and OFF time of the driving pulse can be adjusted accordingly.

This proposed modulation is different from PWM which has a constant frequency and only can adjust pulse width, or PFM which has a constant ON time or OFF time and only can adjust OFF time or ON time respectively.

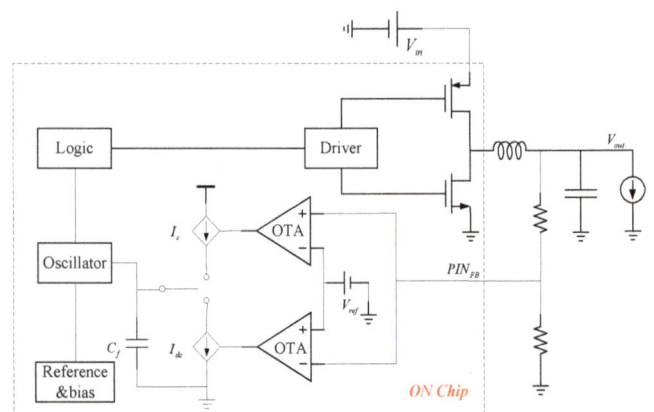

Fig. 1. Schematic of self-adaptive control modulation buck DC-DC converter with synchronous rectifier.

2. Self-Adaptive Modulation

2.1. Modulation Process

By the voltage sampling circuit with divided resistors, a feedback voltage $V_{feebback}$ can be generated according the varying output voltage V_{out} and transfer into the chip.

The operational transconduction amplify (OTA) can identify and amplify the difference between $V_{feebback}$ and reference voltage V_{ref}, As depicted in Fig.02 self-adaptive modulation schematic.

Two identical OTAs are used in the schematic. The noninverting input of one of OTAs is $V_{feebback}$, while the other one OTA's inverting input is $V_{feebback}$. The output of OTAs will be across on resistors R_{ON} and R_{OFF} through voltage followers and adjust the current. As a result, the two identical OTAs will generate two completely opposite output signals to control the charging and discharging current accordingly.

Two completely opposite outputs of OTAs will convert into relevant current which charge or discharge the capacitor C_F respectively.

The range of voltage of C_F is restrained between reference voltages V_{high} and V_{low}. Combining with logic circuit, with opposite changing currents can adjust both T_{on} and T_{off} of pulse.

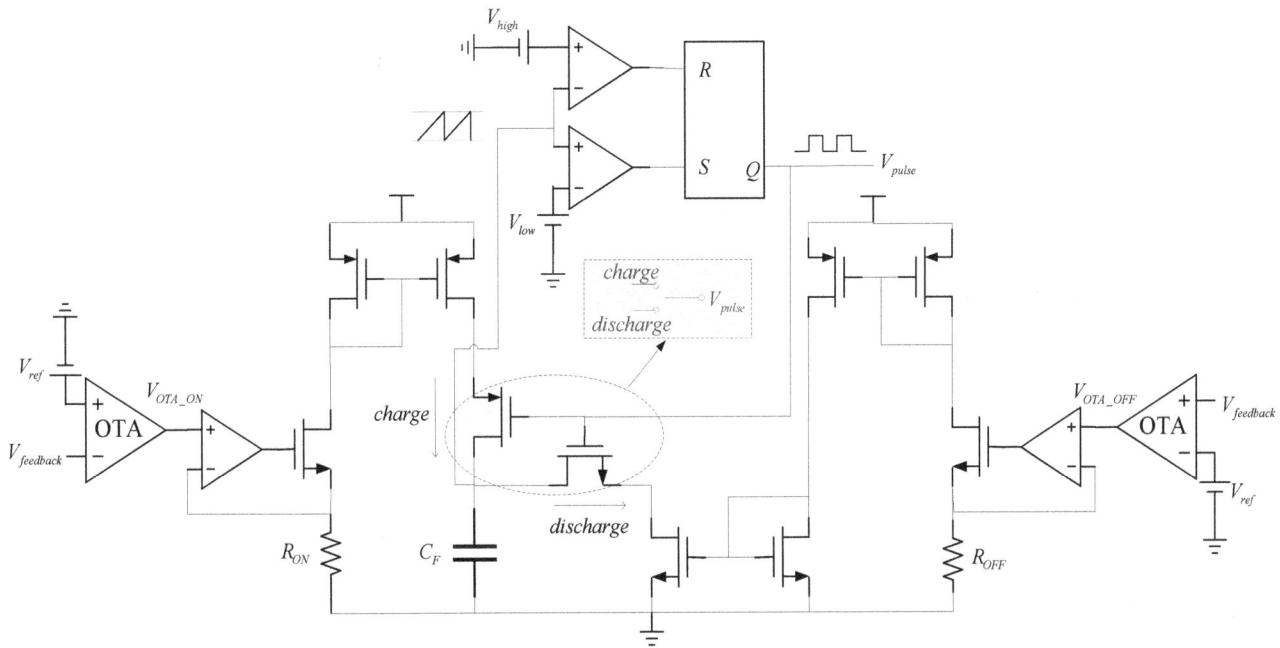

Fig. 2. Schematic for specifying ON time and OFF of driving pulse.

2.2. Quantitative Analysis

As depicted in Fig.02, the voltage followers can keep the voltages V_{OTA_ON} and V_{OTA_OFF} across on the resistors R_{ON} and R_{OFF} respectively. The charge current of C_F is,

$$I_{charge} = \frac{V_{OTA\_ON}}{R_{ON}} \tag{1}$$

And the discharge current of C_F is,

$$I_{discharge} = \frac{V_{OTA\_OFF}}{R_{OFF}} \tag{2}$$

As depicted in the Fig.02, the voltage across capacitor C_F rises and falls between V_{high} and V_{low}.

The switching PMOS and NMOS be turned on or turned off respectively in accordance with output voltage pulse of Set-Reset Latch. In addition, the output voltage pulse of Set-Reset Latch can control the switch of charge or discharge of capacitor C_F. Consequently, both T_{ON} and T_{OFF} will change based on varying V_{out}.

$$T_{ON} = \frac{V_{high} - V_{low}}{V_{OTA\_ON}} R_{ON} C_F \tag{3}$$

$$T_{OFF} = \frac{V_{high} - V_{low}}{V_{OTA\_OFF}} R_{OFF} C_F \tag{4}$$

$$f = \frac{1}{T_{ON} + T_{OFF}} \tag{5}$$

As depicted in Fig.03, the waveform of OFF time and ON time changing. A negative feedback for stabilizing output voltage can be achieved by adjusting the OFF time and ON time of the driving pulse at the same time.

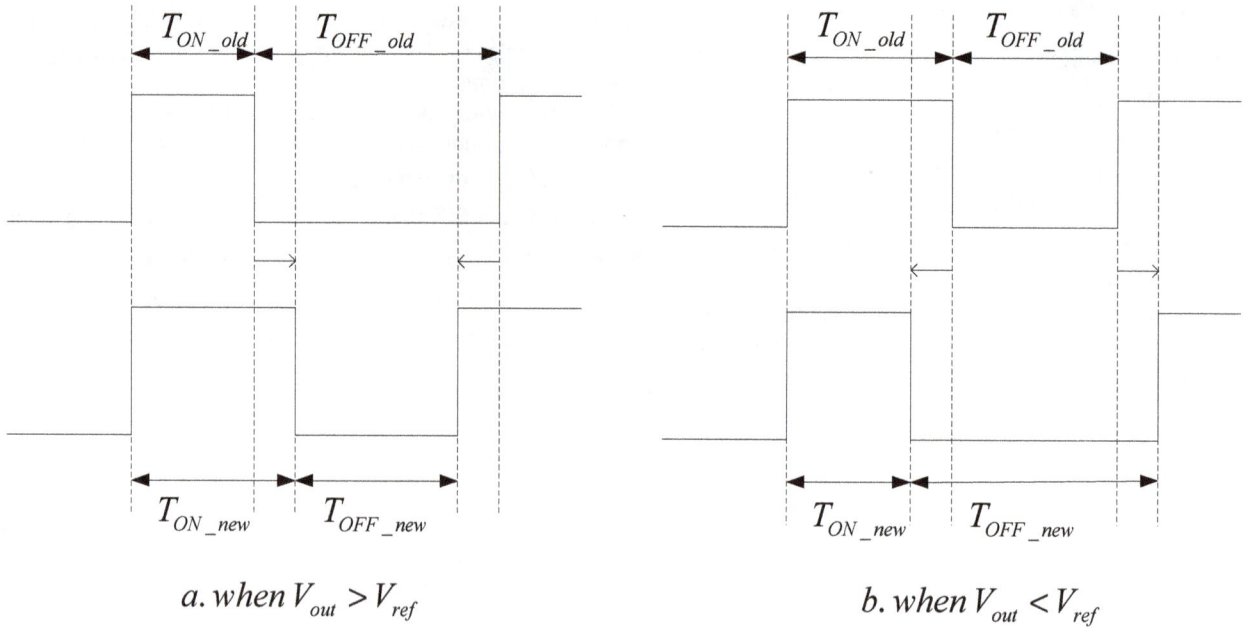

$$a.\ when\ V_{out} > V_{ref}$$

$$b.\ when\ V_{out} < V_{ref}$$

Fig. 3. *Waveform of OFF time and ON time changing according feedback voltage.*

Negative feedback control process as described following:

$$V_{out} > V_{ref} => V_{feebback} \uparrow => V_{OTA\_ON} \downarrow\ and\ V_{OTA\_OFF} \uparrow => T_{ON} \uparrow\ and\ T_{OFF} \downarrow => V_{out} \downarrow\ ;$$

$$V_{out} < V_{ref} => V_{feebback} \downarrow => V_{OTA\_ON} \uparrow\ and\ V_{OTA\_OFF} \downarrow => T_{ON} \downarrow\ and\ T_{OFF} \uparrow => V_{out} \uparrow\ .$$

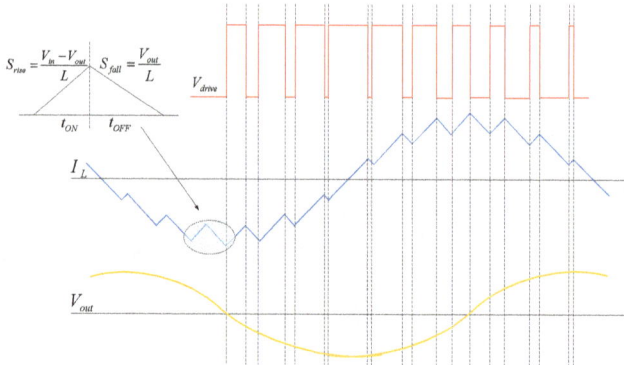

Fig. 4. *Waveform of inductor current and self-adaptive driving pulse.*

As depicted in Fig.04, the wave shape of V_{drive} has a varying ON time and OFF time. From the vertex of V_{out} to the bottom of V_{out}, the duty cycle is larger and larger until reaching to the minimum value, vice versa.

During each driving period,

When $0 < t < t_{ON}$,

$$i_{L(n)} = i_{L(n-1)} + S_{rise} \cdot t$$

When $t_{ON} \leq t \leq T$,

$$i_{L(n)} = i_{L(n-1)} + S_{rise} \cdot t_{ON} - S_{fall} \cdot (t - t_{ON})$$

Under the affection of filter capacitor, the output voltage can be given as,

$$V_{out(n)} = \frac{\sum_{n=0}^{n} \int_{t=0}^{t=T} i_{L(n)}(t)\,dt}{C_{fiter}} \quad (6)$$

3. Operational Transconductance Amplify

The operational transconductance amplifier (OTA) can be defined as an amplifier where all nodes are low impedance except the input and output nodes [8]. Low-impedance node always is connected to the source or gate-drain of MOS. As depicted in Fig.05, it is the schematic of operational transconductance amplifier without buffer. The resistive load is very large, almost 2Mohm. Otherwise, a resistive load will kill the gain of the OTA.

The OTA used in the schematic with a cascode low-voltage (wide-swing) current mirrors structure for increasing the magnification ability [8-10]. The terminology "1: m" indicates that N2 and N4 are sized m times wider than the N1 and N2.

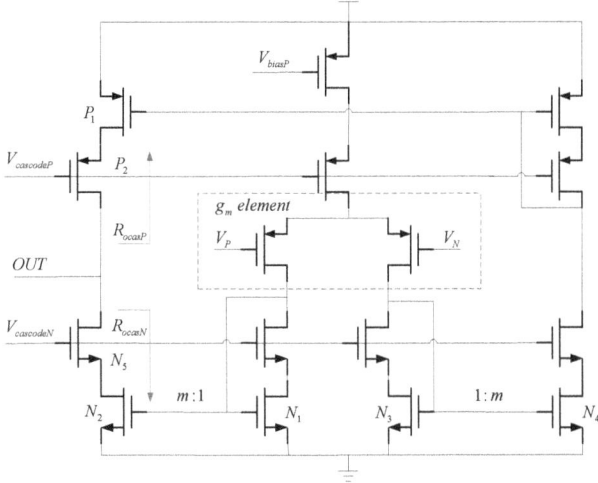

Fig. 5. Schematic of operational transconductance amplifier.

The low-frequency voltage gain is given by

$$A_V = \frac{v_{out}}{v_P - v_N} \tag{7}$$
$$= m \cdot g_{mP} \cdot (R_{ocasP} \parallel R_{ocasN})$$

Where the cascade resistors can be de given as

$$R_{ocasP} = (2 + g_{mP1} \cdot R_{oP1}) R_{oP2}$$

And

$$R_{ocasP} = (2 + g_{mN5} \cdot R_{oN5}) R_{oN2}.$$

The 3-dB frequency is now

$$f_{3dB} = \frac{1}{2\pi (R_{ocasN} \parallel R_{ocasP}) \cdot C_L} \tag{8}$$

For biasing the operational transconductance amplifier and other schematic, a beta-multiplier reference (BMR) is elaborately designed. The BMR is supply independent biasing for two constraint conditions without VDD specifying reference current [8].

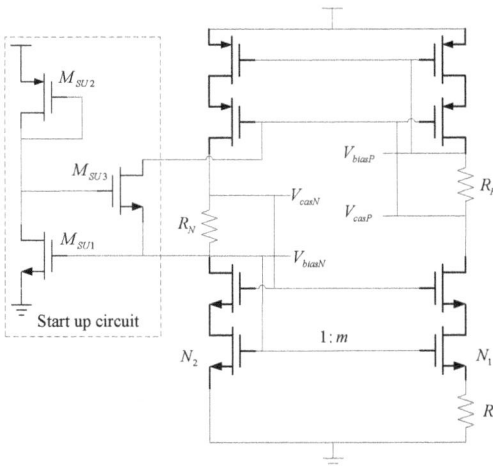

Fig. 6. Schematic of cascode beta-multiplier reference.

As depicted in Fig.06, PMOS cascode current mirror makes sure that both sides flow the same current. N1 is m times than N2 for smaller gate-source voltage. One of the constraint conditions is given by

$$V_{biasN} = V_{gsN2} = V_{gsN1} + I_{ref} \cdot R \tag{9}$$

Based on typical MOSFET square-law equation, there is another constraint condition given by

$$V_{gs} = \sqrt{\frac{2I_d}{\mu_n \cdot C_{ox}' \cdot \frac{W}{L}}} + V_{thN} \tag{10}$$

Obviously, these two constraint condition equations don't include VDD. Combining the two constraint conditions, the biasing current of the proposed design can be expressed in the following form,

$$I_{ref} = \frac{2}{R^2 \mu_n C_{ox}' \cdot \frac{W_1}{L_1}} (1 - \frac{1}{\sqrt{m}})^2 \tag{11}$$

A start-up circuit is added in the beta-multiplier reference for avoiding unwanted operating point that zero current flows in the circuit. If unfortunately, this unwanted state happens with the V_{biasN} closing to ground and V_{biasP} closing to VDD. When in this state, M_{SU1} is off and M_{SU2} turns on. A leaks current flows into N_2 from V_{biasP}. This causes the operating point converte into normal current point. After that, start-up circuit turns off and doesn't affect the beta-multiplier reference operation.

4. Current Mode Bandgap Reference

In voltage mode bandgap reference, for matching the coefficients between PTAT (Proportional to Absolute Temperature) voltage and CTAT (Complementary to Absolute Temperature) voltage, only a specified reference voltage, usually around 1.2V, can be generated. However, based on current mode bandgap reference, it has a flexible reference voltage [5, 11].

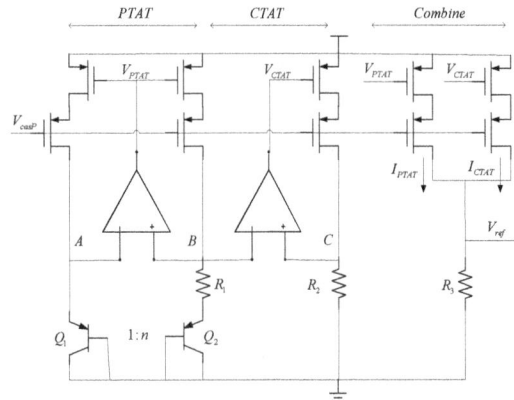

Fig. 7. Schematic of current-mode bandgap reference.

The two operational amplifiers, as depicted in Fig.07, forces nodes A, B and C to be at the same potential. The diode Q_2 is n time than Q_1. The PTAT current can be described as

$$I_{PTAT} = \frac{\Delta V_{BE}}{R_1} = \frac{kT \ln n}{qR_1} \qquad (12)$$

Where

$$\begin{aligned}
\Delta V_{BE} &= V_{BE1} - V_{BE2} \\
&= V_T \cdot \ln \frac{nI_0}{I_{s1}} - V_T \cdot \ln \frac{I_0}{I_{s2}} \\
&= V_T \cdot \ln n \\
&= \frac{kT}{q} \cdot \ln n
\end{aligned} \qquad (13)$$

The change in the PTAT current with temperature is

$$\frac{\partial I_{PTAT}}{\partial T} = \frac{k}{qR_1} \cdot \ln n \qquad (14)$$

The CTAT current can be given as

$$I_{CTAT} = \frac{V_{BE,Q_1}}{R_2} \qquad (15)$$

The change in the CTAT current with temperature is

$$\frac{\partial I_{CTAT}}{\partial T} = \frac{V_{BE} - (4+m)V_T - E_g/q}{T \cdot R_2} \qquad (16)$$

The reference voltage is then the sum of the PTAT and CTAT current multiplied by resistor,

$$V_{ref} = (I_{PTAT} + I_{CTAT}) \cdot R_3 \qquad (17)$$

For eliminating the resistor temperature effect, the resistors R_1, R_2 and R_3 are fabricated by same poly. From the Eq.18, the temperature behavior of the resistor can be fell out from the reference voltage.

5. Simulations Result

The proposed self-adaptive buck DC/DC converter driving chip is implemented in LFoundry GmbH, Germany LF150 CMOS process. It used Cadence Virtuoso to simulate the schematic with L=500n H and C=1m F, with a 500mA 500KHZ load.

As shown in Fig.08, the simulation result meets the analysis in Fig.04. Form the simulation result, it is obvious that the DC-DC operation in current continuous mode. But driving current is considerable large. Driving PMOS and NMOS should endure so much current. It is really a challenge to MOSFETs. More deliberate design should avoid it and over current protection is needed.

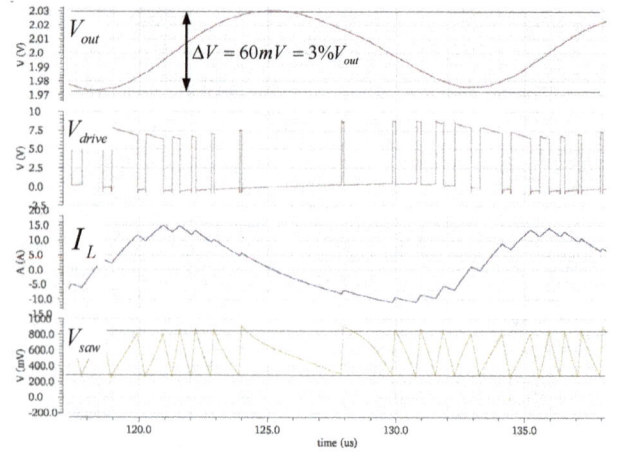

Fig. 8. Self-adaptive DC-DC converter simulation result.

Table 1. Specification of the proposed self-adaptive mode control.

Description	Value	Unit
input voltage	8	V
output voltage	2	V
load	500	mA
ripple	60	mV
Min frequency	130	KHZ
Max frequency	562	KHZ

The proposed self-adaptive buck DC/DC converter modulation can change the ON time and OFF time of the pulse simultaneously according the feedback signal. The frequency is not a constant value, but can be adaptive to the application condition. The simulation meets the design specification.

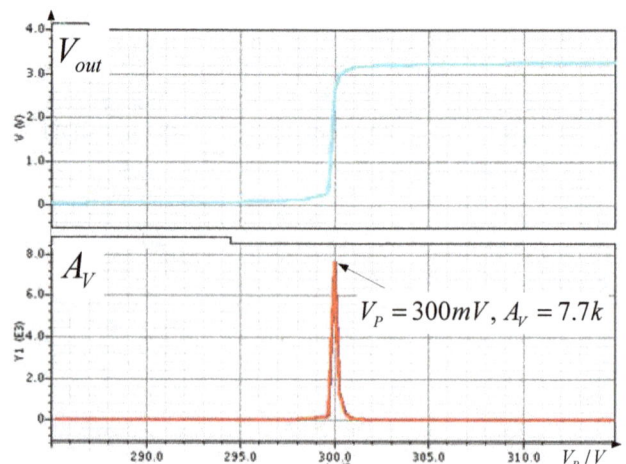

Fig. 9. Operational transconductance amplifier simulation result.

As shown in Fig.09, with inverting input $V_N = 300mV$, the largest voltage gain occurs at noninverting input equal to inverting input.

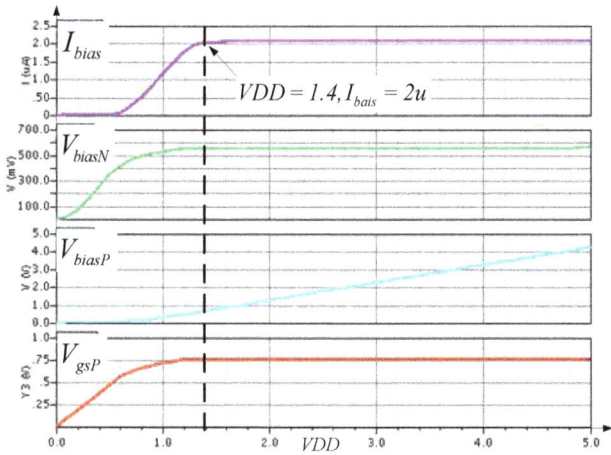

Fig. 10. Beta-multiplier bias simulation result.

The supply independent characteristor can be shown in Fig.10. After the circuit operating in saturation region, $VDD > 1.4V$, biasing current will be constant value 2uA. The NMOS gate-source voltage is equal to V_{biasN} being with constant. Although V_{biasP} proportional to power supply VDD, the difference between VDD and V_{biasP}, namely the PMOS gate-source voltage, also is constant.

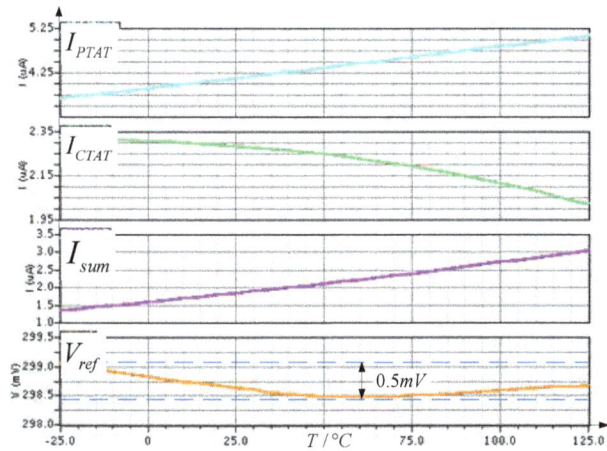

Fig. 11. Current-mode bandgap reference simulation result.

As depicted in Fig.11, the PTAT current has a better linearity than CTAT current. The nonlinear should be compensated in the high precise application [11]. The summing up of PTAT and CTAT current is proportional to absolute temperature. This can be counteracted by poly resistors' negative relation with temperature characteristic.

6. Conclusion

The simulations results have demonstrated the feasibility to change both pule ON time and OFF time on DC-DC converter with comparative low output ripple. It is normal that the final experimental implementation of the on-chip DC-DC converter will degrade some specifications comparative with simulation results. And also stability analysis and load transient analysis must be processed in the future before its expansion into commercial integrated circuits.

Acknowledgment

The author would like to acknowledge Prof. Dr.-Ing. Robert Weigel and M. Sc. Andreas Bänisch in Erlangen-Nurnberg University for supervising my semester project on CMOS DC-DC converter design, and the anonymous reviewers for their constructive comments.

References

[1] B. Sahu, G.A. Rincon-Mora, A accurate, low-voltage,CMOS switching power supply with adaptive on-time pulse-frequency modulation (PFM) control, IEEE Trans. Circuits and Systems, 54 (2) (2007) 312-321.

[2] Pui-Sun Lei, Chang, R.C.-H. A high-efficiency PWM DC-DC buck converter with a novel DCM control under light-load, Circuits and Systems (ISCAS), 2011 IEEE International Symposium on, pp237-240.

[3] B. Arbetter, R. Erickson, and D. Maksimovic´, "DC–DC converter design for battery-operated systems," in Proc. IEEE Power Electron. Specialist Conf., 1995, pp. 103–109.

[4] Wan-Rone Liou, , Mei-Ling Yeh, and Yueh Lung Kuo, A High Efficiency Dual-Mode Buck Converter IC For Portable Applications. IEEE TRANSACTIONS ON POWER ELECTRONICS, VOL. 23, NO. 2, MARCH 2008.

[5] Biranchinath Sahu, Gabriel A. Rincón-Mora, An Accurate, Low-Voltage, CMOS Switching Power Supply With Adaptive On-Time Pulse-Frequency Modulation (PFM) Control, IEEE TRANSACTIONS ON CIRCUITS AND SYSTEMS—I: REGULAR PAPERS, VOL. 54, NO. 2, FEBRUARY 2007.

[6] Abraham I. Pressman, Keith Billings, Taylor Morey. Switching Power Supply Design. Third Edition. McGraw-Hill Professional.2009.

[7] X. Zhou, M. Donati, L. Amoroso, and F. C. Lee, "Improved Light-Load Efficiency for Synchronous Rectifier Voltage Regulator Module", IEEE Transactions on Power Electronics, Vol. 15, NO.5, pp. 826-834, September, 2000.

[8] R.JACOB BAKER. Circuit Design, Layout, and Simulation, Third Edition [M]. United States of America, IEEE Press, 2010:116-118

[9] Phillip E. Allen, Douglas R. Holberg, CMOS analog Circuit Design, 2nd ed. OXFORD UNIVERSITY PRESS. 2002.

[10] Behzad Razavi, Design of Analog CMOS Integrated Circuits, McGraw-Hill, Inc. New York, NY, USA. 2001.

[11] Charalambos M. Andreou, Savvas Koudounas, and Julius Georgiou, A Novel Wide-Temperature-Range, 3.9 ppm/ C CMOS Bandgap Reference Circuit [J]. IEEE JOURNAL OF SOLID-STATE CIRCUITS, VOL. 47, NO. 2, FEBRUARY 2012: 574-580.

20

Analysis on the Basis of Volterra Series Signal–To–Noise Ratio of Nonlinear Device in the Conditions of the Stochastic Resonance Effect

Okcana Kharchenko[1], Vladislav Tyutyunnik[2]

[1]National Technical University "Kharkiv Polytechnic Institute», Kharkiv, Ukraine
[2]Scientific Center, Kharkiv Air Force University named I. Kozhedub, Kharkiv, Ukraine

Email address:
okcana1304@mail.ru (O. Kharchenko), tvlad70@mail.ru (V. Tyutyunnik)

Abstract: In this paper, the stochastic resonance effect is considered. It is shown that the stochastic resonance effect appears in the conditions of operating on the nonlinear system of additive mixture of desired signal and noise. The numerical simulation of the output signal when exposed to the input of the system of additive mixture harmonic signal and noise with a uniform distribution is given. Analytical relational expressions for signal-to-noise ratio on the output of the nonlinear system are got. The analysis of signal-to-noise ratio is conducted on the output of the nonlinear system depending on the parameters of the input signal and noise. In this paper we have shown the stochastic resonance effect occurs mainly at low frequencies.

Keywords: Stochastic Resonance (SR), Nonlinear System, Signal-to-Noise Ratio (SNR), Volterra Series

1. Introduction

In modern conditions the problem of providing of reliable communication of data at presence of interferences is considered one of major problems. Error-correcting codes, optimal filters are created, and are used by the detection method of accumulation, the probabilistic approach to suppressing random disturbances. etc. [1,2,3].

At the same time, researches by physicists at the end of the XX century [2], resulted in paradoxical conclusions. The noise on the input of the nonlinear systems possessing the effect of the so-called stochastic resonance (SR), allows to stand out a weak (as compared to the noise) signal from additive signal–noise mixture. The SR effect characterizes the response of the nonlinear system on a weak input signal. Thus data-output of the nonlinear system, such as an amplification factor and signal-to-noise ratio, at certain terms have the distinctly expressed maximum [4,5].

2. Concept of Stochastic Resonance

Stochastic resonance is a nonlinear physical phenomenon in which the output signals of some nonlinear systems can be enhanced by adding suitable noise under certain conditions

[4,5]. SR is a universal effect, to many inherent nonlinear systems being under external influence simultaneously of chaotic and weak periodic influences. The response of the nonlinear system on a weak external signal in case of SR noticeably increases with the height of the noise intensity in the system and arrives at a certain maximum at some level.

Consider the nonlinear system described by the equation [6]

$$\dot{y}(t) = -y(t) - y^3(t) + x(t) \tag{1}$$

where $x(t) = s(t) + n(t)$ is an input process being additive mixture of desired signal and normal noise;

$y(t)$ is a process on the output of the nonlinear device.

As shown in [6] this system has the SR effect. The numeral simulation of response at affecting input of the system of additive mixture of harmonic signal $s(t)$ and noise with even distribution $n(t)$, illustrating model (1), resulted on Fig.1.

It is seen that as a result of processing in accordance with expression (1) can significantly reduce the noise component fluctuations, although the form of the output signal differs from the harmonic. However, Equation (1) is an Abel equation of the second kind and has no exact analytical solution.

3. The Analysis of Nonlinear Systems using Volterra Series

The principle of superposition, which is the basis of the linear theory, is not applicable for the investigation of phenomena in nonlinear systems. Forced abandonment of the principle of superposition complicates research opportunities. Unfortunately, such powerful and versatile methods of research, which has the linear theory, is absent in the nonlinear theory. The study of nonlinear problems comprises a number of specialized techniques and methods that have different power and scope[[7]

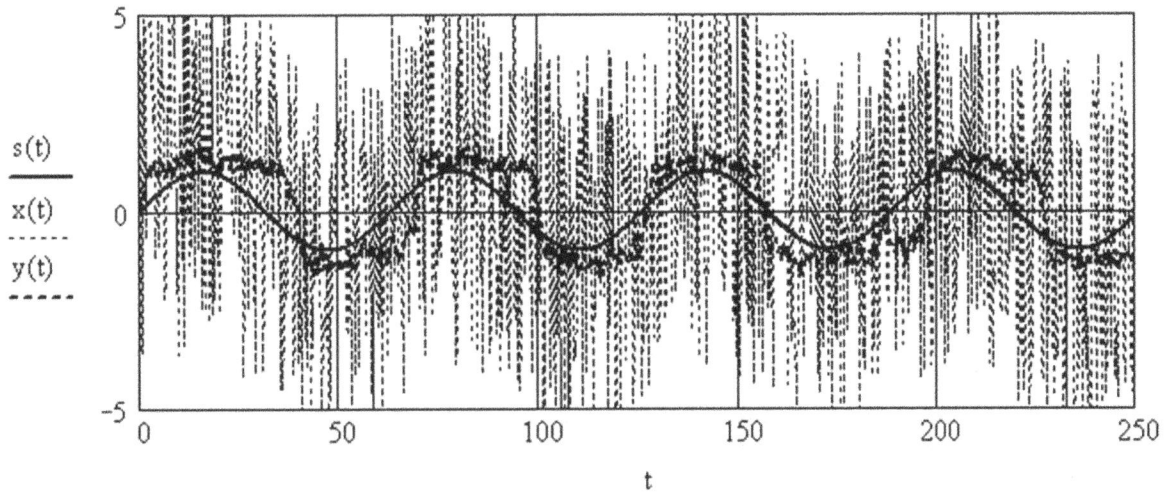

Figure 1. Standing out of signal from additive signal–noise mixture.

To receive the signal at the output of a nonlinear system in the analytical form Volterra series are usually used. They are described as "power series with memory" for displaying an output signal of a nonlinear system in the form of degrees of the input signal [8]. For the nonlinear system described equation (1) will use the following notation series.

$$y(t) = \sum_{n=1}^{\infty} \frac{1}{n!} \int_{-\infty}^{\infty} du_1 ... \int_{-\infty}^{\infty} du_n g_n(u_1,...,u_n) \prod_{r=1}^{n} x(t-u_r) \quad (2)$$

where $y(t)$ is the output signal; $x(t)$ is the input signal; $g_n(u_1,...,u_n)$ are the n-th order Volterra kernels, which describe the system.

It should be noted that the kernel of the first order $g_1(u_1)$ is the pulse response of a linear circuit. Thus, the kernel of a higher order can be viewed as the pulse response of a high order, which are used to describe different orders of nonlinearity.

Appearing in the [8] coefficient $\frac{1}{n!}$ introduced A. Bedrosian and D. Rice to simplify these expressions.

Important for in this analysis is n-order Fourier transform, which looks as follows

$$G_n(f_1,...,f_n) = \int_{-\infty}^{\infty} du_1 ... \int_{-\infty}^{\infty} du_n g_n(u_1,...,u_n) \exp\left[-i(f_1 u_1 +...+ +f_n u_n)\right] \quad (3)$$

Value G_0 equal to zero, as in the present case, the Volterra series begins with $n=1$. The member of the row with number $n=0$ corresponds to the active system (when there is an output signal when no input, which is contrary to the principle of causality).

Component $G_1(f_1)$ is a transfer function of a linear circuit.

Thus, the transformation kernel of Volterra-th n-order is similar to the transfer function of the n-order. The complete formula of the form (2) are infinite series, and the complexity of computing the n-th member with increase of n increases rapidly. However, the analysis of communication systems by members of Volterra series above the second or third order can be neglected [8].

4. Calculation and Analysis of the Spectral Power Density of the Output Signal

Let us consider a nonlinear system described by equation (1). A harmonic signal is at the input of a nonlinear system. As shown in [6] this nonlinear system has the SR effect. Bedrosian A. and Rice D. [8] derived analytical expressions for the spectral power density of the components of the output signal of a nonlinear system using a series of Voltaire. The transfer functions for a given nonlinear system are defined in [6]. The results of the calculations of the output power spectrum main members are given in table 1.

Here A is the amplitude of the harmonic input signal,

W_1 is the spectral power of the noise component in the input signal. Further, to enhance the visibility and convenience of calculations in the resulting formula is used circular frequency ω_0 instead f_0.

If the input signal is an additive mixture of the unmodulated carrier and noise then the spectral power density of the output signal includes three main parts [9,10]:

$F_{SxS}(\omega)$ corresponds to the beatings between signal components and its harmonics (discrete part of the spectrum);

$F_{NxN}(\omega)$ corresponds to the beatings noise components (continuous part of the spectrum);

$F_{SxN}(\omega)$ corresponds to the beatings signal and noise components (continuous component of the spectrum).

Therefore, the energy spectrum of the output of the nonlinear system is defined as

$$F(\omega) = F_{SxS}(\omega) + F_{SxN}(\omega) + F_{NxN}(\omega) \qquad (4)$$

In order to determine the signal-to-noise ratio (SNR) at the output of a nonlinear system, you must decide to signal or noise attributed to the part of the output spectrum $F_{SxN}(\omega)$. Thus, we get two expressions:

a) when the beating between the signal components and noise related to the noise:

$$SNR = \frac{\int\limits_{-\infty}^{\infty} F_{SxS}(\omega)d\omega}{\int\limits_{-\infty}^{\infty} [F_{SxN0}(\omega) + F_{NxN}(\omega)]d\omega} \qquad (5)$$

b) when the beating between the signal components and noise related to the signal

$$SNR = \frac{\int\limits_{-\infty}^{\infty} [F_{SxS}(\omega) + F_{SxN}(\omega)]d\omega}{\int\limits_{-\infty}^{\infty} F_{NxN}(\omega)d\omega} \qquad (6)$$

In our case, the signal will include components of the first and third lines (Table 1). Other components refer to the output noise. The resulting formula for the signal-to-noise ratio is quite cumbersome, so we present the results of calculations in Fig. 2. The signal-to-noise ratio is a function of three variables: A, ω_0 and W.

Fig. 2. shows that SNR has a strongly pronounced maximum at a certain frequency at a given amplitude A. Note that the SNR has a minimum too. In addition, the dependence of the SNR on the spectral power density of input noise is non-monotonic.

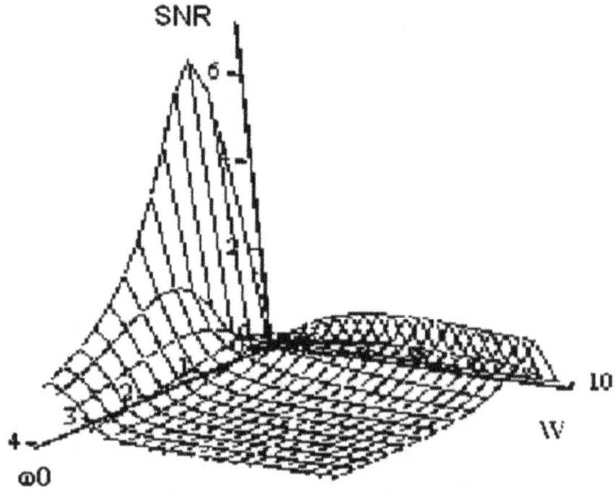

Figure 2. *SNR dependence on the frequency ω_0 of the input harmonic signal (A=1 – amplitude of input signal, W – spectral power density of input noise)*

Table 1. *The calculation results of the output power spectrum.*

The main members of the output power spectrum introduced by Bedrosian and Rrice	The main members of the output power spectrum of a nonlinear system described by equation (1)
The pulses due to the harmonic components	
$\delta(f-f_0)\left\|\dfrac{A}{2}G_1(f_0)+\dfrac{A^3}{16}G_3(f_0,f_0,-f_0)\right\|^2$	$\delta(f-f_0)A^2 \times \times \dfrac{16\omega_0^2(1+\omega_0^2)^2+(2(1+\omega_0^2)(2-3W_n)-3A^2)^2}{64(1+\omega_0^2)^4}$
$\delta(f-3f_0)\left\|\dfrac{A^3}{48}G_3(f_0,f_0,f_0)\right\|^2$	$\delta(f-3f_0)\dfrac{A^6}{64(1+9\omega_0^2)(1+\omega_0^2)^3}$
Members of the negative frequency range when replacing f_0 for $-f_0$	
$\delta(f+f_0)\left\|\dfrac{A}{2}G_1(-f_0)+\dfrac{A^3}{16}G_3(-f_0,-f_0,+f_0)\right\|^2$	$\delta(f+f_0)A^2 \times \times \dfrac{16\omega_0^2(1+\omega_0^2)^2+(2(1+\omega_0^2)(2-3W_n)-3A^2)^2}{64(1+\omega_0^2)^4}$
$\delta(f+3f_0)\left\|\dfrac{A^3}{48}G_3(-f_0,-f_0,-f_0)\right\|^2$	$\delta(f+3f_0)\dfrac{A^6}{64(1+9\omega_0^2)(1+\omega_0^2)^3}$
Continuous part of the spectrum	
$W(f)\left\|G_1(f)+\dfrac{A^2}{4}G_3(f_0,-f_0,f)+\dfrac{1}{2}\int\limits_{-\infty}^{\infty}df_1W(f_1)G_3(f_1,-f_1,f)\right\|^2$	$\dfrac{W_n[(2-3W_n)(1+\omega_0^2)-3A^2]^2+4\omega^2(1+\omega_0^2)^2}{4(1+\omega_0^2)^2(1+\omega^2)^2}$
$W(f-2f_0)\left\|\dfrac{A^2}{8}G_3(f_0,-f_0,f-2f_0)\right\|^2$	$\dfrac{9W_nA^4}{16(1+\omega_0^2)^2[1+(\omega-2\omega_0)^2](1+\omega^2)}$

The main members of the output power spectrum introduced by Bedrosian and Rrice	The main members of the output power spectrum of a nonlinear system described by equation (1)
$W(f+2f_0)\left\|\dfrac{A^2}{8}G_3(-f_0,f_0,f+2f_0)\right\|^2$	$\dfrac{9W_n A^4}{16(1+\omega_0^2)^2[1+(\omega+2\omega_0)^2](1+\omega^2)}$
$\dfrac{1}{2!}\displaystyle\int_{-\infty}^{\infty} df_1 W(f_1)W(f-f_1-f_0)\times\times\left\|\dfrac{A}{2}G_3(f_1,f_0,f-f_1-f_0)\right\|^2$	$\dfrac{9W_n^2 A^2}{2(1+\omega_0^2)[4+(\omega-\omega_0)^2](1+\omega^2)}$
$\dfrac{1}{2!}\displaystyle\int_{-\infty}^{\infty} df_1 W(f_1)W(f-f_1+f_0)\times\times\left\|\dfrac{A}{2}G_3(f_1,-f_0,f-f_1+f_0)\right\|^2$	$\dfrac{9W_n^2 A^2}{2(1+\omega_0^2)[4+(\omega+\omega_0)^2](1+\omega^2)}$
$\dfrac{1}{3!}\displaystyle\int_{-\infty}^{\infty} df_1 \int_{-\infty}^{\infty} df_2 W(f_1)W(f_2)W(f-f_1-f_2)\times\times\left\|G_3(f_1,f_2,f-f_1-f_2)\right\|^2$	$\dfrac{3W_n^3}{2\pi^2(9+\omega^2)(1+\omega^2)}$

As it is shown in Fig. 3, the SNR (for a given frequency of the first harmonic ω_0) is only a maximum value at certain value W.

With increasing frequency of the harmonic signal ω_0 the SNR maximum is shifted in the direction of increasing of the spectral power density of input noise W. From the Figures it is seen that the SR effect is manifested mainly at low frequencies. Thus, at relatively low values of frequency using nonlinear equations (3) to solve the problem of the selection signal is at a high level of noise.

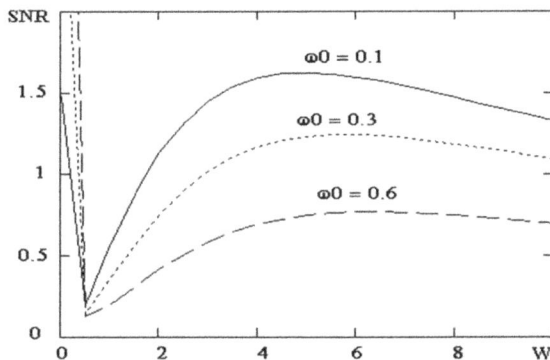

Figure 3. The dependence of the SNR from the spectral power density of input noise W of the different values of frequency of the harmonic signal ω_0 (the amplitude of the input sine signal $A=1$)

5. Conclusion

In this paper, both by the numeral simulation and by the analytical form Volterra series simulation, the possibility of improving the output signal in the nonlinear system by adding noise has been established. An input process consisting of the additive mixture of the harmonic signal and the normal noise has been considered. Abel equation of the second kind in which an effect of stochastic resonance takes place is represented.

The analytical expressions for the components of the spectral power density of the nonlinear system output signal using Volterra series are deduced. Based on the method of decomposition in the Volterra series analysis of nonlinear equations describing the response of a nonlinear system, which has the SR effect.

Components of the power spectral density of the output signal of a nonlinear system are entered and calculated.

Signal-to-noise ratio at the output of the nonlinear device has been obtained and analyzed.

The values of the parameters of the input oscillations, corresponding to the SR effect have been shown. In this paper, we have established that the output signal amplitude increases with increasing noise intensity, passes through a maximum, and then decreases again.

It should be pointed out that the results obtained in this paper make it possible to calculate the optimum level of spectral power density of additive noise at which the SNR maximum is reached. One can see that the SNR dependence shows non-monotonic behavior with the increase of the spectral power density of input noise W and frequency of the harmonic signal ω_0.

It has been shown that by choosing the optimal level of the noise power, we can achieve the maximum level of the signal at a given frequency.

The analytical expressions deduced here can be the basis for the design of radio systems operating in conditions of high noise levels.

References

[1] Kharkevich A.A. Borba s pomekhami [Interference Contro]l, M: Nauka, 1965, 275 p. (in Russian)

[2] Sklar B. Digital Communication. Fundamentals and Applications, Second Edition, Prentice Hall PTR, 2003, 1099 p. (in Russian)

[3] Tikhonov V.I. Statisticheskay rediotechnika [Statistical Radioengineering], M: Sov. Radio, 1966, 678 p. (in Russian)

[4] Anishenko V.S., Neman V.B., F., Gayer L. Stokhasticheskiy resonans kak indyshirovanay shumom effect uvelicheniy stepeni poraydka. [Stochastic resonance as induced by noise the effect of increasing the degree of order]. – Uspekhi fizicheskikh nayk, 1999, vol. 169, № 1, pp. 7–37. (in Russian)

[5] Kharchenko O.I., Chumakov V.I., Sklyaruk V.L. K raschotu otklika nelineynoi sistemy s pemyatu s primeneniem rjaydov Volterra. [On the calculation response of the nonlinear system with memory response using the Voltaire series]. Zb. Nauk. prats AMVS,2010, vip. 3(3), pp. 110-115. . (in Russian

[6] Geraschenko O.V. Stokhasticheskiy resonans v asimmetrichnoi bistabilnoy sisteme. [Stochastic resonance in an asymmetric bistable system]. Pisma v GTF, 2003, vol. 29, vyp. 6, pp.82-86. (in Russian)

[7] Kharkevich A.A. Nelineynye i parametricheskiy ustroyctva. [Nonlinear and parametric devices]. M.: Gosudarstvenoe izd. tekhniko-teoreticheskoi literatury, 1955, 184 p. (in Russian)

[8] Bedrosian E., Rice S. The Output Properties of Volterra Systems (Nonlinear Systems with Memory) Driven by Harmonic and Gaussian Inputs // IEEE.- Vol. 59, No 12.- 1971. – pp.58–82.

[9] Levin B.R. Teoreticheskie osnovy statisticheskoi radiotekhniki. [Theoretical foundations of statistical radio engineering]. M: Sov. Radio, 1969, 752 p. (in Russian)

[10] David Middleton, An Introduction to Statistical Communication Theory: An IEEE Press Classic Reissue, Wiley-IEEE Press, 1996.–1184 p. (in Russian)

21

Smart Inverter with Active Power Control and Reactive Power Compensation

Zaiming Fan[1], Xiongwei Liu[2]

[1]Faculty of Health and Science, University of Cumbria, Lancaster, United Kingdom
[2]Entrust, The Innovation Centre, Science Technology Daresbury, Cheshire, United Kingdom

Email address:
z.fan@lancaster.ac.uk (Zaiming Fan), xiongwei@en-trust.co.uk (Xiongwei Liu)

Abstract: Conventional grid-tied single phase inverters of renewable power generators (solar PV systems typical) have limited reactive power compensation capability and do not have active power control. This paper presents a novel control strategy which provides active power control with reactive power compensation for a DC/AC inverter connected to a single-phase AC grid, which supplies electricity to local loads. By sampling the instantaneous current on the grid at the local load side, which represents the domestic load current, an orthogonal signal is constructed using second order generalised integrator. The active and reactive current of the local loads are then rapidly detached from the orthogonal signal through specific trigonometric calculation. The reference current for the inverter output are produced by combining the active current and the reactive current which is detached from the domestic load current. Comparing the reference current with the inverter output current generates the PWM signals which are used to control the IGBT devices of the inverter bridge with capacitive impedance output to achieve domestic reactive compensation for inductive loads. The output current remarkably improves the load capacity of the grid and reduces the demand of reactive power from the gerid.

Keywords: Grid-Connected, Inverter, Active, Reactive, SOGI

1. Introduction

The power grid provides both active power and reactive power to satisfy the requirement of application of non-linear loads, such as induction motors driving domestic appliances, which require a large amount of reactive power. Therefore the fluctuation of reactive power demand impacts the balance of grid voltage. With high penetration of distributed power sources such as grid-tied single-phase photovoltaic power generation system, unbalanced voltage can easily go beyond the limit which forces the inverter stop injecting power into the grid even with strong solar irradiation [1] [2]. The feeder randomly and geographically extends out that brings huge challenge to find an appropriate approach to achieve reactive power compensation to stabilise the voltage of local electricity grid. Static compensators installed at the residential terminals will bring extra costs to the residents. Utilities are considering charging reactive power demand. For example, Italy's leading power distributor has decided to install more than 20 million household electricity meters with active and reactive power measurement [3]. For the

safe operation of the power grid and improving power quality, the grid infrastructure must be configured reasonably, and also have optimal control methodologies, which can smoothly adjust the amplitude, current, frequency and phase angle. Secure and economic operation of power grid depends largely on its controllability, i.e. the power control, including active and reactive power control. Electromechanical oscillations have been observed widely in many power systems worldwide[4]. There are a number of situations which could cause the power system oscillation, such as a fast exciter for a wind turbine generator, fluctuation of the loads and intermittence of distributed power generation, such as fluctuated renewable power generation. The problems of oscillation may lead to power system blackout or power interruption [5]. The loadability of a power grid depends on the demands of active and reactive power that is received from distributed lines. As the capacity of electricity system load approaches its maximum critical point, the demands of both active and reactive power increase rapidly. Therefore, the reactive power supports have to be locally available [6].

The current mainstream approach to solve reactive power compensation is to connect a Flexible AC Transmission System (FACTS) device in parallel with the power line. There are a number of FACTS controllers [7], which are developed by engineers to damp the power oscillation [4] [8], such as Static Var Compensators (SVC), Thyristor Controlled Series Capacitor (TCSC), Static Synchronous Series Compensator (SSC), Unified Power Flow Controller (UPFC), and Static Synchronous Compensator (STATCOM). Among the available FACTS devices, the STATCOM is a good one to be introduced in renewable generation to improve dynamic stability, steady state stability and transient stability [9] [10].

Micro renewable power generation interfaces with the power grid by inverters, which inject power to the grid as much as it can. Sometimes the loads of the micro-grid may require a sinusoidal current which is not in phase with the micro-grid voltage [11]. Conventional inverter rather considers how to inject the maximum power to the grid as much as it can whatever circumstance, even operating in the situation of the voltage magnitude of grid approaches collapse due to a plenty of renewable power generation is installed in the community. As the result of that the inverter stops output in order to meet the standard of Engineering Recommendation G83 [2].

The second order generalised integrators (SOGI) based on Phase Locked Loop (PLL) are well-known. Several researchers have successfully extended application in various purpose, such as synchronous signal with grid-converted in the application of active rectifier, active filters, uninterruptible power supplies and distributed generation etc.[12].

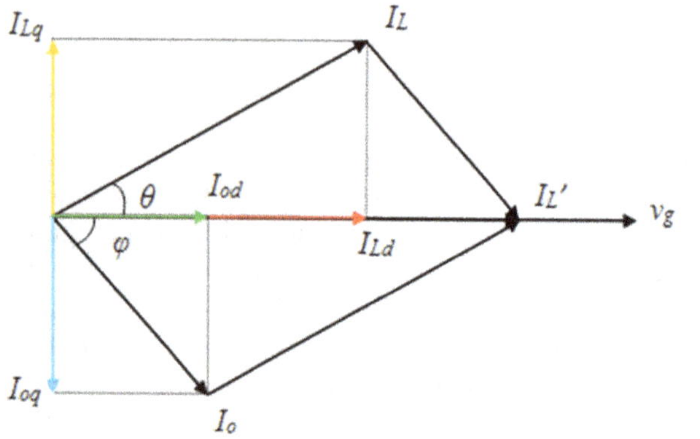

Figure 1. *Envisaged the vector relationships between currents and voltage.*

Figure 1 illustrates the vector relationships between the inverter output current, the load current I_L and the grid voltage v_g. The load current I_L can be decomposed into active current I_{Ld} and reactive current I_{Lq} respectively. As shown in Figure 2, the inverter output current I_o is decomposed by an active component I_{od} and a reactive component I_{oq}. If I_{oq} has the same magnitude of the load current's reactive component I_{Lq} and opposite direction, the whole system reactive power can be completely compensated by the inverter, then it can inject active power to the grid with reactive power compensation.

2. Control Strategies

2.1. Overall of Control Methodology

Through the above analysis, the domestic reactive power compensation of single phase grid can be achieved by adjusting instantaneous reactive current which is decomposed out from the output current of single phase inverter according to the instantaneous active and reactive power theory.

The instantaneous active and reactive power theory well-known as d-q theory is the mostly widely used time reference current generation technique[13]. The second order generalised integrator (SOGI) is widely exploited to achieve Phase Locked Loop (PLL) [13] [14] [15] and to eliminate or reduce the instantaneous noise level in many fields, such as control theory, relaying protection, signal processing, radio frequency, power systems, etc. [15]. The present strategy is shown in Figure 2, SOGI is used to construct a pair of the orthogonal trigonometric function for the current of loads which is sampled from the grid. Afterwards yielding the reactive current of loads i_q through decomposes load's active and reactive current by means of d-q theory, which is combined with the active current of demand to produce an instantaneous expected reference current compares with the actual output current of inverter i_o. The comparative error is exploited to generate PWM signal that is applied to control the inverter bridge.

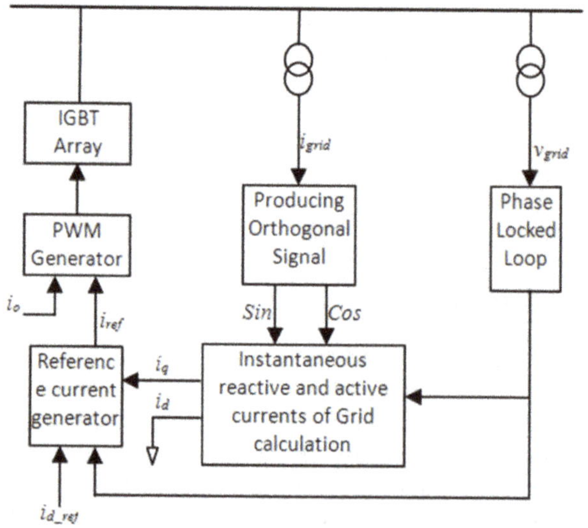

Figure 2. *Method of reactive power compensation.*

The single phase instantaneous voltage of the power grid and the load current are given by:

$$U_g = U_A \cos \omega t \qquad (1)$$

$$I_\alpha = I_A \cos(\omega t - \theta) \qquad (2)$$

Where

U_A is the root mean square (rms) value of the grid voltage,

I_A is the rms value of the load current,

I_α is the instantaneous current of the grid, and

θ is the phase angle.

The orthogonal equation of (2) the instantaneous current of loads can be obtained as:

$$I_\beta = I_A \sin(\omega t - \theta) \tag{3}$$

Applying trigonometric calculation for equation (2) and (3) yields:

$$I_\alpha = I_A \cos \omega t \cos \theta + I_A \sin \omega t \sin \theta \tag{4}$$

$$I_\beta = I_A \cos \omega t \sin \theta - I_A \sin \omega t \cos \theta \tag{5}$$

$$\begin{bmatrix} I_\alpha \\ I_\beta \end{bmatrix} = \begin{bmatrix} \cos \omega t & \sin \omega t \\ -\sin \omega t & \cos \omega t \end{bmatrix} \begin{bmatrix} I_A \cos \theta \\ I_A \sin \theta \end{bmatrix} \tag{6}$$

To obtain the active and reactive current of the load by multiplying matrix $\begin{bmatrix} \cos \omega t & -\sin \omega t \\ \sin \omega t & \cos \omega t \end{bmatrix}$ at both side of the equation (6)

$$\begin{bmatrix} I_A \cos \theta \\ I_A \sin \theta \end{bmatrix} = \begin{bmatrix} \cos \omega t & -\sin \omega t \\ \sin \omega t & \cos \omega t \end{bmatrix} \begin{bmatrix} I_\alpha \\ I_\beta \end{bmatrix} \tag{7}$$

Obviously the active and reactive current I_d, I_q have been apart from the sampling load current by equation (7), which are $I_A\cos\theta$, $I_A\sin\theta$ respectively.

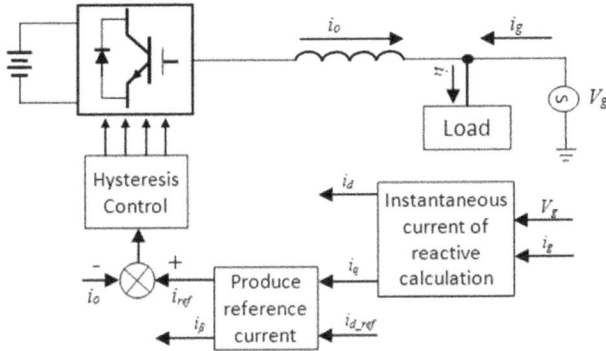

Figure 3. Decomposes reactive and active load current.

In the equation 7, sine and cosine functions can be acquired by a phase-locked loop.

Figure 3 is a module of to obtain instantaneous active and reactive current from electric utility.

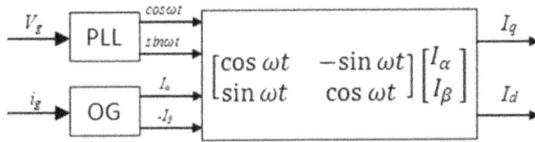

Figure 4. Equation of obtaining active and reactive current.

The system samples the voltage of grid to ensure the output power from inverter is completely synchronous with the grid in the term of frequency, phase and the room mean square (RMS) voltage.

Figure 4 implements equation (7), which illustrates that the module produces the output reference current of the inverter according to the demanded reactive power of the grid, the

ability of inverter provides active power and PLL signals.

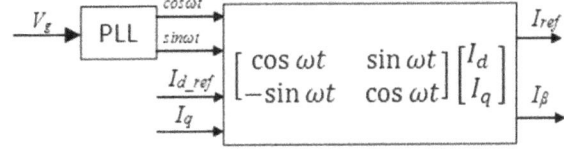

Figure 5. Producing the reference current.

2.2. Orthogonal System Generation

Assuming that an input signal is a sinusoidal signal $v_i=sin\omega t$, the orthogonal function of sinusoidal function is cosine function that can be obtained by generalized integration as the following equation:

$$y = \int_{-\infty}^{+\infty} \sin \omega t \, dt = -\frac{1}{\omega}.\cos \omega t + C \tag{8}$$

And the input signal can be retrieved by means of derivative of above equation as below.

$$\frac{dy}{dt} = \sin \omega t = v_i \tag{9}$$

If only considers that the input signal v_i start from time 0, which means the function $sin\omega t$ is defined for all $t \geq 0$, then Laplace transfer of sine trigonometric function is:

$$F(s) = \mathcal{L}(y) = \frac{\omega}{s^2 + \omega^2} \tag{10}$$

And the Laplace transfer of the derivative of function y' can be granted as:

$$\mathcal{L}(y') = \frac{s\omega}{s^2 + \omega^2} \tag{11}$$

So the equation above is the response function of system H_s that needs to be constructed.

$$H_s = \frac{v_o}{v_i} = \frac{s\omega}{s^2 + \omega^2} \tag{12}$$

$$=> v_o(s^2 + \omega^2) = v_i.s\omega \tag{13}$$

$$=> v_o = \left(v_i - v_o \frac{\omega}{s}\right)\frac{\omega}{s} \tag{14}$$

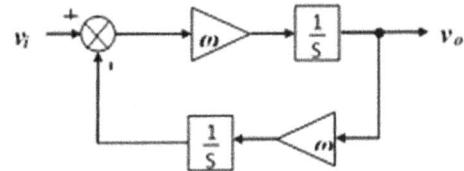

Figure 6. The response of open-loop system of orthogonal generation.

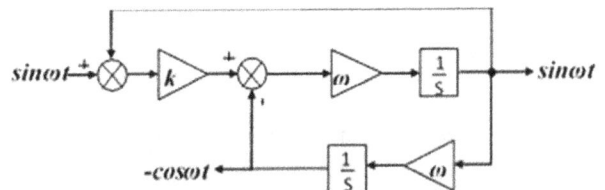

Figure 7. Closed-loop system of orthogonal generation.

Figure 6 illustrates the open-loop structure of system response according to Equation 14, and Figure 7 is the structure of closed-loop system. In Figure 7, the gain parameter k affects the bandwidth of the closed-loop system.

3. Simulated Results

Figure 8 illustrates the results of orthogonal system output compare with input signal. As the appearance of figure, the two output signal accurately formed a pair of quadrature. Although there is some unstable at the time starting, the system rapidly operated at steady-state and the produced sinusoidal signal overlaps with input signal.

Obviously, the loads current of the grid leads certain phase angle with respect to the voltage of the grid in Figure 9 (a). There is considerable contrast with Figure 9 (b) which produced by the inverter utilizing the presented method of reactive compensation to coincide with output of active power. The phase angel of loads current perfectly follows the voltage of the grid after delivered reactive power to the grid by the inverter.

The simulation of inverter embedded the presented approach demonstrates that the reactive power supported by the grid significantly fall down as Figure 10 (b) illustrated.

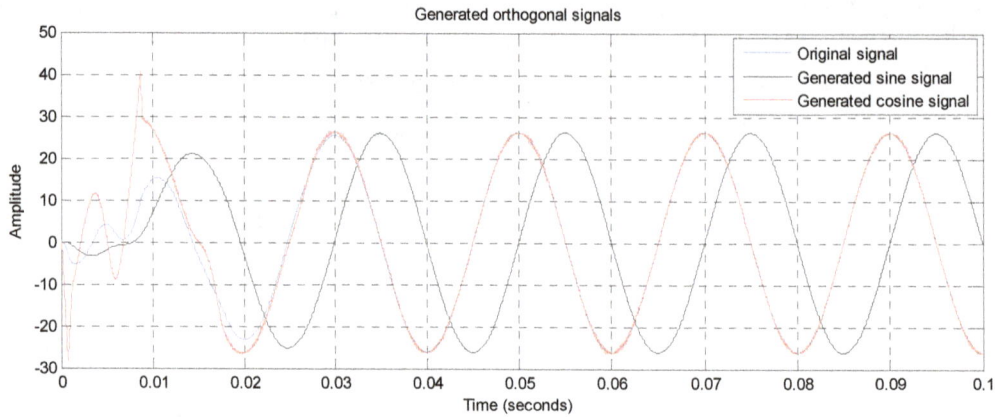

Figure 8. Orthogonal system output.

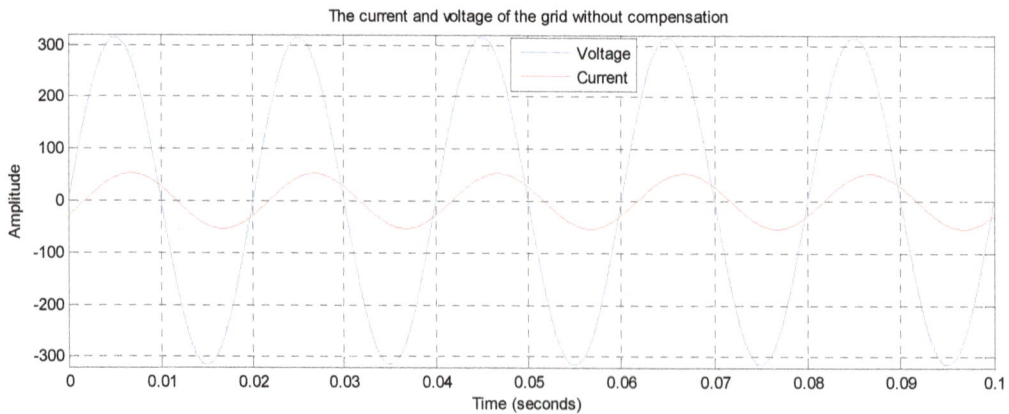

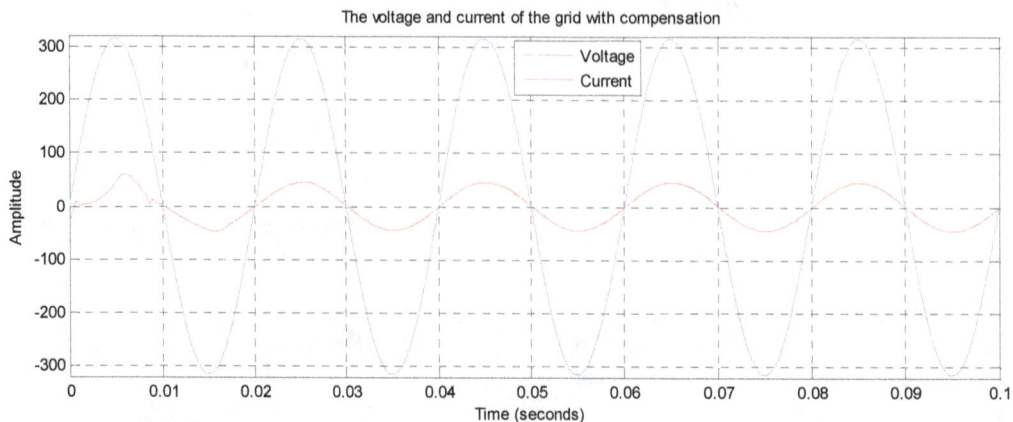

(a) The current and voltage of grid without compensation (b) The current and voltage of grid after compensation

Figure 9. The current and voltage of the grid.

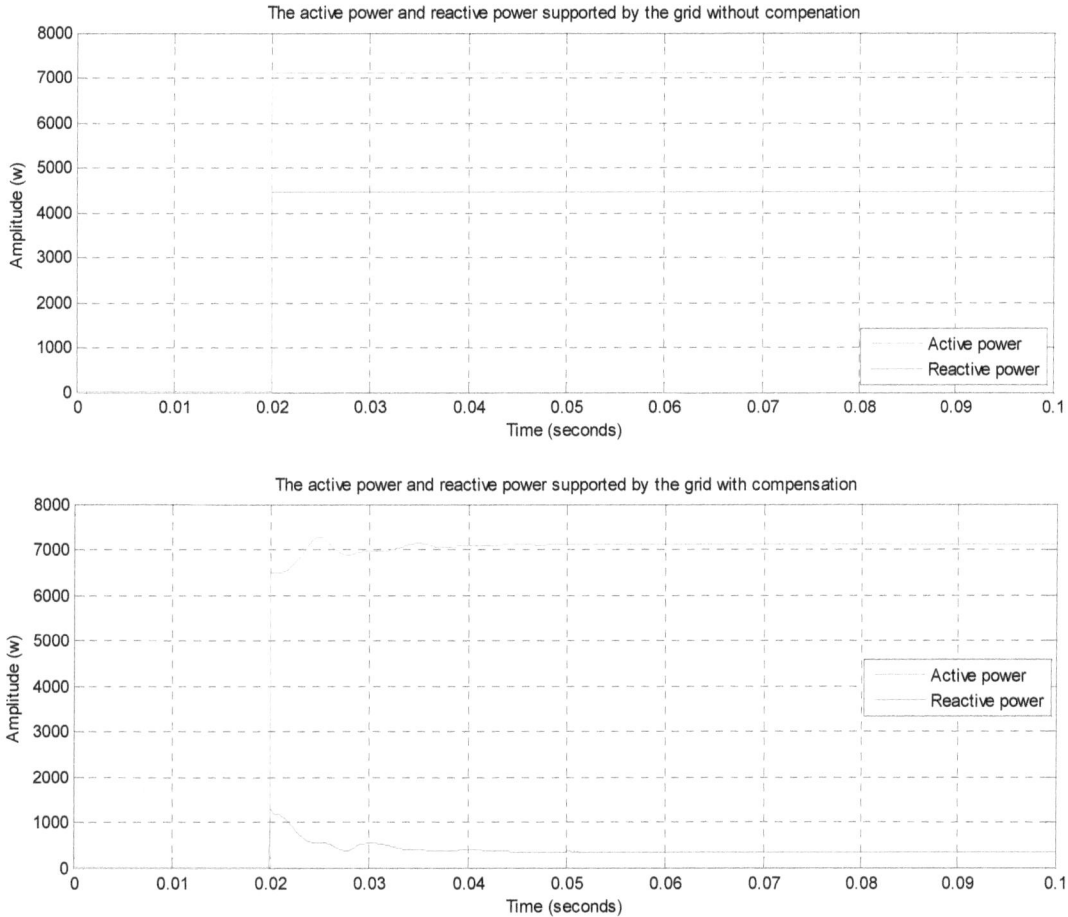

(a) The results of without compensation (b) The results of after compensation

Figure 10. *The comparison between the inverter with the presented method of compensation and without compensation.*

In Figure 10, the simulated results obtained in the situation of that the inverter only injects reactive power to the power grid. As Figure10 (b) illustrates that the active power supported by the grid maintains at same level contrast with the system without compensation.

Figure 11 shows the simulate model can output active power, and in the meantime also can produce reactive power. Figures 10 and 11 demonstrate that the proposed measure can separately control the active power and reactive power according to the demands of customs.

Figures 12 (a) and (b) elucidate that the power factor has been considerably improved, increased from approximate 0.85 to nearly 1. Improving power factor can reduce power system loses to relatively increase load carrying capabilities in the local power system.

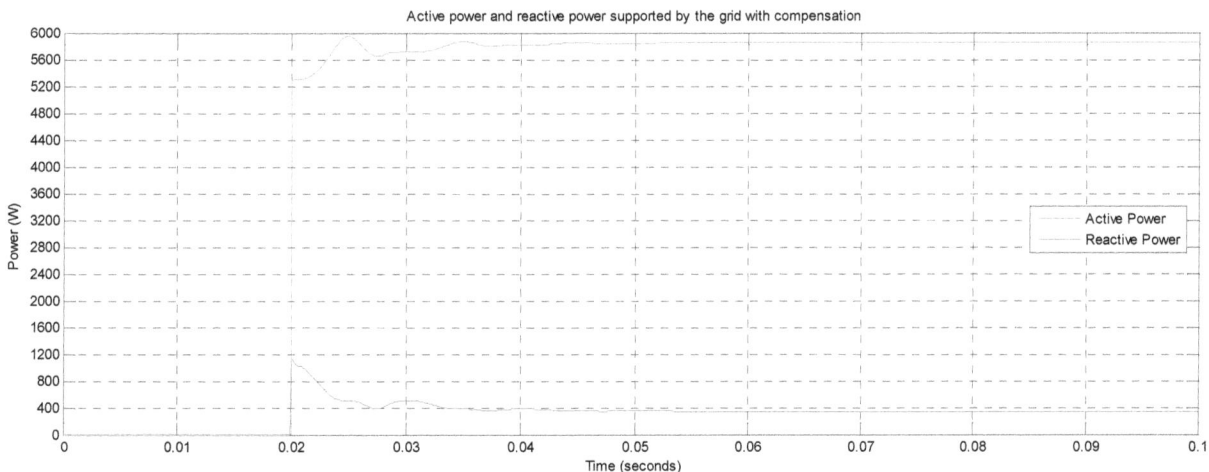

Figure 11. *The results of the inverter model output active and reactive power.*

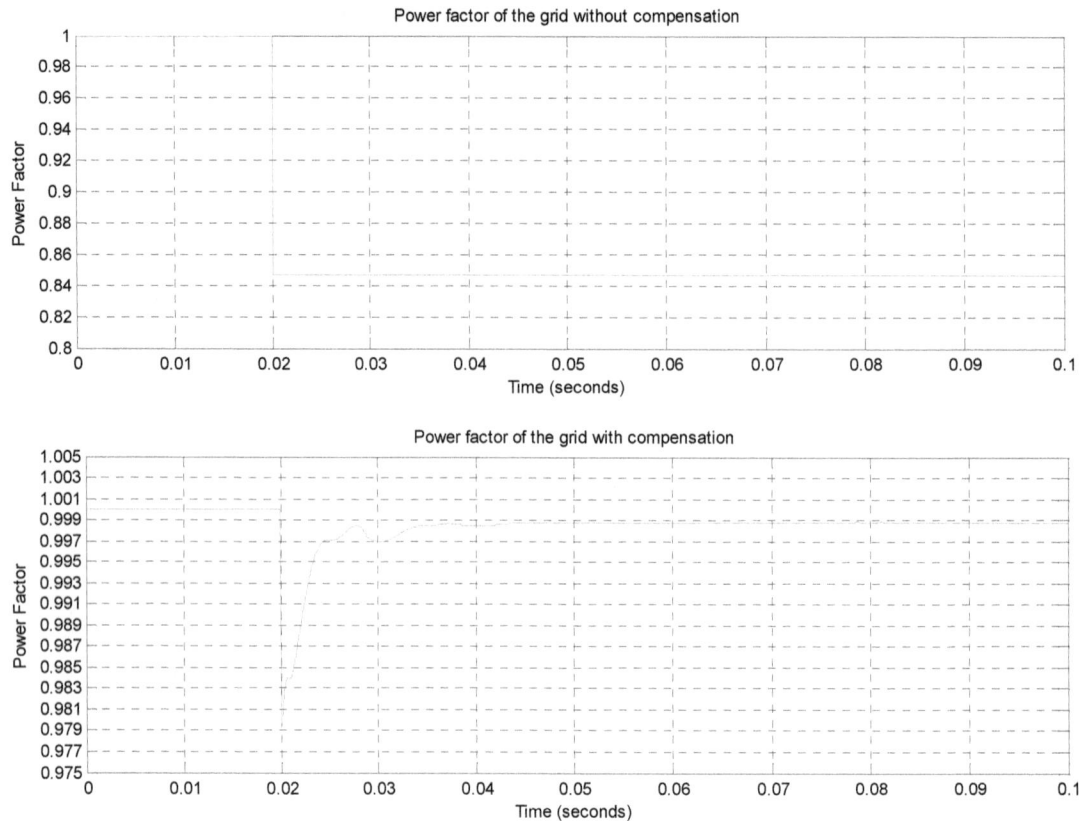

(a) The power factor of the grid before the model provided compensation (b) The power factor of the grid after compensation

Figure 12. The comparison of power factor.

4. Conclusion

The simulation results demonstrate that the proposed method by separating active and reactive current of the grid using SOGI to construct an orthogonal system is successfully approved by means of simulated consequence. It means that the inverter can inject active power into the grid to coincide with certain amount of reactive power compensation.

References

[1]   R. Majumder, "Reactive Power Compensation in Single-Phase Operation of Microgrid," *IEEE TRANSACTIONS ON INDUSTRIAL ELECTRONICS,* vol. 60, no. 4, pp. 1403 - 1414, 2013.

[2]   Enery Networks Association, "Engineering Recommendation G83," Energy Nerworks Association, London, 2012.

[3]   E. Moulin, "Measurring Reactive Power in Energy Meters," *Metering International,* no. 1, pp. 52 - 54, 2002.

[4]   M. Mithulananthan, C. A. Canizares, J. Reeeve and G. J. Rogers, "Comparison of PSS,SVC and STATCOM Controllers for Damping Power System Oscillations," *IEEE Trans. Power Systems,* pp. 1-8, Octber 2002.

[5]   M. Jaiswal, P. Kaur and P. Jaiswal, "Review and Analysis of Voltage Collapse in Power System," *International Journal of Scientific and Reasearch Publications,* January 2012.

[6]   A. Sode-Yome and N. Mithulananthan, "Comparison of shunt capacitor, SVC and STATCOM in static voltage stability margin enhancement," *International Journal of Electrical Engineering Education,* vol. 41/2, no. 0020-7209, pp. 158-171, 2004.

[7]   N. G. Hingorani, "Flexible AC Transmission Systems," *IEEE Spectrum,* pp. 40-45, Apr. 1993.

[8]   D. Murali, M. Rajaram and N. Reka, "Comparison of FACTS Devices for Power System Stability Enhancement," *International Journal of Computer Applications,* pp. 30-35, October 2010.

[9]   A. Ganesh, R. Dahiya and K. G. Singh, "A Novel Robust STATCOM Control Scheme for Stability Enhancement in Multimachine Power System," *ELEKTRONIKA IR ELEKTROTECHNIKA,* pp. 22-28, 26 January 2014.

[10] T.-L. Lee, S.-H. Hu and Y.-H. Chan, "D-STATCOM with Positive-Sequence Admittance and Negtive-Sequence Conductance to Mitigate Voltage Fluctuations in High-Level Penetration of Distributed-Generation Systems," *IEEE TRANSCATION ON INDUSTRIAL ELECTRONICS,* pp. 1417-1428, April 2013.

[11] S. Bolognani and S. Zampieri, "A Distributed Control Strategy for Reactive Power Compensation in Smart Microgrids," *Automatic Control, IEEE Transactions on,* pp. 2818-2833, 20 June 2013.

[12] J. F. Rodriguez, E. Bueno, M. Aredes, B. L. Rolim, S. F. Neves and C. M. Cavalcanti, "Discrete-time implementation of second order generalised integrators for grid converters," in *Industrial Electronics,2008. IECON 2008. 34th Annual Conference of IEEE,* Orlando, FL, 2008.

[13] S. Golestan, M. Monfared and J. M. Guerrero, "Second Order Generalized Integrator Based Reference Current Heneration Method for Single-Phase Shunt Active Power Filters Under Adverse Grid Conditions," in *Power Electronics, Drive Systems and Technologies Conference*, Tehran, 2013.

[14] P. Rodriguez, A. Luna, I. Etxeberria, J. Hermoso and R. Teodorescu, "Multiple Second Order Generalized Integrators for Harmonic Synchronization of Power Converters," in *Energy Conversion Congress and Exposition*, San Jose, CA, 2009.

[15] G. Fedele, "Non-adaptive Second-order generalized integrator for sinusoidal parameters estimation," *Electrical Power and Energy Systems,* vol. 42, no. 0142-0615, pp. 314-320, November 2012.

22

Study on Transformer Fault Diagnosis Based on Dynamic Fault Tree

Fei Peng[1, 2], Lin Cheng[1, 2], Kaikai Gu[2], Zhenbo Du[2], Jiang Guo[3]

[1]School of Electrical and Electronic Engineering, Huazhong University of Science and Technology, Wuhan, China
[2]Wuhan Nari Group Corporation of State Grid Electric Power Research Institute, Wuhan, Hubei, China
[3]School of Power and Mechanical Engineering, Wuhan University, Wuhan, Hubei, China

Email address:

759733212@qq.com (Fei Peng), chenglin@nari-ge.com (Lin Cheng), gkkyf@126.com (Kaikai Gu), 52781821@qq.com (Zhenbo Du),
guo.river@163.com (Jiang Guo)

Abstract: In this paper, according to theoretical diagnosis of fault tree, the author builds a diagnosis model based on dynamic fault tree and illustrates the model's construction method and diagnosis logic in detail. According to case analysis, compared with conventional fault tree diagnosis, the above-mentioned method is advanced in fault-tolerant ability. Plus, the diagnosis results record some intermediate processes of the diagnosis, with relevant information being returned to the researchers as ideas facilitating further analysis in the event of incomplete information.

Keywords: Transformer, Dynamic Fault Tree, Fault-Tolerant Ability, Fault Diagnosis

1. Introduction

Since the operation state of the transformer has a direct bearing on the security and stability of the whole power grid, it is necessary to prevent and minimize the occurrence of transformer faults and accidents to the largest degree. In recent years, having been used for transformer fault diagnosis, with certain achievements being achieved, various intelligent techniques [1-5] have provided bases for transformer fault diagnosis. Among many types of intelligent techniques, fuzzy set and fault tree analysis are the two categories applied and researched most. But mostly in transformer fault diagnosis, simplex intelligent techniques are used for this purpose, the final judgments of which lack guiding significance with respect to the formulation of maintenance strategies. In actual operating process, transformer faults are usually characterized by complexity, uncertainty and concurrency of multiple faults, in which, the use of simplex intelligent fault diagnosis will cause such problems as low accuracy, poor reasoning, and so on, difficult to obtain satisfactory diagnostic effect, especially for fault tree analysis and fuzzy mathematics theory. In fault diagnosis process, fault tree is poor in interpretation, hard to accurately and quantitatively describe the occurrence possibility of fault tree nodes, with a transition zone existing between "healthy" state and "faulty" state, and in the case of

incomplete information, fuzzy set theory is not possible be determined.

Aiming at using the appropriate combination of fuzzy set and fault tree, study on transformer fault diagnosis based on the two concepts mutually makes up their deficiencies with their advantages, in order to overcome the respective disadvantages of fuzzy set or fault tree and improve the accuracy and efficiency of transformer fault diagnosis.

2. Construction of Transformer Fault Tree

Study on fault diagnosis based on the fault tree principle has made a large number of research results, with many successful cases occurring in industrial field. However, through years of application researches, following defects or deficiencies are also exposed gradually: 1) Precise fault positioning is not possible; 2) Fault tree node lacks relevant information required in fault detection; 3) It is hard to accurately and quantitatively describe the possibility of the occurrence of fault tree nodes.

2.1. Dynamic Fault Tree Nodes

In this paper, to overcome deficiencies of fault tree applied in transformer fault diagnosis, the author proposes dynamic fault tree nodes, as shown in Figure 1, based on a comprehensive reference to conventional fault tree theory and the actual situation of transformer fault diagnosis.

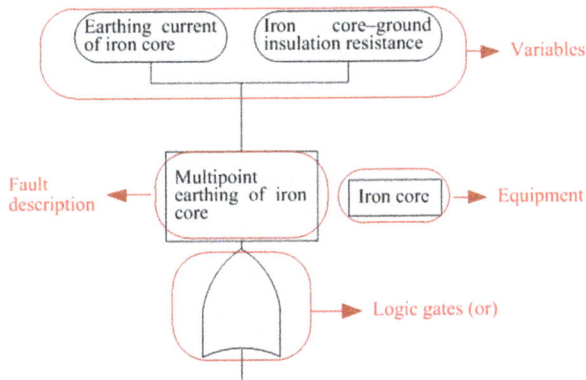

Fig. 1. A dynamic fault tree node with variables.

Compared with conventional fault tree nodes, dynamic fault tree nodes have introduced 2 parts, namely, "variables" and "equipment".

1) Variables refer to the relevant information needed in fault diagnosis, and these characteristics that have shown are the "variables" of this system. Along with the occurrence of each fault, there will be some phenomenon, such as "temperature rise", "overcurrent", and "partial discharge volume exceeding the standard". The introduction of variables is to connect the system in series successfully. When received by the system, original-state data of the transformer (inspection, online monitoring, test data, and so on) will be first processed through a series of procedures (eliminating algorithm, regression algorithm, trend analysis, and so on) to turn from original data into state information, which, then, will be processed and transferred into variables of fault tree for serving the diagnosis of the whole fault tree. Secondly, another advantage of introducing variables is that: in the event that the conclusion obtained from fault tree diagnosis is indeterminate, the corresponding state quantity can be traced according to the fault tree node variables obtained through diagnosis, with further analysis being performed based on respective test recommendations of the state quantity. Variables are also a key to make fuzzy input of fault tree possible, which will be discussed hereinafter.

2) Equipment refers to components or sub-components of the fault tree node which are connected to the transformer. Eventually, fault diagnosis is a troubleshooting service, and it will achieve precise fault positioning by connecting the fault tree node to the equipment. Troubleshooting personnel can locate the specific parts requiring repair or replacement according to this information.

Table 1. List of variables for dynamic fault tree node.

Fault name				Variables
Faults of iron core			-	
	Local overheating			Overheat fault is detected by oil chromatographic analysis
				Loss of idle load is too large
				Temperature of top-level oil is too high
				Iron core's earthing current is too large
		Multipoint earthing of iron core		Thermal fault is detected by oil chromatographic analysis
				Loss of idle load is too large
				Temperature of top-level oil is too high
				Total hydrocarbon is abnormal
			Foreign objects exist in the box	Ground insulation resistance of iron core is abnormal
			Core touches the shell or the clamping piece	Clamping-piece insulation resistance of iron core is abnormal
			Paperboard of the base damps	Trace water volume exceeds the standard level
				Ground insulation resistance of iron core is abnormal
				Dielectric loss exceeds the limit
				Over-excitation protection actions
		Magnetic saturation		Thermal fault is detected by oil chromatographic analysis
				Loss of idle load is too large
				Temperature of top-level oil is too high
				Total hydrocarbon is abnormal
		Partial short circuit of iron core		Dielectric loss exceeds the limit
				Thermal fault is detected by oil chromatographic analysis
				Loss of idle load is too large
				Total hydrocarbon is abnormal
				Dielectric loss is abnormal
			Insulation layer between laminates is damaged	Electric capacity is abnormal
				Loss of idle load is too large
				Temperature of top-level oil is too high
				Total hydrocarbon is abnormal
	Suspended electro-discharge			Partial discharge volume exceeds the standard
				Electro-discharge fault is detected by oil chromatographic analysis
				Acetylene exceeds the standard

2.2. Construction of Transformer Fault Tree

Systematic research analysis is the key to build a transformer fault tree. Small components or faults of high frequency should not be ignored simply, because some small faults are probably the root causes of serious system faults. Since transformer is of a complex structure with many sub-components, "modular decomposition" should be used. This method is to divide transformer into 8 subsystems first, namely, iron core, winding, insulation oil, casing, cooling system, tap switch, and non-electricity protection, and analyze each sub-system and build subtrees according to principles of building fault tree, and then combine these subtrees together. After constructing the fault tree completely, determine for each fault tree node the relevant state quantity, namely, "variables", which is as shown in Table 1.

3. Fault Diagnosis Based on Dynamic Fault Tree

3.1. Establishment of State Set

Carry out state evaluation for indicators, which can be divided into 4 levels, with their respective possible words being "healthy", "attentive", "abnormal", and "faulty". Given that the state set is:

$$V = \{V_1, V_2, V_3, V_4\} \tag{1}$$

In this formula: $V_i (i = 1, 2, 3, 4)$ refers to State Level i. To make full use of information provided by the state evaluation, set the corresponding point of the given state levels 1-4 as 1.00, 0.75, 0.5, and 0.25.

3.2. Determination of Membership Degree

Since membership function is the basis of applying fuzzy set, whether the membership function is constructed correctly [6] is one of the key factors to make good use of fuzzy set. Most of the variables of transformer fault tree nodes are measurable values, namely quantitative variables, which can be divided into 3 categories: the smaller the better, the nearer to the middle the better, and the bigger the better. For variables of the smaller the better, its membership function adopts formula (2). For variables of the bigger the better, its membership function usually adopts formula (3).

$$A(x) = \begin{cases} 1 & x \leq a \\ f(x) & x > a \end{cases} \tag{2}$$

In this formula: $f(x)$ is a nonincreasing function, with $0 \leq f(x) \leq 1$.

$$A(x) = \begin{cases} 1 & x \leq a \\ g(x) & x > a \end{cases} \tag{3}$$

In this formula: $g(x)$ is a nondecreasing function, with $0 \leq g(x) \leq 1$.

For variables of the nearer to the middle the better, its membership function can be demonstrated by the distribution of middle-type fuzzy functions.

3.3. Fuzzy State Matrix of Dynamic Fault Tree Nodes

State of dynamic fault tree nodes is directly related to variables, and Dynamic fault tree node F contains k (a number) variables. Thus, F can be divided into the following k sets of divisors and factors:

$$F = \{F_1, F_2, \cdots, F_k\} \tag{4}$$

Among them, the simplex-factor state matrix of subset i is as follows:

$$F_i = [F_{i1}, F_{i2}, F_{i3}, F_{i4}] \tag{5}$$

In this formula: F_{ij} (i=1,2,...,k; j=1,2,3,4) refers to the corresponding membership of Node F's Variable i to State Level V_j.

According to the above analysis, the fuzzy state matrix of Node F is obtained as follows:

$$R_{k \times 4} = \begin{bmatrix} F_{11} & F_{12} & F_{13} & F_{14} \\ F_{21} & F_{22} & F_{23} & F_{24} \\ \cdots & \cdots & \cdots & \cdots \\ F_{k1} & F_{k2} & F_{k3} & F_{k4} \end{bmatrix} \tag{6}$$

When the fuzzy state matrix of dynamic fault tree nodes needs to be obtained based on the simplex-factor state matrix of variables, the weight set of variables is required to be known. Fuzzy mathematical operation is performed based on the fuzzy matrix of variables and its weight set, which will lead the fuzzy state matrix of dynamic fault tree nodes being obtained.

3.4. Determination of Variables' Weights by Analytic Hierarchy Process

To get variables' weights of dynamic fault tree nodes, it is necessary to establish a judgment matrix. For example, to get the corresponding weight of such variables as that "overheat fault is detected by oil chromatographic analysis", "loss of idle load is too large", and "temperature of top-level oil is too high" to the dynamic fault tree node "local overheating", it is necessary to compare the significance of these three kinds of variables for "local overheating" respectively, with the compared results being based to form a 3x3 judgment matrix [6-7]. By finding the corresponding eigenvector to the largest eigenvalue of the matrix, it is available to get the corresponding weights of the three variables to the dynamic fault tree node "local overheating".

In formula (4), the general procedures of getting the

corresponding weight of Variable $F_i(i=1,2,\cdots,k)$ to dynamic fault tree node F are as follows:

1) Invite an expert to compare the significance of F_i for F respectively, the basis of which is shown in Table 2[14];

Table 2. List of scale.

Scale	Degree Description
1	Equally important
3	Somewhat more important
5	Obviously more important

2) Make judgment according to the rules given by Table 2, and establish a judgment matrix $H_{k\times k}$ based on the respectively compared results, among which, Factor $a_{ij}(i,j=1,2,\cdots,k)$ refers to the result of comparing variable F_i with variable F_j, with $a_{ii}=1, a_{ij}=1/a_{ji}$. For example, when $a_{12}=1$, it means F_1 and F_2 are equally important.

3) In the consistency test for the judgment matrix, the information of analytic hierarchy process comes from the judgment of experts on the relative importance between each two factors on one level. After quantization by appropriate scale, the information will form a judgment matrix. When any factor in Matrix $H_{k\times k}$ meets the condition of $a_{ij}=a_{il}a_{lj}(i,j,l=1,2,\cdots,k)$, it means that the matrix has certain consistency. As for a consistent matrix, it is available to obtain the weight of each indicator by getting the corresponding characteristic quantity to its maximum feature root. For those inconsistent cases, it is necessary to adjust the judgment matrix so as to achieve consistency, the specific adjustment method of which is shown in Reference [8] and Reference [9].

4) Establish weight set for indicators. After getting the consistent Judgment Matrix H, take its corresponding eigenvectors $[w_1,w_2,\cdots,w_k]$ to its Maximum Feature Root λ as weight coefficients, with $\sum_{i=1}^{k}w_i=1$, where w_i refers to the weight of variable i with respect to the influence on dynamic fault tree node. If H is not a consistent one, the corresponding normalized eigenvectors to its maximum feature root should be taken as Weight Set w, with the following formula coming into being:

$$Hw^T=\lambda w^T \qquad (7)$$

3.5. State Evaluation on Dynamic Fault Tree Nodes

After obtaining the simplex-factor state matrix of each variable and determining the weight vectors of each layer's indicators, it is available to adopt fuzzy mathematical operation to get the fuzzy state set of the dynamic fault tree node. Max-min compositional operation is performed:

$$F=w_{1\times k}\circ R_{k\times 4} \qquad (8)$$

In this formula: F is the fuzzy matrix on the judgment level of the dynamic fault tree node; $w_{1\times k}$ is the corresponding

membership degree of the variable to dynamic fault tree node F; and $R_{k\times 4}$ is the fuzzy state matrix of the variable.

In F, the computational formula of each Factor F_i is as follows:

$$F_j=\bigvee_{i=1}^{k}(w_i\wedge r_{ij})\ \text{i=1,2,..,k; j=1,2,3,4} \qquad (9)$$

In this formula: w_i and r_{ij} are factors in Matrixes $w_{1\times k}$ and $R_{k\times 4}$, with $\vee$ and $\wedge$ representing supremum and infimum respectively.

According to the rules of highest membership degree, the corresponding state level to the largest element in the matrix is the state evaluation level of the dynamic fault tree node. Put the value of the state level in the following form of computational vectors:

$$Q=[1\quad 0.75\quad 0.5\quad 0.25]$$

Normalize the Fuzzy Judgement Matrix F obtained based on the formula, and get Matrix F'. Multiply F' with Q, which will get the evaluation score S of this dynamic fault tree node:

$$S=F'Q^T \qquad (10)$$

3.6. Fault Reasoning Strategies

The specific processes and strategies of transformer fault diagnosis based on compact fusion of fuzzy set and fault tree are as follows:

1) Build transformer fault tree, set membership function for each variable, and then establish weight relationship for variables of dynamic fault tree nodes according to the actual conditions.

2) As for fuzzy state matrix of state quantity, input the test value for the state quantity, and get its fuzzy state matrix based on its membership function.

3) As for the state judgment for dynamic fault tree node, based on the corresponding Weight Set w of the state quantity to the dynamic fault tree node, the state score of the dynamic fault tree node is obtained according to Formula (8) and Formula (9), which, thus, will lead the state of the dynamic fault tree node being obtained.

4) As for fault tree diagnosis, a variable will be used in more than one dynamic fault tree node. As shown in Table 1, the variable "temperature of top oil layer" is associated to such dynamic fault tree nodes as "local overheating", "multipoint earthing of iron core", "magnetic saturation", and "insulation layer between laminates being damaged". It is available to get dynamic fault tree node states caused by a set of variables through Step 2) and Step 3), with several diagnosis branches being formed on the fault tree.

5) As for fault positioning, in fault tree diagnosis, if all branches can be positioned to the root causes, then they should be traced back to the corresponding root causes. However, for transformer, such a complex system, which is often

impossible to realize precise fault positioning through one diagnosis mode, needs to be supported by other detection and experiment methods, with branches of root causes not positioned being returned to unidentified variables of the next layer of the dynamic fault tree node. Then, put values of these variables and initial state quantities into Step 2), and repeat Step 3), Step 4), and Step 5) until all branches are positioned to root causes.

4. Diagnosis Examples and Result Analysis

A real diagnosis example is given below. The state information of a transformer is that oil chromatographic analysis shows overheating fault, excess loss of idle load, and extremely high temperature of the top oil level. According to Table 1, the current fault is local overheating, which is that the dynamic fault tree node "local overheating" is lightened. Now, it is required to identify the root cause of the fault. The specific steps are as follows:

1) Focus on the variables on the next level of the dynamic fault tree node "local overheating", and test its value by being supported by other means. The processes of calculating the state matrix for the iron core earthing current are given below, and the type of the iron core earthing current is the indicator of the smaller the better. Theoretically, completely healthy iron core earthing current tends to 0, and for transformers of small capacity, generally it is regarded as a healthy one when the iron core earthing current is lower than 90mA. Roughly, it can be defined that the iron core earthing currents of 90,100 and 110 correspond to 1-4 grade respectively. So that, it is available to get the Membership Function $A_i(\mathrm{x})$ of iron core earthing current belonging to the Fuzzy Set $V_i(i=1,2,3,4)$ as follows:

$$A_1(\mathrm{x})=\begin{cases} 1 & 0\leq x\leq 90 \\ \dfrac{100-x}{100-90} & 90\leq x\leq 100 \\ 0 & x\geq 100 \end{cases}$$

$$A_2(\mathrm{x})=\begin{cases} \dfrac{x}{90} & 0\leq x\leq 90 \\ 1 & 90\leq x\leq 100 \\ \dfrac{110-x}{110-100} & 100\leq x\leq 110 \\ 0 & x\geq 110 \end{cases}$$

$$A_3(\mathrm{x})=\begin{cases} 0 & 0\leq x\leq 90 \\ \dfrac{100-x}{100-90} & 90\leq x\leq 100 \\ 1 & 100\leq x\leq 110 \\ \dfrac{110}{x} & x\geq 110 \end{cases}$$

$$A_4(\mathrm{x})=\begin{cases} 0 & 0\leq x\leq 100 \\ \dfrac{110-x}{110-100} & 100\leq x\leq 110 \\ 1 & x\geq 110 \end{cases}$$

The measured value of iron core earthing current is 45mA, after which is put into these 4 functions, the state matrix of iron core earthing current will be [1,0.5,0,0].

2) Measure the score of the lower-level node ("multipoint earthing of iron core", "magnetic saturation", "partial short circuit of iron core" and "poor heat dissipation"). Take the calculation of the score of "multipoint earthing of iron core" as an example. The state matrixes of iron core earthing current, oil chromatographic analysis, idle-load loss, top-level oil temperature, and total hydrocarbon are [1,0.5,0,0], [0,0,0.5,1], [0,0,0.5,1], [0,0,0.25,1], and [0,0.5,1,0.5] respectively, with the Fuzzy Matrix $R_{5\times4}$ based on Formula (6) being as follows:

$$R_{5\times4}=\begin{bmatrix} 1 & 0.5 & 0 & 0 \\ 0 & 0 & 0.5 & 1 \\ 0 & 0 & 0.5 & 1 \\ 0 & 0 & 0.25 & 1 \\ 0 & 0.5 & 1 & 0.5 \end{bmatrix}$$

The corresponding Weight Judgment Matrix $H_{5\times5}$ of five variables to the dynamic fault tree node "multipoint earthing of iron core" is as follows:

$$H_{5\times5}=\begin{bmatrix} 1 & 2 & 2 & 2 & 2 \\ 1/2 & 1 & 1 & 1 & 1 \\ 1/2 & 1 & 1 & 1 & 1 \\ 1/2 & 1 & 1 & 1 & 1 \\ 1/2 & 1 & 1 & 1 & 1 \end{bmatrix}$$

It is shown that $H_{5\times5}$ has consistency, and Weight Set w is obtained based on Formula (7):

$$w=\begin{bmatrix} 0.36 & 0.16 & 0.16 & 0.16 & 0.16 \end{bmatrix}$$

The Fuzzy Judgment Matrix F of dynamic fault tree node "multipoint earthing of iron core":

$$F=\begin{bmatrix} 0.36 & 0.16 & 0.16 & 0.16 & 0.16 \end{bmatrix}\circ\begin{bmatrix} 1 & 0.5 & 0 & 0 \\ 0 & 0 & 0.5 & 1 \\ 0 & 0 & 0.5 & 1 \\ 0 & 0 & 0.25 & 1 \\ 0 & 0.5 & 1 & 0.5 \end{bmatrix}$$

$$= \begin{bmatrix} 0.36 & 0.16 & 0.16 & 0.16 \end{bmatrix}$$

After normalization, $F'=\begin{bmatrix} 0.43 & 0.19 & 0.19 & 0.19 \end{bmatrix}$
According to Formula (10), score S is:

$$S=\begin{bmatrix} 0.43 & 0.19 & 0.19 & 0.19 \end{bmatrix}\begin{bmatrix} 1 & 0.75 & 0.5 & 0.25 \end{bmatrix}^T$$
$$=0.715$$

It is shown that the state of dynamic fault tree node "multipoint earthing of iron core" is attentive. Similarly, the states of dynamic fault tree nodes "magnetic saturation", "partial short circuit of iron core", and "poor heat dissipation" are "attentive", "faulty", and "faulty" respectively.

3) As for unidentified variables the returned diagnosis results of which are root causes "poor heat dissipation" and "partial short circuit of iron core", since the partial circuit of the iron core is the fault node in the middle, it requires further observation on the state of its next-level node, which means to repeat Step (1) and Step (2) until the root causes are identified.

Discussions on Results: The interpretation ability of conventional fault tree diagnosis is low, with the results being normal or abnormal only. Meanwhile, the diagnostic capacity is low when information is incomplete. According to the above calculation example, conventional fault tree diagnosis shows the result of "poor heat dissipation". Compared with conventional fault diagnosis, the transformer fault diagnosis based on compact fusion of fuzzy set and fault tree is stronger in fault-tolerant ability, with the diagnosis results recording some intermediate processes of the diagnosis, which will return the relevant variables to the researchers as ideas for their further analysis in the case of incomplete information.

5. Conclusion

In this paper, based on respective advantages of fuzzy set and fault tree, the author presents the transformer fault diagnosis based on compact fusion of fuzzy set and fault tree, with case analysis demonstrating the accuracy being enhanced by this method for the transformer diagnosis.

1) Equipment and variables are introduced to the conventional dynamic fault tree node, with the precise positioning of fault diagnosis results and the combination of equipment state information and fault tree being achieved.

2) Construction method of the improved fault tree is described, with the flexibility and scalability of constructing fault tree being improved.

3) The compact fusion of the fuzzy set theory and dynamic fault tree nodes is realized, which brings fuzzy input to dynamic fault tree nodes, together with more states (healthy, attentive, abnormal, faulty), more consistent with the actual process.

4) Diagnosis capacity of the fault tree in the case of incomplete information gets improved. When information does not support the fault tree to locate the root cause, the variable will be returned to the lower-level dynamic fault tree node directly for providing ideas for researchers' further analysis.

As for the application of fuzzy set and fault in transformer diagnosis, the research methods and conclusions of this paper play important role both in terms of reference and practical engineering.

References

[1] FU Qiang, CHEN Tefang, ZHU Jiaojiao. Transformer Fault Diag-nosis Using Self-adaptive RBF Neural Network Algorithm[J]. High Voltage Engineering, 2012, 38(6): 1368-1375.

[2] CHEN Xiaoqing, LIU Juemin, HUANG Yingwei, FU Bo. Transformer Fault Diagnosis Using Improved Artificial Fish Swarm with Rough Set Algorithm[J]. High Voltage Engineering, 2012, 38(6): 1403-1408.

[3] BAI Cuifen, GAO Wensheng, JIN Lei, YU Wenxuan, ZHU Wenjun. Integrated Diagnosis of Transformer Faults Based on Three-layer Bayesian Network[J]. High Voltage Engineering, 2013, 39(2): 330-335.

[4] Wang Jianyuan, Ji Yanchao. Appliacation of fuzzy petri nets knowledge representation in electric power transformer fault diag-nosis[J]. Proceedings of the CSEE, 2003, 23(3): 121-125.

[5] ZHAO Wenqing, ZHANG Shenglong, NIU Dongxiao. Transformer fault diagnosis based on multi-Agent[J]. Electric Power Automation Equipment, 2011, 31(3): 23-26.

[6] Lei Yaguo. Research on hybrid intelligent technique and its applications in fault diagnosis[D]. Xi'an, China: Xi'an Jiaotong University, 2007.

[7] Xie Jijian, Liu Chengping. Fuzzy mathematics method and its application[M]. Wuhan, China: Huazhong University of Science and Technology Press, 2000.

[8] Lan Jibin, Xu Yang, Huo Liangan, Liu Jiazhong. Research on the Priorities of Fuzzy Analytical Hierarchy Process [J]. Systems Engineering – Theory & Practice, 2006, 26(9):107-112.

[9] SATTY T.L. The analytic hierarchy process.New York, NY, USA: McGraw-Hill Inc, 1980.

[10] Wang Yingluo. Systems engineering[M]. Beijing, China: Machinery Industry Press, 2001.

23

Microcontroller Based Electrical Parameter Monitoring System of Electronic Load Controller Used in Micro Hydro Power Plant

Nan Win Aung, Aung Ze Ya

Department of Electrical Power Engineering, Mandalay Technological University, Mandalay, Myanmar

Email address:
nanwinaung.ep@gmail.com (N. W. Aung), dr.aungzeya010@gmail.com (A. Z. Ya)

Abstract: In stand-alone micro-hydro power system, water turbine will vary in speed due to the variation of consumer load. This speed variation will cause in fluctuation in both voltage and frequency output from a generator. To solve this problem, electronic load controllers were invented and used in micro-hydro power system. The objective of this paper is to monitor the electrical parameters such as voltage, current, power and frequency of electronic load controller (ELC) by using microcontroller and liquid crystal display (LCD). It explains how to monitor and sense the above parameters and isolate between the power line and microcontroller. The voltage is sensed by using the step down transformer and voltage divider circuit. The current is measured with the help of an ACS 712 current sensor. Frequency signal is obtained by using a frequency signal converter circuit. Microcontroller 16F887 and liquid crystal display are used as the main devices to monitor the above parameters according to the values obtaining from the sensing circuits.

Keywords: Electronic Load Controller (ELC), Electrical Parameters Such as Voltage, Current, Power and Frequency, Liquid Crystal Display (LCD), Monitoring System, Microcontroller, Micro Hydro Power Plant

1. Introduction

The problem in micro-hydro power system is fluctuation in frequency and voltage generated by the generator under consumer load variation. The fluctuation in frequency and voltage cause adverse affect in various electrical appliances. Electronic Load Controller (ELC) is used to solve that problem. Control is done by diverting the unused power to the ballast load. An ELC is a solid state electronic device designed to regulate output power of a micro-hydro power system and maintaining a near-constant load on the turbine. Monitoring system is required to inform and check the present parameter values to an operator when micro hydro power system is running on load.

Electrical parameters in the system are voltage, current, frequency, power and power factor. There are three types of power in the system such as apparent power (S) in VA, active power (P) in W and reactive power (Q) in var. Apparent power can be measured by using the voltage and current values. However, power factor measurement is needed to measure the active and reactive power when the load is not the pure resistive loads. But the main parameters of ELC are the system voltage and frequency. Because the purpose of ELC control is to be stable the system voltage and frequency. Therefore, only apparent power (S) is measured and simulated in this paper. Microcontroller based electronic load controller (ELC) should be added with microcontroller based monitoring system to be a perfect and more advance the controller.

2. Principle Operation of ELC

The synchronous generator–ELC system consists of a three-phase star-connected generator driven by a micro hydro turbine and an ELC. Since the input power is nearly constant, the output power of synchronous generator is held constant at varying consumer loads. The power in surplus of the consumer load is dumped in a dump load through the ELC. Thus, synchronous generator feeds two loads in parallel such that the total power is constant, that is,

$$PG = PC + PD$$

Where,

PG = Generated power of the generator (which should be kept constant),

PC = Consumer load power, and

PD = Dump load power

The power dissipated in the dump load can be used for battery charging, water heating, cooking, etc[13].

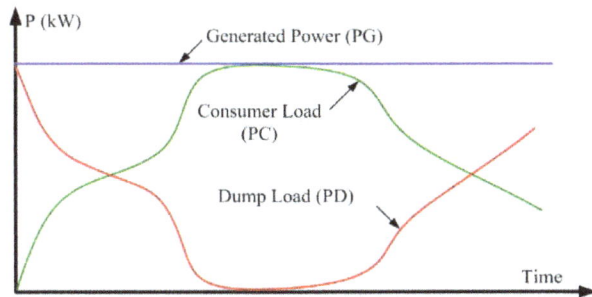

Figure 1. *Principle operation of ELC.*

The main type of ELC designs that are prevalent are:
- Binary load regulation
- Phase angle regulation
- Pulse width regulation
- Controlled bridge rectifier
- Uncontrolled bridge rectifier with a chopper

2.1. Binary Load Regulation

In binary load regulation the ballast load is made up from a switched combination of binary arrangement of separate resistive loads. In response to a change in the consumer load, a switching selection is made to connect the appropriate combination of load steps. This switching operation occurs during the transient period only, thereafter full system voltage is applied to the new fraction of the ballast load and hence harmonics are not produced by this method in the steady-state. In addition, it is usually the practice to adopt solid-state switching relays which include a zero-voltage switching circuit that reduces the harmonic distortion associated with the transient switching period [3]. The number of dump loads and the associated wiring is high and to achieve smooth regulation, these dump loads should all have exactly the right capacity. With a low number of dump loads, steps between dump load combinations remain too large and the system cannot regulate smoothly[4].

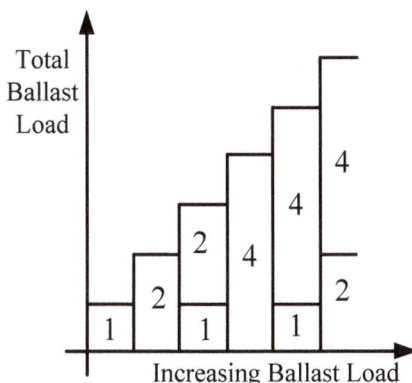

Figure 2. *Binary weighted ballast load.*

2.2. Phase Angle Regulation

In phase angle regulation, the ballast load comprises of a permanently connected single resistive load circuit of magnitude equal to (or slightly greater than) the full load rated output of the generator. As a result of the detection of a change in the consumer load, the firing angle of present power electronic switching device, such as a triac, is adjusted, thus altering the average voltage applied to and hence the power dissipated by, the ballast load.

As with all power electronic switching of this nature, this technique introduces harmonics onto the electrical system. In phase angle regulation method some of the shortcomings present are the presence of harmonics and that effectiveness limited by timing accuracy of trigger pulse. Phase angle modulation also seriously distorts the generator and this leads to the increasing of size of the generator to almost 25%.

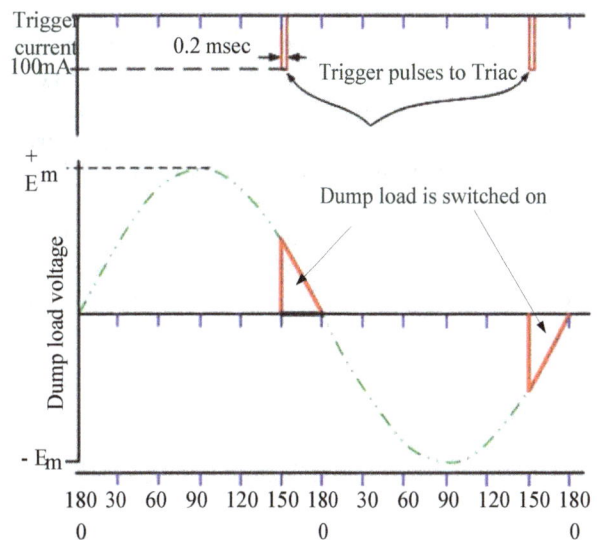

Figure 3. *Firing angle of phase angle regulation.*

2.3. Pulse Width Regulation

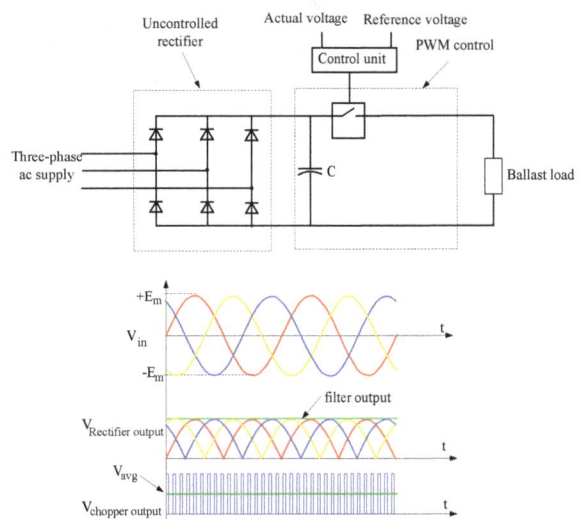

Figure 4. *Circuit diagram of uncontrolled rectifier with pulse width regulation control.*

In pulse width regulation, AC voltage is first rectified and dump load is switched on and off with a variable duty cycle. Duty cycle is the ratio of switch on time of a cycle to the time for a cycle. Control is done by varying the on-time of a cycle when the time of a cycle is constant for fixed frequency. PWM control can have fast response and compared to other schemes they usually have very smooth speed control, but total heat that is produced in this type of ELC is high and this is due to current problems with both rectifier and the transistor switching losses is really noticeable and significant in high frequency [3].

2.4. Controlled Bridge Rectifier

In controlled bridge rectifier, AC voltage is not only rectified to DC voltage but also controlled to variable DC output voltage whose magnitude is varied by phase control. So, a controlled bridge rectifier involves both conversion and control of electrical power. To achieve both conversion and control of electrical power, silicon controlled rectifiers (SCR) which are also called thyristors are used in power circuit as the power electronic devices. To turn SCR on, gate pulse must be provided to the gate of SCR. To achieve the dump load control, the rectifier output voltage is controlled by varying the delay time of gate pulse called delay angle (α).

As phase angle regulation, this technique introduces harmonics onto the electrical system. Moreover the timing accuracy of trigger pulses is very complex and limits the effectiveness of the system.

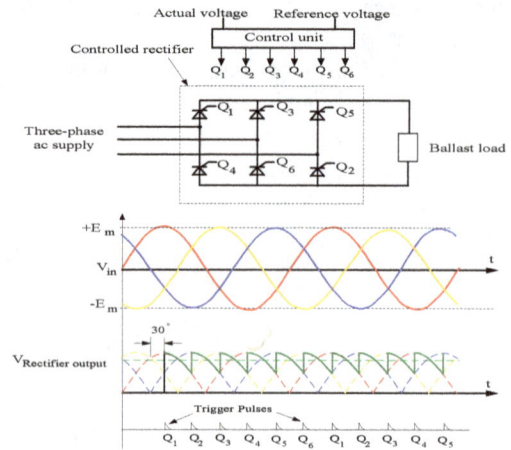

Figure 5. *Circuit diagram of controlled rectifier regulation.*

2.5. Uncontrolled Bridge Rectifier with a Chopper

In uncontrolled bridge rectifier with a chopper, AC voltage is first rectified to DC voltage and then a chopper controls it to variable DC voltage by varying the chopper duty cycle (D) for dump load control. Dump load power is controlled by adjusting the duty cycle (D). It is very similar to pulse width regulation method except the chopper design. But it has two control methods such as pulse width modulation and pulse frequency modulation. In pulse frequency modulation, the time of a cycle must be varied for frequency modulation when the switch on time is constant [4].

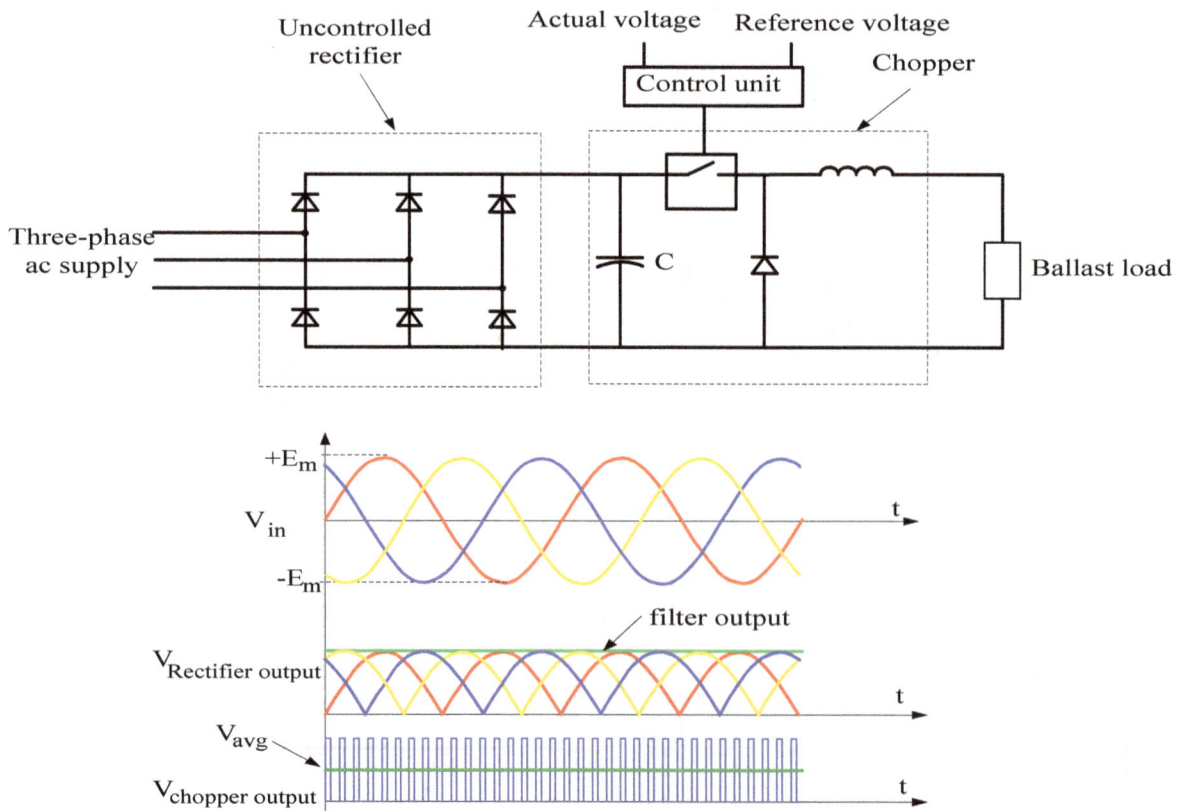

Figure 6. *Circuit diagram of uncontrolled rectifier with a chopper.*

Table I. *Advantages and disadvantages of control methods.*

Method	Advantages	disadvantages
Binary load regulation	-Minimal harmonics	-Fixed damp load size -Requires large number of dump load -Effectiveness limited by number of dump loads.
Phase angle regulation	-Can use any number/size combination of dump load	-Harmonics -Effectiveness is limited by timing accuracy of trigger pulse. -Control system is complex.
Pulse width regulation	-Can use any number/size combination of dump load -Control system is simplest.	-Harmonics -Effectiveness is limited by timing accuracy of duty cycle.
Controlled bridge rectifier	-Can use any number/size combination of dump load	-Harmonics -Effectiveness is limited by timing accuracy of trigger pulse. -Control system is complex.
Uncontrolled bridge rectifier with a chopper	-Can use any number/size combination of dump load	-Harmonics -Effectiveness is limited by timing accuracy of duty cycle.

3. Proposed ELC Design and System Configuration

The selected method for proposed 15 kW ELC is the combination of binary load regulation and pulse width regulation. Because binary load regulation is minimal harmonics and pulse width regulation is fast respond compared to others schemes.

Proposed ELC design can be divided into two circuits such as control circuit and power circuit.

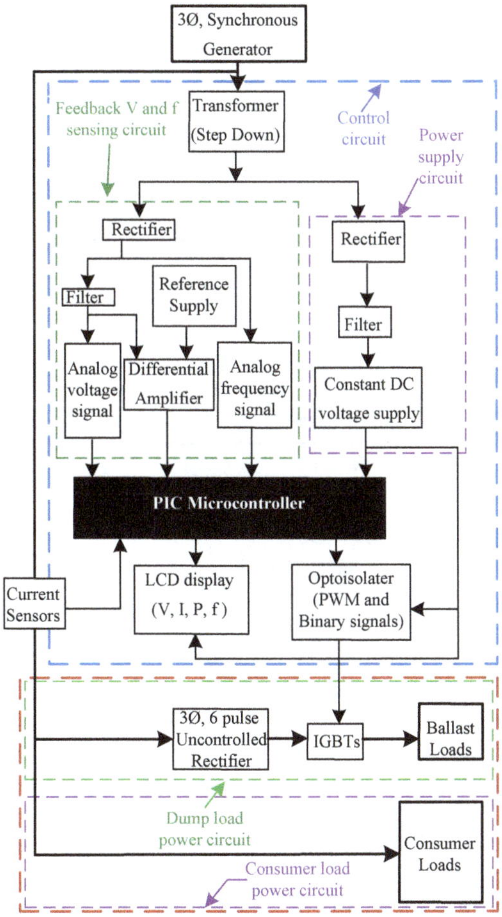

Control circuit consists of:
- Feedback voltage and frequency sensing circuit
- Power supply circuit and
- Microcontroller, LCD display and optoisolator.

Power circuit consists of:
- Dump or ballast load power circuit and
- Consumer load power circuit.

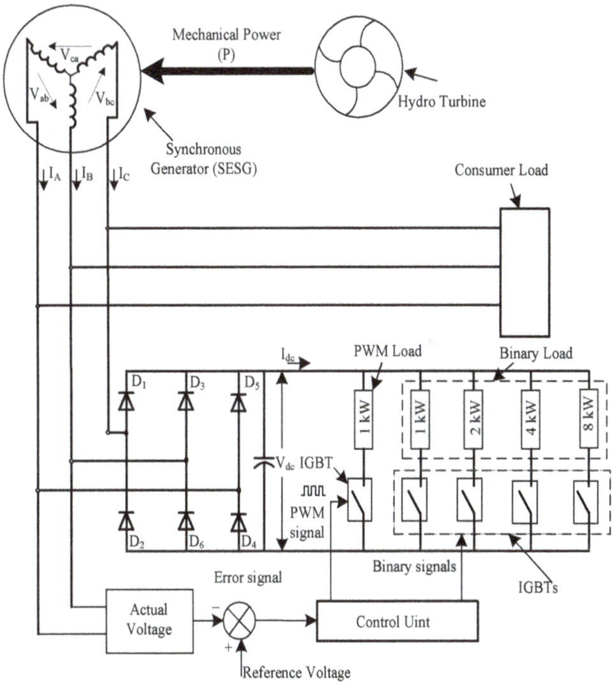

Figure 8. *Power line circuit diagram of proposed ELC design.*

The rating of bridge rectifier and PWM switch depends on the rated voltage and power of the synchronous generator. For 400V system voltage, the DC output voltage of uncontrolled bridge rectifier is given as below:

$$V_{dc} = (3\sqrt{2}\, V_L)/\pi$$

$$V_{dc} = (3\sqrt{2} \times 400)/\pi = 540 \text{ V}$$

The PWM load resistance is calculated as :

$$R_{PWM} = V_{dc}^2 / P_{PWM} = 540^2 / 1000 = 291.6\ \Omega$$

Figure 7. *Block diagram of proposed ELC design.*

Table II. Duty Cycle Calculation of PWM Load for 100 W Range.

P_{PWM} (W)	$V_o = \sqrt{(P_{PWM}R_{PWM})}$(V)	$I_o = V_o / R_{PWM}$ (A)	$D = V_o / V_{dc}$
100	171	0.59	0.31
200	241	0.83	0.45
300	296	1.02	0.55
400	342	1.17	0.63
500	382	1.31	0.71
600	418	1.44	0.77
700	452	1.55	0.84
800	483	1.65	0.89
900	512	1.76	0.95
1000	540	1.85	1

Table III. Sample Load Control Sharing of Combination System.

P_C(kW)	$P_D=P_B+P_{PWM}$ (kW)	P_B(kW)	Binary Load Status(8, 4, 2,1)	P_{PWM}(W)	D
15.0	0	0	0000	0	0
12.5	2.50	2	0001	500	0.71
11.6	3.40	3	0011	400	0.63
9.15	5.85	5	0101	850	0.92
8.75	6.25	6	0110	250	0.50
7.90	7.10	7	0111	100	0.31
4.45	10.55	10	1010	550	0.74
2.70	12.30	12	1100	300	0.55
0	15	15	1111	0	0

4. Microcontroller

4.1. PIC 16F887 Microcontroller

The PIC16f887 is the one of the latest products from microchip. It features all the components which modern microcontrollers normally have. For its low price, wide range of application, high quality and easy availability. Ithas 35 I/O pin and 14 analog channels. It is an ideal solution in applications such as: the control of different process in industry, machine control devices, measurement of different values etc [9].

4.2. Analog to Digital Converter (ADC) in PIC Microcontroller

ADC module of PIC microcontroller have usually 5 input for 28 pin devices and 8 inputs for 40 pin devices. The conversion of analog signal to PIC ADC module results in corresponding 10 bit digital number. PIC ADC module has software selectable high and low voltage reference input to some combination of VDD, Vss, RA2 and RA3 [9].

5. Liquid Crystal Display (LCD)

Liquid crystal cell displays (LCDs) are used in similar applications where LEDs are used. These applications are display of numeric and alphanumeric characters in dot matrix and segmental displays.

LCDs are of two types:
1. Dynamic scattering type
2. Field effect type

5.1. Dynamic Scattering Type

The liquid crystal material may be one of the several components, which exhibit optical properties of a crystal though they remain in liquid form. Liquid crystal is layered between glass sheets with transparent electrodes deposited on the inside faces. When a potential is applied across the cell, charge carriers flowing through the liquid disrupt the molecular alignment and produce turbulence. When the liquid is not activated, it is transparent. When the liquid is activated the molecular turbulence causes light to be scattered in all directions and the cell appears to be bright. This phenomenon is called dynamic scattering [8].

5.2. Field Effect Type

The construction of a field effect liquid crystal display is similar to that of the dynamic scattering type, with the exception that two thin polarizing optical filters are placed at the inside of is also of different type from employed in the dynamic scattering cell [8].

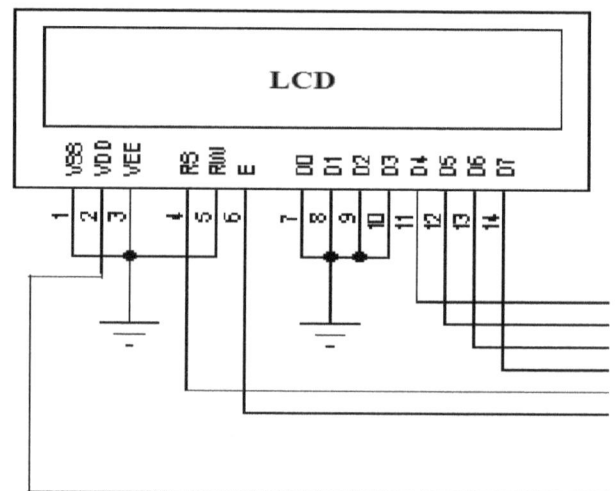

Figure 9. LCD Interfacing with microcontroller.

Table IV. Pin Definitions of Liquid Crystal Display (LCD).

No.	Symbols	Function
1	VSS	To power
2	VDD	power positive
3	VEE	Liquid crystal display bias
4	RS	Data/ command options
5	R/W	Read/write choice
6	E	By using the signal
7	D0	Data
8	D1	Data
9	D2	Data
10	D3	Data
11	D4	Data
12	D5	Data
13	D6	Data
14	D7	Data

6. Measurement of Alternating Voltage

6.1. Voltage Sensor

Voltage transformers (VTs), also referred to as "Potential Transformers" (PTs), are designed to have an accurately known transformation ratio in both magnitude and phase, A voltage transformer is intended to present a negligible load to the supply being measured. Voltage transformer downs the AC voltage from 440 V to 6V according to transformer ratio.The low secondary voltage is rectified by full wave bridge rectifier to convert AC voltage to DC voltage. The output voltage of the rectifier contains both DC and AC (ripple) components. A filter is required to filter out ripple component to obtain the smooth DC voltage. Voltage divider circuit, variable resistor, divides the DC voltage to be an acceptable DC voltage (0- 5V) for microcontroller. 5.1 V zener diode is connected across at voltage divider output to protect the microcontroller from over voltage.

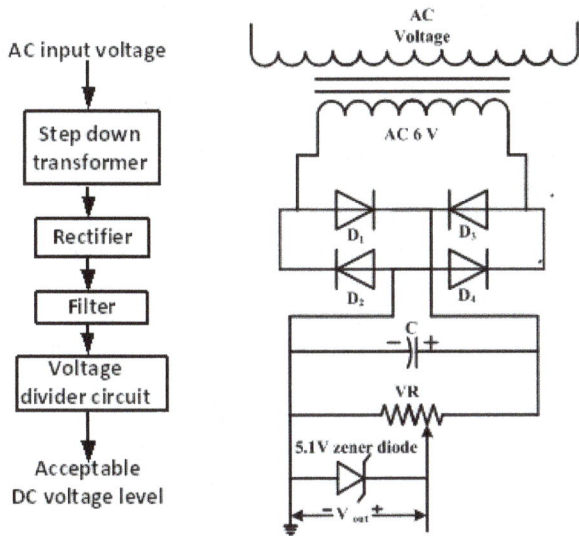

Figure 10. *Voltage sensor circuit for voltage monitoring system.*

6.2. Simulation of AC Voltage Sensing and Measurement

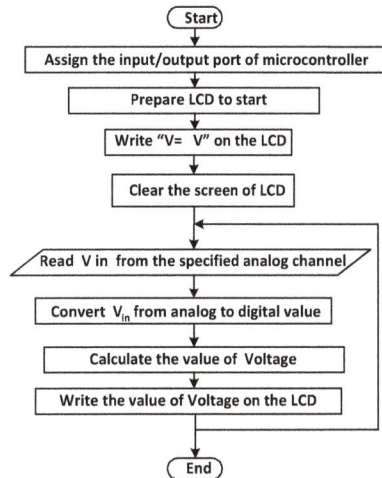

Figure 11. *Flow chart program of voltage monitoring system.*

The firmware program for the microcontroller is compiled with the MikroC software and simulation is done by Proteus professional software 8.0. The source code is written in MikroC software. The flow of program as shown in figure 11 goes like this: after assigning the input/output port of microcontroller, LCD is prepared to start for the system. The character "V=" and "V" are written on the LCD at the specified locations. Then the screen of LCD must be cleared to remove the previous system. Microcontroller read the input voltage from the specified analog channel and the input voltage is converted into digital value from analog value by analog-digital converter (ADC). This value has to be multiplied with voltage ratio gain and ADC gain to obtain the actual RMS value. For example, the voltage ratio gain for 500 V RMS voltage is 500/5 = 100. The ADC gain for 10 bit digital number is 5/1023= 0.004888. After that, the value of voltage is calculated by written program and the result is monitored on the LCD by changing into the numerical character according to American Standard Code for Information Interchange (ASCII).

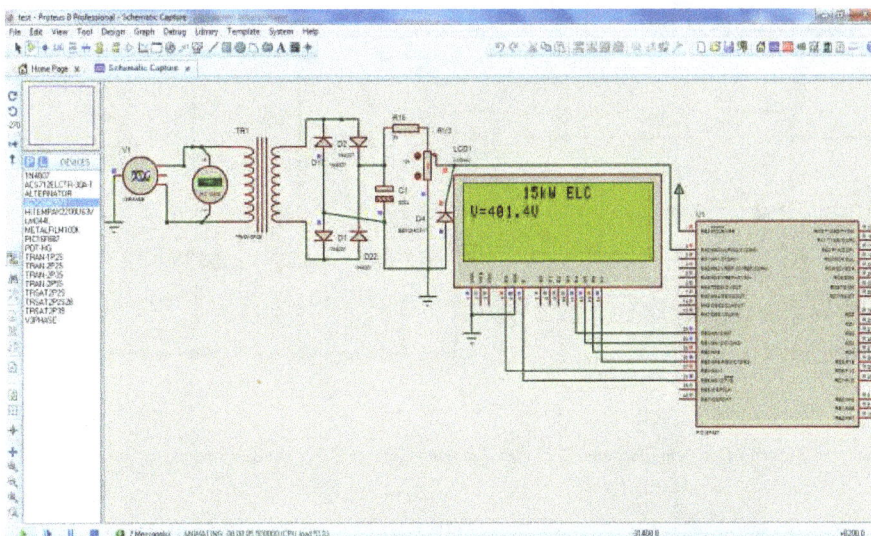

Figure 12. *Simulation result of voltage monitoring system.*

7. Measurement of Alternating Current

7.1. Current Sensor

In electrical engineering, a current transformer (CT) is used for measurement of electrical currents when the current in a circuit is too high to directly apply to measuring instruments. Now, AC or DC current sensors for microcontroller based sensing are available in the market cheaply.

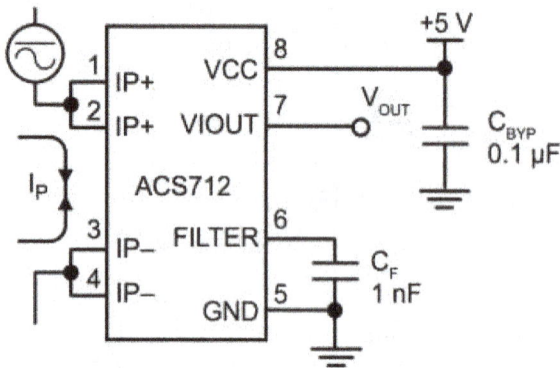

In this research, Allegro™ACS712 current sensor is used to measure the AC current. The device consists of a precise, low-offset, linear Hall circuit with a copper conduction path located near the surface of the die. Applied current flowing through this copper conduction path generates a magnetic field which the Hall IC converts into a proportional voltage. Device accuracy is optimized through the close proximity of the magnetic signal to the Hall transducer.

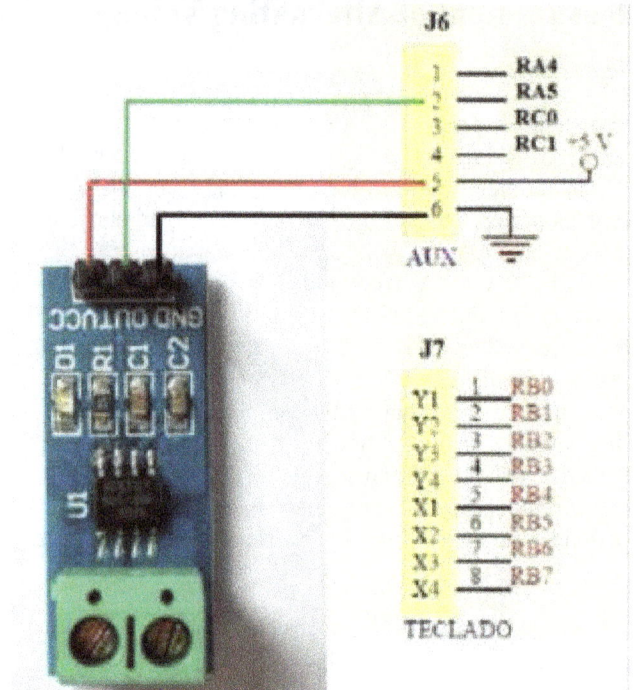

Figure 13. Typical application of ACS 712 current sensor.

The output of the device has a positive slope (>VIOUT(Q)) when an increasing current flows through the primary copper conduction path (from pins 1 and 2, to pins 3 and 4), which is the path used for current sampling. The internal resistance of this conductive path is 1.2 mΩ typical, providing low power loss. The thickness of the copper conductor allows survival of the device at up to 5× over current conditions. The terminals of the conductive path are electrically isolated from the signal leads (pins 5 through 8). This allows the ACS712 to be used in applications requiring electrical isolation without the use of opto-isolators or other costly isolation techniques.

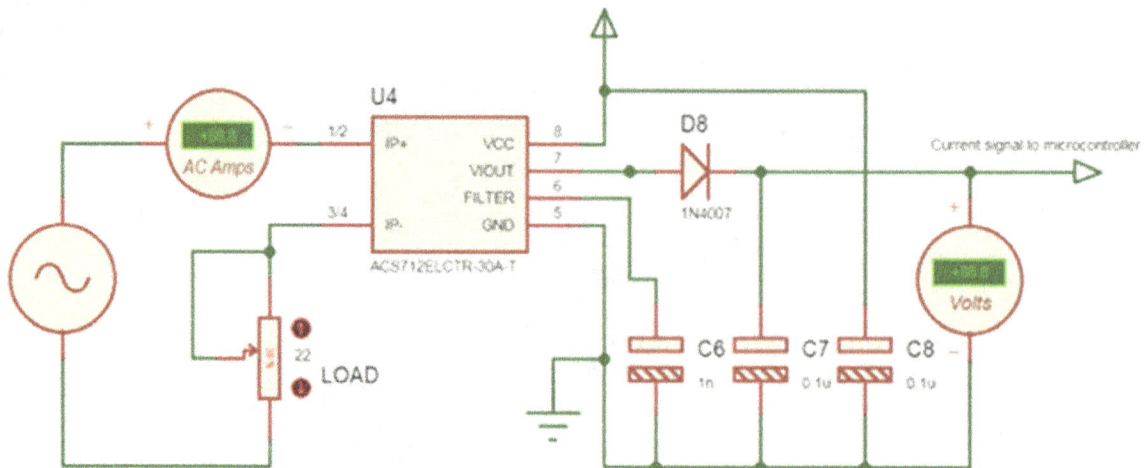

Figure 14. AC current sensing circuit diagram using ACS 712 current sensor.

The ACS712 outputs an analog signal, VOUT, that varies linearly with the uni- or bi-directional AC or DC primary sampled current, IP, within the range specified. CF is recommended for noise management, with values that depend on the application [10].

7.2. Simulation of AC Current Sensing and Measurement

The flow of program as shown in figure 15 goes like this: after assigning the input/output port of microcontroller, LCD is prepared to start for the system. The character "I=" and "A" are written on the LCD at the specified locations. Then the screen of LCD must be cleared to remove the previous system. When alternating current flows through the sensor device, the output voltage of sensor also vary between 0V and 5V proportionally with AC current. So, to obtain the steady output voltage of sensor, a diode and a capacitor are connected to output terminal. But diode has the forward voltage drop across the anode and cathode. It reduces the output voltage of sensor. To solve that problem, various gains which refer to diode voltage drop are added to the output voltage in the program whenever load current increases or decreases one ampere. Then the value of current is calculated by written program and the result is monitored on the LCD by changing into the numerical character according to American Standard Code Information Interchange (ASCII).

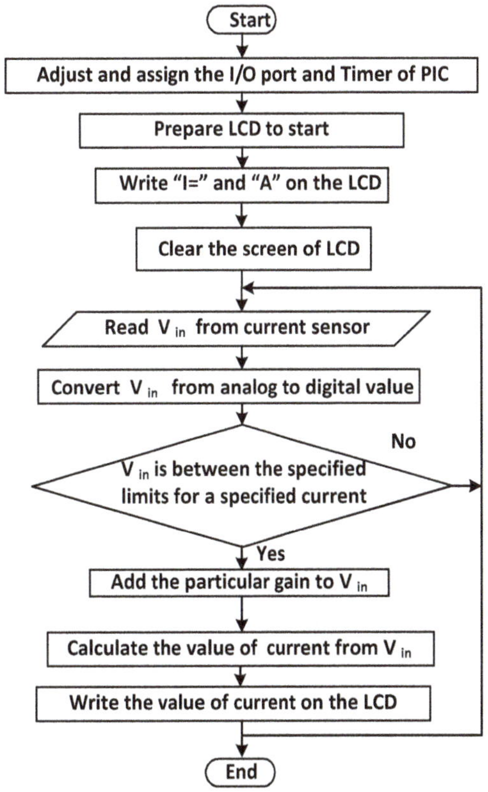

Figure 15. *Flow chart program of current monitoring system.*

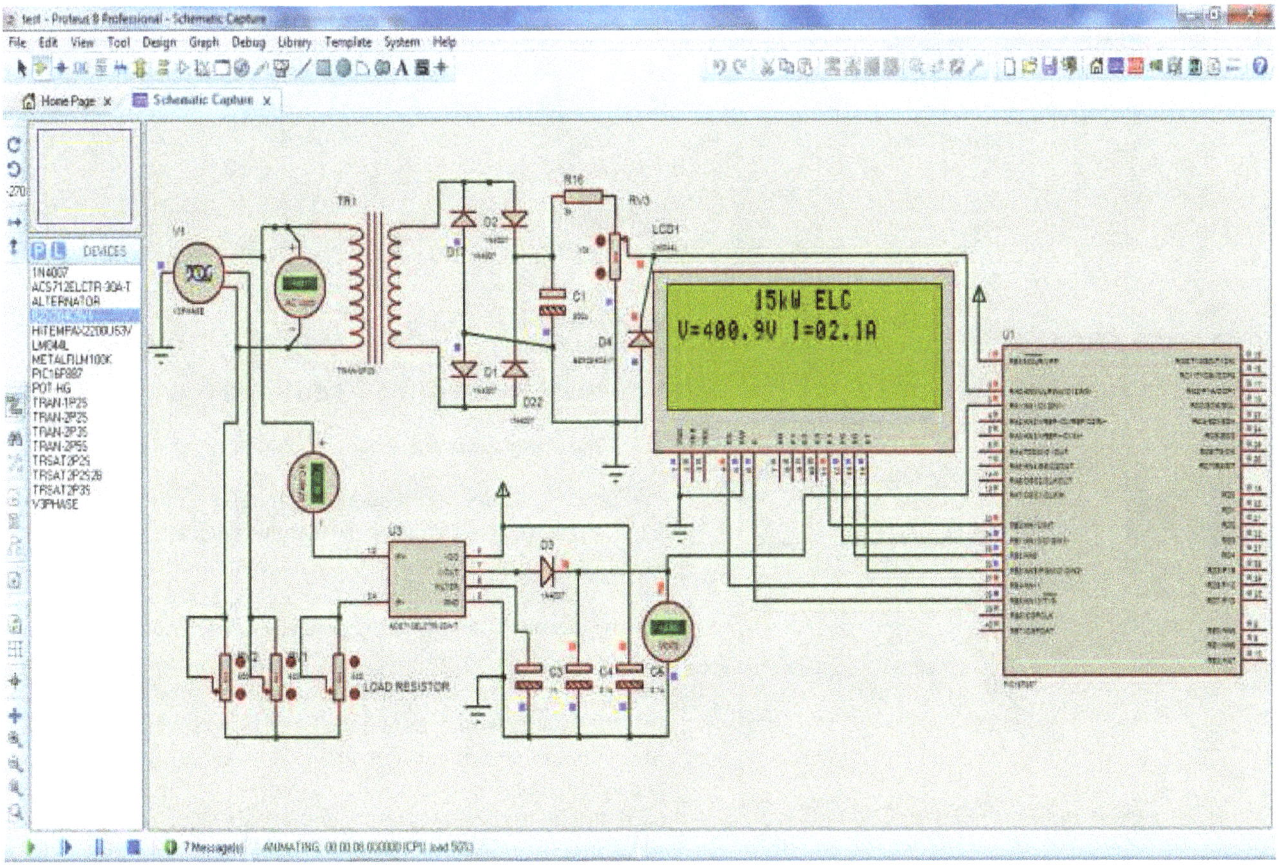

Figure 16. *Simulation result of current monitoring system.*

8. Measurement of Generated Power

Power measurement is easily and simply done by using microcontroller and LCD after achieving the voltage and current sensing. Generated power (PG), apparent power, is measured by multiplying the voltage and current values from each sensing circuit according to the following equation.

For three phase,

$$PG \text{ or } S = \sqrt{3}VI \times 10^{-3} KVA$$

For single phase,

$$PG \text{ or } S = \sqrt{2}VI \times 10^{-3} KVA$$

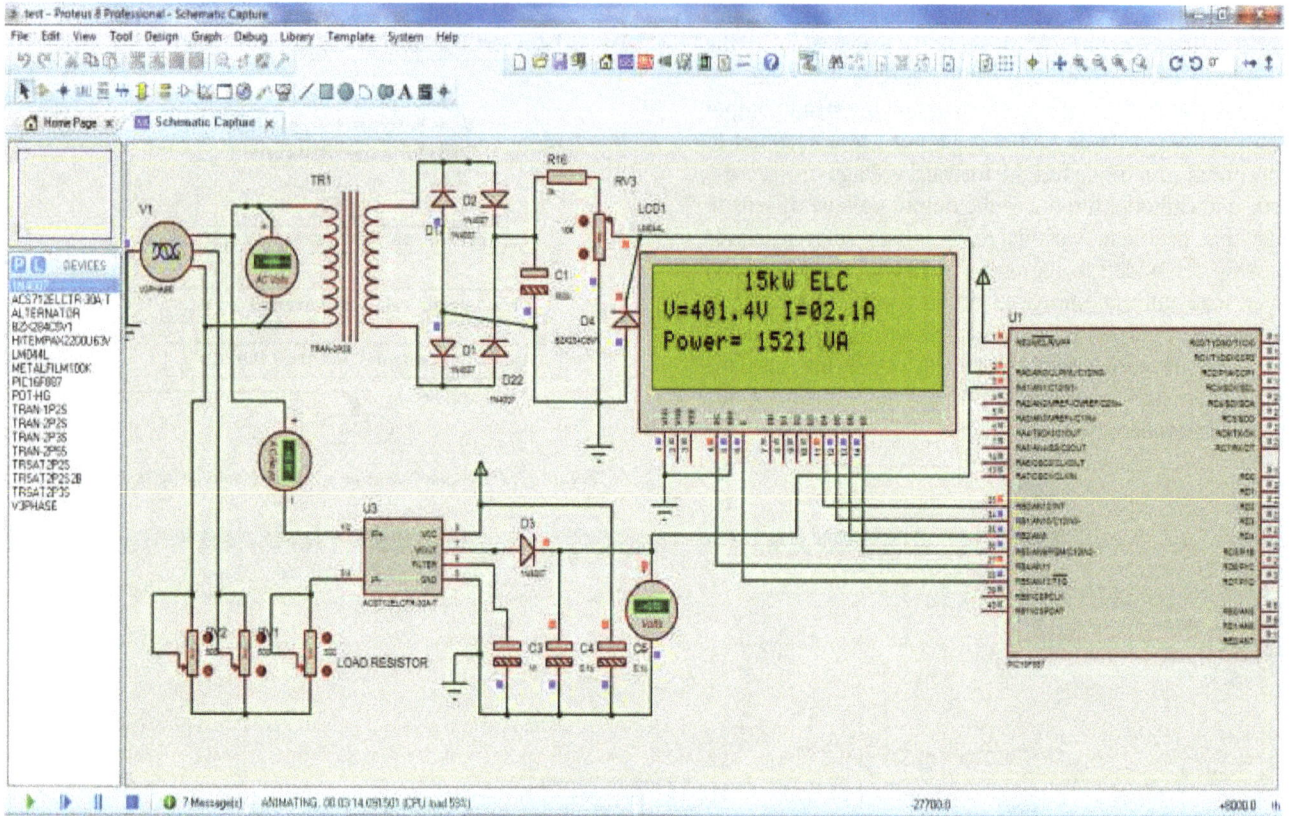

Figure 17. Simulation result of power monitoring system.

9. Measurement of Frequency

9.1. Frequency Sensor

Figure 13 shows the frequency converter circuit to convert from AC frequency to binary frequency signal with acceptable magnitude to be able to read by microcontroller. In circuit, series connection of resistor R_1 and R_2 operate as a voltage divider circuit. Transistor Q is used as a switching device. The purpose of R_s is to limit the base current (I_B). In the positive half cycle of the AC voltage, tansistor Q obtain the base current with forward bias and then turn on like a closed switch. Microcontroller will accept the binary signal as 0. In the negative half cycle of AC voltage, transistor Q obtain the base current with reverse bias and then turn off like a open switch. Now, microcontroller will accept the binary signal as 1. The binary frequency output of the circuit is shown in figure19.

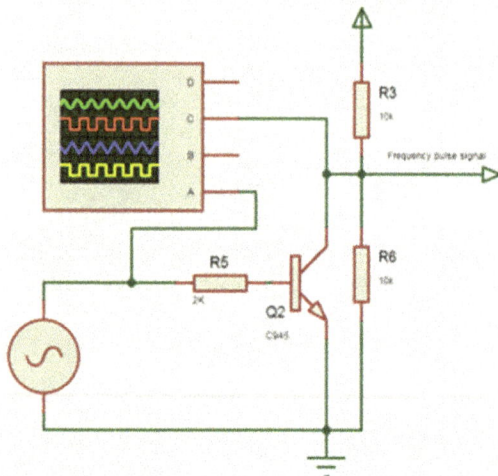

Figure 18. Frequency converter circuit diagram.

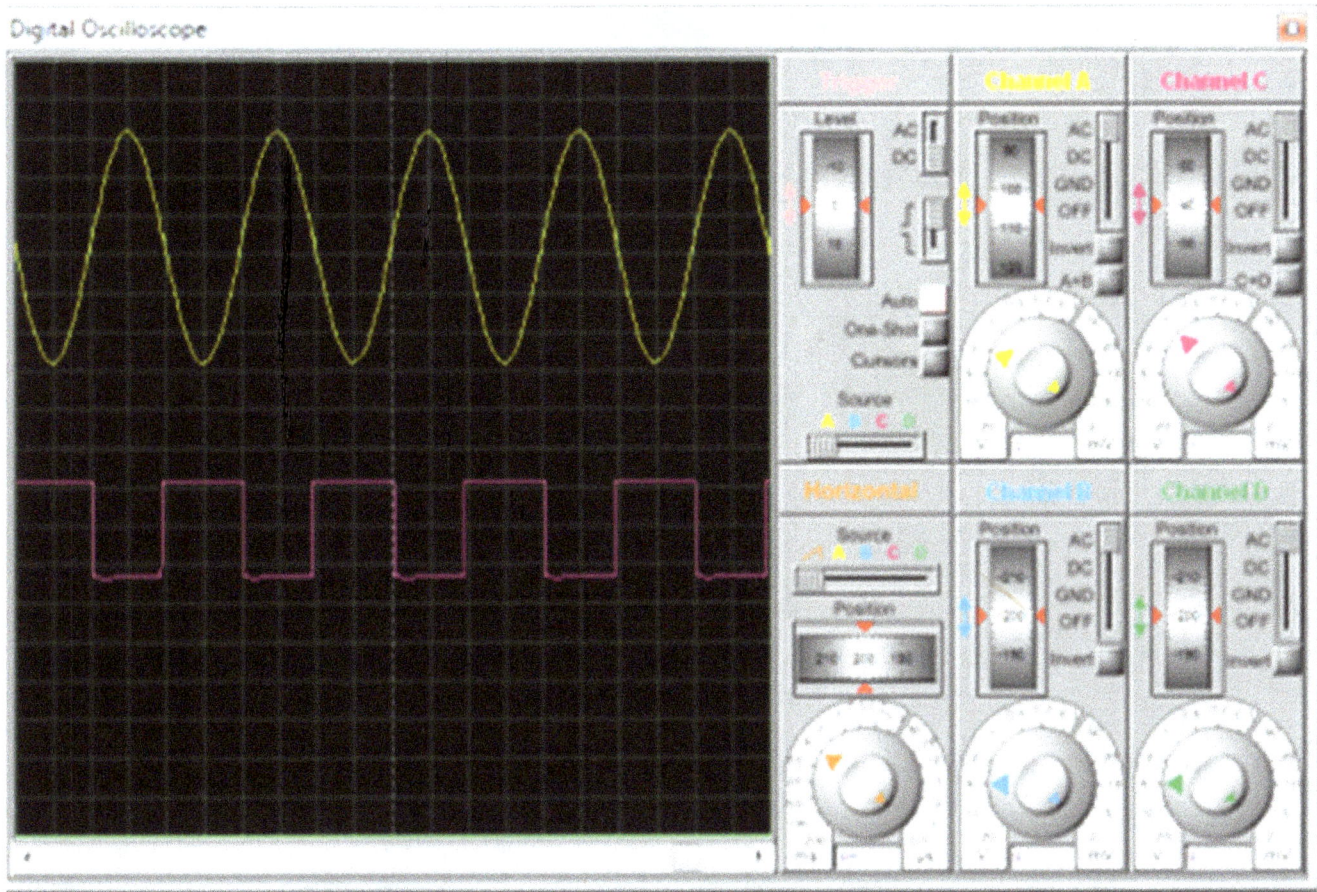

Figure 19. Binary frequency output of the frequency converter circuit.

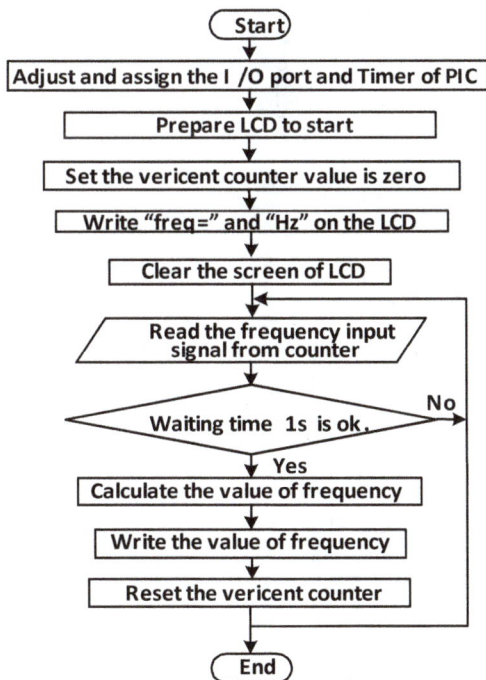

Figure 20. Flow chart program of frequency monitoring system.

9.2. Simulation of Frequency Sensing and Measurement

The flow of program as shown in figure 20 goes like this: after assigning the input/output port of microcontroller, LCD is prepared to start for the system. The character "freq=" and "Hz" are written on the LCD at the specified locations. Then the screen of LCD must be cleared to remove the previous system. After initialization the code enters an endless loop where it continuously performs a measurement and display operation. After an accurate 1 second delay the counter result is processed and displayed on the LCD. The main operation of this code is within the interrupt routine that both counts the input edges and obtains an accurate 1s time by counting the edges of the internal oscillator clock (Fosc/8).The most important part of frequency counter is the interrupt() routine. This is where all the action and decisions are made. The interrupt code for Timer1 is very simple and all it does is increment a long variable for counting multiple input events. The more tricky interrupt code, for Timer 0, counts time as described above. It counts 3906 overflows followed by a single 64 cycle count to reach a time of 1 second after which it captures the event count and then triggers an update to the LCD to calculate and display the frequency.

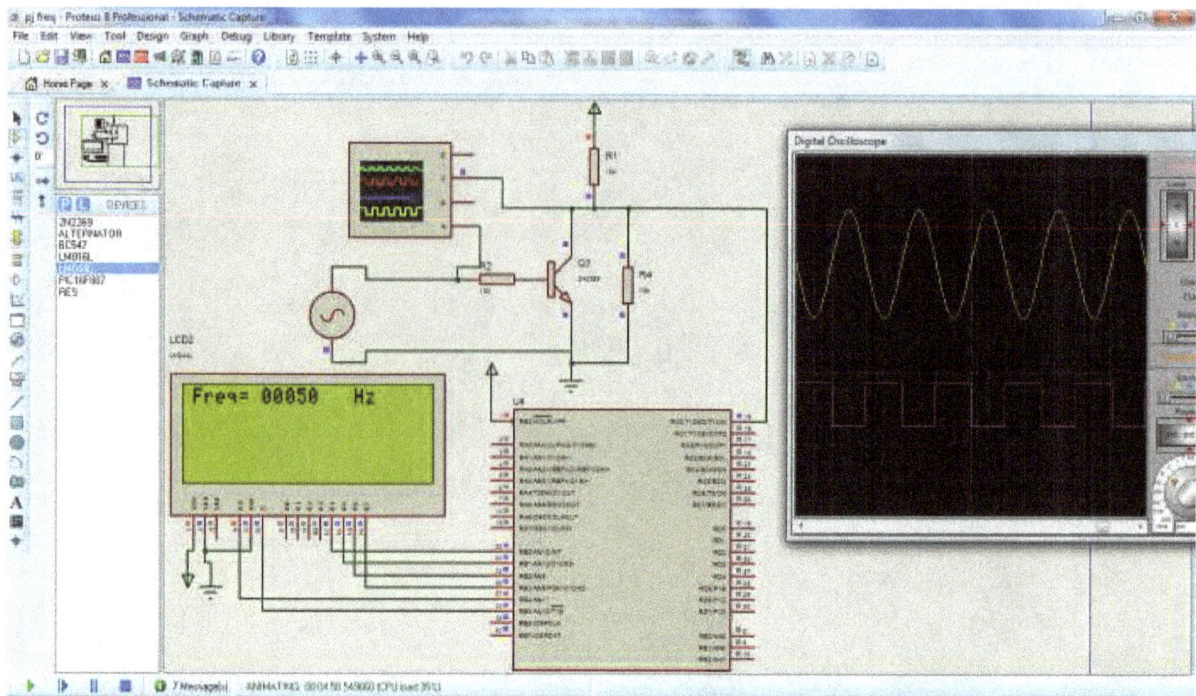

Figure 21. Simulation result of frequency monitoring system.

10. Simulation of Proposed ELC Design

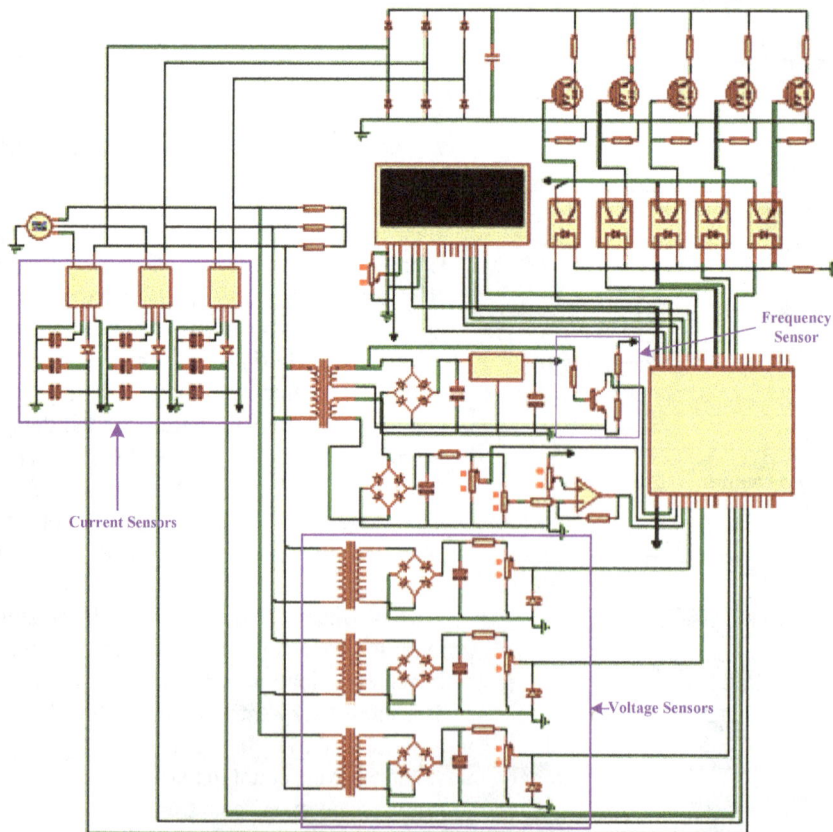

Figure 22. Complete circuit diagram of proposed ELC.

Figure 22 shows the complete circuit diagram of proposed ELC to simulate the monitoring system. In system monitoring, three system line voltages, three line currents, generated power (PG) and system frequency are displayed on the LCD screen. In simulation, both consumer and dump load are pure resistive loads. Therefore, power factor is assumed as unity. Figure 23 shows the simulation result of monitoring system.

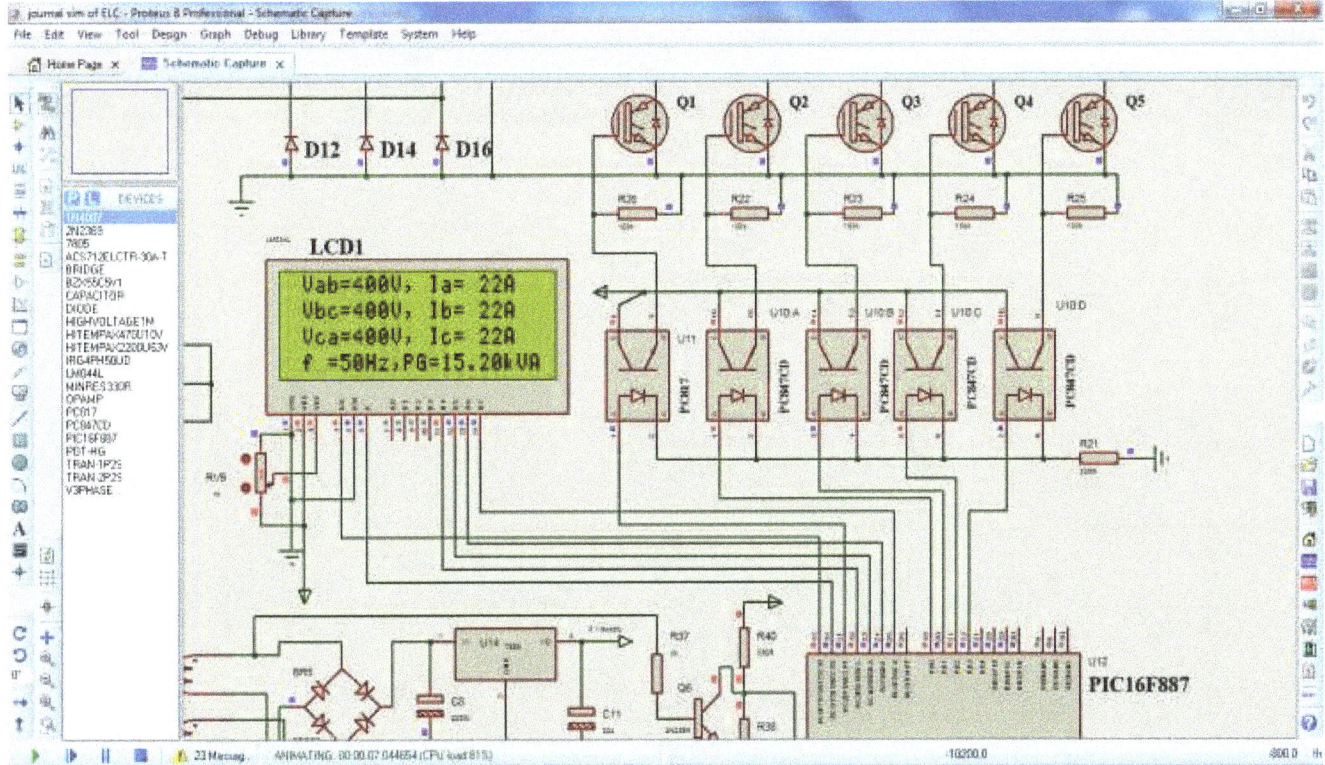

Figure 23. *Overall monitoring system by simulation result.*

11. Conclusion

The micro-hydro power generation is a viable option in remote and rural areas where grid electricity is not available. So, electronic load controllers (ELC) were being invented to be advance more and more with the wide using of micro hydropower generation. The developed microcontroller based ELC is found to be reliable, compact, cost effective and above all. Moreover, microcontroller based circuit can easily sense the system parameters such as voltage, current, frequency, power and power factor. Three system voltages, three line currents, power and system frequency must be measured and monitored on the ELC panel. In microcontroller based monitoring system, these eight parameters can be monitored on the single LCD. But, in ordinary monitoring system, eight measuring instruments have to be used for the monitoring system of ELC. The cost of these instruments is very much higher than the cost of the LCD. Using microcontroller based monitoring system not only reduces the cost of ELC but also compact the size of ELC.

Acknowledgements

The author is deeply gratitude to Dr. Myint Thein, Rector, Mandalay Technological University, for his guidance and advice. The author would like to thank to Dr. Yan Aung Oo, Professor, Head of Department of Electrical Power Engineering, Mandalay Technological University, for his kind permission, providing encouragement and giving helpful advices and comments. The author would like to express grateful thanks to his supervisor, Dr. Aung Ze Ya, Associated Professor, Department of Electrical Power Engineering, Mandalay Technological University, for thoroughly proof-reading these paper and giving useful remarks on it. Finally, the author wishes to express his special thanks to his parents for their supports and encouragement to attain his destination without any trouble throughout his life.

References

[1]   Vimal Singh Bisht, Y.R Sood, Nikhil Kushwaha, and Suryakant, Review On Electronic Load Controller,International Journal of Scientific Engineering and Technology, www.ijset.com, Volume No.1, Issue No.2 pg: 93-102, 2012

[2]   Ned Mohan, Tore M. Undeland and William P. Robbins, Power Electronics, Converters, Applications, Design, 2003

[3]   H.Ludens, Electronic Load Controllerfor micro-hydro system, 2010

[4]   J.Portegijs, The 'Humming Bird' Electronic Load Controller/ Induction Generator Controller, 2000, 6 December

[5]   D. Henderson, An Advanced Electronic Load Controller for Control of Micro Hydroelectric Generation, IEEE TRANSACTIONS ON ENERGY CONVERSION, vol.13, pp.300-304, 1998

[6]   Renerconsys, Digital load Controller for Synchronous Generator: Manual Instruction, 2010

[7]   J. M. Jacob, Power Electronics: Principles and Applications, Vikas Publishing House, 2002.

[8]   Dogan Ibrahim, Advanced PIC Microcontroller projects in C, 2008

[9]   Datasheet and Guide Book ofJ204ALiquid CrystalDisplay(LCD),www.alldatasheet.com/J204A

[10] DatasheetandGuideBook ofPIC 16F887Microcontroller, www.alldatasheet.com/PIC

[11] DatasheetandGuideBookofACS 712 ACcurrentsensor, www.alldatasheet.com/Acs712.

24

Dynamic Comparator with Using Negative Resistance and CMOS Input Pair Strategies in F_S=4MHz-10GHz

M. Dashtbayazi[1], M. Sabaghi[2, *], S. Marjani[1]

[1]Department of Electrical Engineering, Ferdowsi University of Mashhad, Mashhad, Iran
[2]Laser and Optics Research School, Nuclear Science and Technology Research Institute (NSTRI), Tehran, Iran

Email address:
msabaghi@aeoi.org.ir (M. Sabaghi)

Abstract: A 4MHz-10GHz, 10ps/dec dynamic comparator with using negative resistance and CMOS input differential pair is proposed and designed in IBM 130nm CMOS process technology. In this design, we effort that taking maximum sampling frequency from CMOS technology and the proposed comparator consumes 110nw-146μW at 1.5V supply.

Keywords: Comparator, Negative Resistance, Optical Communication Systems, Transconductance Boosting, Dual-Rail Differential Input

1. Introduction

Generally, comparator is an important part of electronic systems, like Multi-GHz sample rate Analog-to-Digital convertors (ADCs) [1-4] as crucial components in optical communication systems. Due to limited accuracy, comparison speed and power consumption, comparison of the input data is regarded as one of the limiting factor of the high speed VLSI data conversion system. There is a growing need for the development of low power, low voltage and high speed circuit techniques and system building blocks because of increasing demand for portable and hand held high speed devices. On the one hand, low voltage operation is required to use as few batteries as possible for low weight and small size. On the other hand at the high frequency of operation, Low power consumption is demanded to prolong battery lifetime as much as possible [5, 6]. In the recent decade, due to the rapid advancement in Analog-to-Digital conversion devices, the rapid advancement of VLSI technology causes the evolution of digital integrated circuit technologies for signal processing systems that operate on a wide variety of continuous-time signals including speech, medical imaging, sonar, radar, consumer electronics and telecommunications.

Already with voltage scaling, sub-threshold voltage region demands require technology for new architecture and innovative circuits. Previous advances in MOS technologies meet it for higher speed applications, however due to MOS device mismatches, it is difficult to achieve a high speed and accuracy at the same time [7].

One of the key components in ADC as the fundamental block for the A/D conversion, are the comparators. In fact they are the link between analog domain and digital domain for compare a set of variable or unknown values against that of a constant or known reference value [8]. Many high speed ADCs demands high-speed, low-power comparators with small chip area. For low-voltage operation, developing new circuit structures is preferable in order to avoid stacking too many transistors between the supply rails [9]. The critical parameters in the ADC circuit design are speed, resolution, and power consumption. The ADC operates continuously, however processing digital signals are dependent on specific demands.

The organization of this paper is as follows. Section 2 described the proposed dynamic comparator. Section 3 illustrates the simulation results. A comparison to the conventional comparators is also presented, followed by the conclusions in section 4.

2. Proposed Dynamic Comparator

2.1. Structure

The comparators are consists of two parts namely pre-amplifier and latch that these parts operate in the same phase. Therefore, comparator can pull the latch. In the second phase, the pre-amplifier output nodes up to the power supply voltage level. Also, the comparator offset can be cancelled in this phase [10-11]. Consequently, the comparator offset effect has not been considered in this paper. Fig. 1 shows the circuit diagram has been used in simulations that has been obtained from [12] with adding Q2 and Q4 that a dual rail pre-amplifier is achieved. For decreasing the loading effect on the comparator output, two CMOS inverters have been used to supply capacitive loads. The comparator consumes dynamic power because the comparator consists of dynamic latch and pre-amplifier [13].

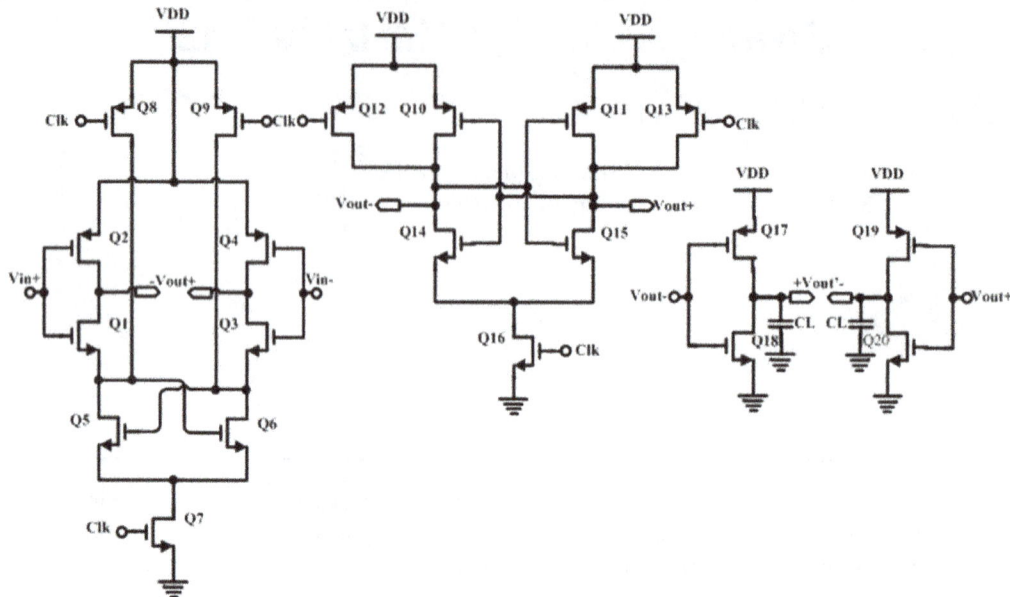

Fig. 1. Schematics of the dynamic comparator and CMOS inverters.

2.2. Transient Behavior

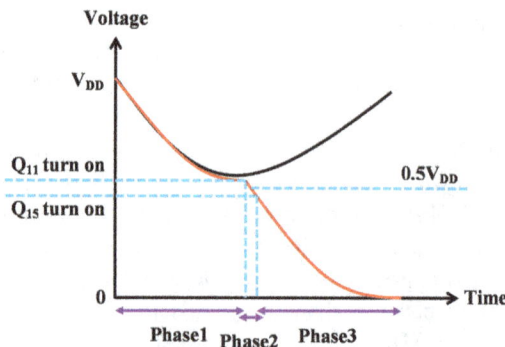

Fig. 2. The sketch map for the transition behavior of the dynamic comparator.

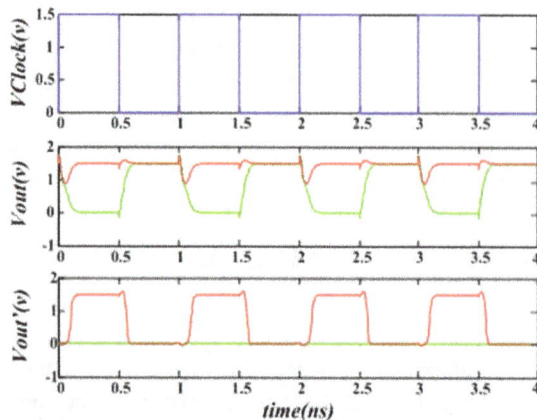

Fig. 3. Simulated transient behavior of the proposed comparator structure.

Fig. 2 shows the sketch map of the transient behavior of the dynamic comparator in the comparison phase. By the switching transistors Q8, Q9, Q12, and Q13 during the reset phase, the output terminals and the drain side voltages of Q5 and Q6 are all pulled high to VDD, respectively. By turning on the two tail transistors Q7 and Q16, the comparison phase begins. Therefore, the CMOS input stage transfers the differential small signal to the cross coupled stage. Q2, Q4 are activating when the input common mode is lower than the Q1 or Q3 threshold voltage. Therefore, the dual-rail differential input obtain since they are supplies the preamplifier output. Based on [14], the comparison transition can be divided into three phases (from phase 1 to phase 3). The output terminals are pulled down by two tail transistors Q7 and Q16 during the phase 1. Until one of the two output terminal voltages decreases to (V_{DD}-V_{thp}), the p-channel transistors Q10 and Q11 remain cut-off. The Q2 and Q4 activate when the cross-coupled stage composed from Q1, Q3, Q5 and Q6 cannot start. Therefore, the speed of proposed idea is more than paper [12]. In order to enhance the voltage difference between the output terminals, the cross-coupled inverters provide strong positive feedback in the phase 2.The transition state changes from phase 2 into phase 3 when one of the transistors Q14 and Q15 is cut-off. There is no static power dissipation since the current flows through these n-channel MOSFETs stop automatically after the transition. Fig. 3 shows transient behavior of the proposed dynamic comparator at 1GHz. As seen, both of the output terminals are pulled low in the beginning of the compare operation. One of the output terminals charge through the p-channel transistor when the transition goes from phase 1 into phase 2.

Therefore, the strong positive feedback provided by these

two cross-coupled inverters separates the output voltages. Against paper [12], this design aren't use from p-wells, thus the cost of comparator chip is lower.

3. Simulation Results

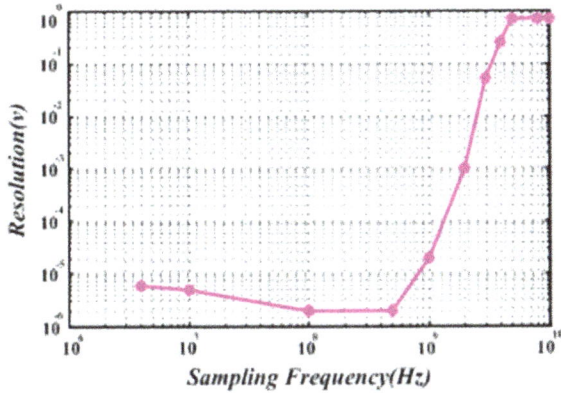

Fig. 4. *Simulated Resolution voltages versus Sampling Frequency at 1.5V supply voltage.*

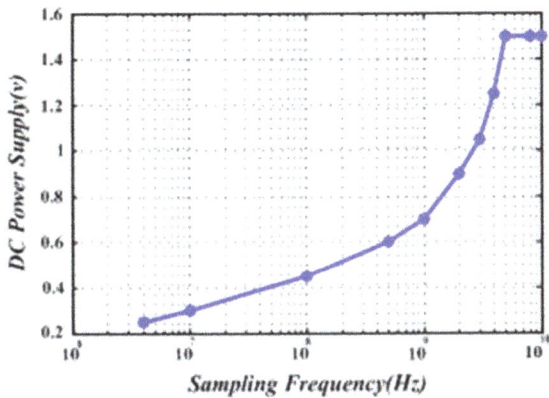

Fig. 5. *Simulated DC Power Supply versus Sampling Frequency.*

The comparator is designed in IBM 130nm CMOS process with 1.5V low threshold voltage device model (Vt≈0.3V). The resolution voltage (difference between input signals) versus sampling frequency is shown in Fig. 4. According to this Figure, resolution voltage is changing in range of 4MHz to 10GHz from 2µv to 700mv. The dc power supply versus sampling frequency is shown in Fig. 5. According to this figure, the minimum supply voltage is 250mv and maximum supply voltage is 1.5V. The power consumption versus sampling frequency is shown in Fig. 6. According to this figure, the minimum power consumption is 110nw at 4-MHz and maximum power consumption is 146µw at 10-GHz. According to simulation results, the delay versus CMOS differential input voltage (resolution) is shown in Fig. 7. The delay/log(ΔVin) in this figure is 10-ps/dec. Fig. 8 and Fig. 9 summarized the simulation results of delay and energy/decision versus input common-mode voltage, respectively. The delay of the proposed structure is also insensitive to input common-mode voltage (Vcm) that is the same as [15]. According to figure 8, the input common voltage is dual-rail because the range of common mode is between 0.1 to 1.1V. Therefore the proposed paper than papers [12] is very

better. The energy dissipation is 2.5fJ/decision at Clk=1GHz and ΔVin=1.5V. That is very better than 71, 88 and 90fJ/decision in [12], [14]-[15], respectively. Table I summarizes the specifications of proposed dynamic comparator. We demonstrate that our comparator structure has the best delay/log (ΔVin) and energy/decision performance in comparison with recent publications [12], [14]-[16].

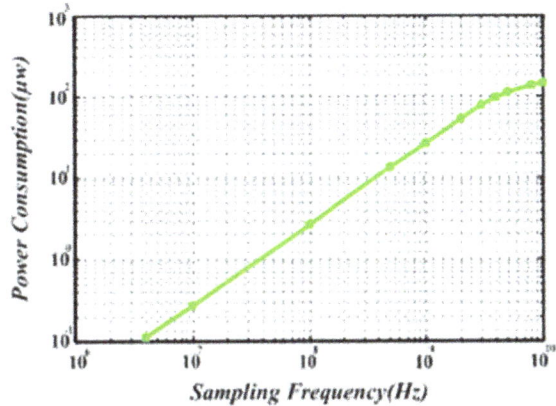

Fig. 6. *Simulated Power Consumption versus Sampling Frequency at 1.5V supply voltage.*

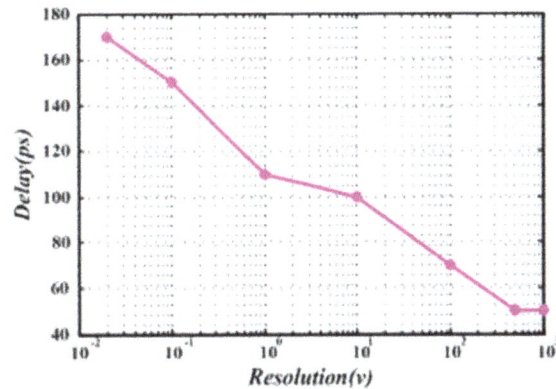

Fig. 7. *Simulated delay of different dynamic comparators versus differential input voltage. The delay is the time between clock edge and differential output signal of comparator crosses at 1/2 supply voltage.*

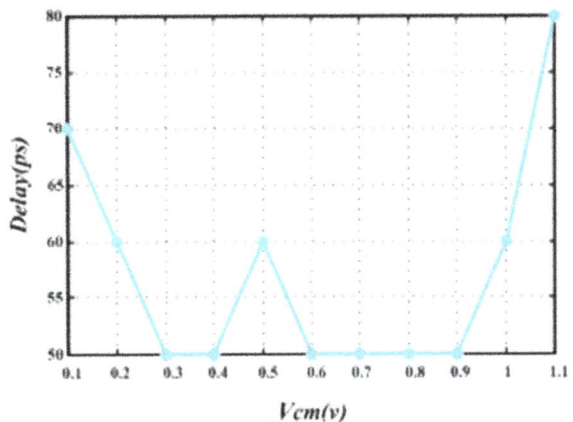

Fig. 8. *Simulated delay versus input common mode voltage at 1-GHz and 1.5V supply voltage.*

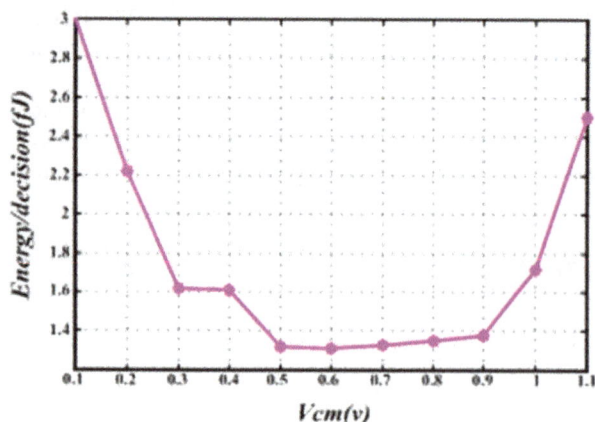

Fig. 9. *Simulated Energy/decision versus input common mode voltage at 1-GHz and 1.5V supply voltage.*

Table 1. *Performance summary and comparison This Work.*

Specifications	[12]	[14]	[15]	[16]	This Work
CMOS Process(nm)	90	Scaled to 90	90	65	130
Sampling Rate(GHz)	3	3	3	7	1
Delay/log(Resolution)	22	47	35	--	10
Supply/ICMV	1.2/1	1.2/1	1.2/1	1.2/1	1.5/1
Energy/decision(fJ)	71	88	90	185	2.5

4. Conclusions

The proposed dynamic comparator structure adds the CMOS input pair to negative resistance that causes the higher speed and range of V_{CM} of dynamic comparator at low power consumption. For receiving to multi-giga Hz in advanced deep sub-micron CMOS process in this design, the fingers of transistors increase that they achieve to minimum parasitic capacitance. Therefore this design received to highest sampling frequency in CMOS technology. In this paper, we received to delay/log (ΔVin) and Energy/decision equal 10 and 2.5fJ at 1GHz Respectively. Also in this paper, we received to sampling Frequency 10GHz at Resolution equal 1v.

References

[1] H. Traf, G. Holmbert and S. Eriksson "Application of switched current technique to algorithmic DA- and AD-converters", *IEEE Intl. Symp. Circuits and Systems ISCAS*, pp.1549-1552, June 1991.

[2] D.G. Nairn and C.A.T. Salama, "Current-mode algorithmic analog-to-digital converters", *IEEE Journal of Solid-State Circuits*, vol. 25, pp.997 – 1004, Aug. 2002.

[3] M. Kijima, K. Ito, K. Kamei and S. Tsukamoto, "A 6b 3GS/s Flash ADC with Background Calibration", *IEEE Custom Integrated Circuits Conference Digest of Technical Papers*, pp.283–286, Sept. 2009.

[4] Y. Nakajima, A. Sakaguchi, T. Ohkido, T. Matsumoto and M. Yotsuyanagi, "A Self-Background Calibrated 6b 2.7GS/s ADC with Cascade-Calibrated Folding-Interpolating Architecture", *IEEE Symposium on VLSI Circuits Digest of Technical Papers*, pp.266–267, June 2009.

[5] V. Peluso, P. Vancorenland, A.M. Marques, M.S.J. Steyaert and W. Sansen, "A 900-mV Low-Power ΣΔ A/D Converter with 77-dB Dynamic Range", *IEEE Journal of Solid-State Circuits*, vol. 33, pp. 1887 – 1897, Aug. 2002.

[6] H. Roh, Y. Choi and J. Roh, "A 89-dB DR 457-W 20-kHz Bandwidth Delta-Sigma Modulator with Gain-Boosting OTAs" *Analog Integrated Circuits and Signals Processing*, vol. 64, pp. 173 – 182, 2010.

[7] G.M. Yin, F. Op't Eynde and W. Sansen, "A High-speed CMOS Comparator with 8-b Resolution", *IEEE Journal of Solid-State Circuits*, vol. 27, pp. 208 – 211, Aug. 2002.

[8] L. Yu, J.Y. Zhang, L. Wang and J.G. Lu, "A 12-bit Fully Differential SAR ADC with Dynamic Latch Comparator for Portable Physiological Monitoring Applications", *4th international Conference on Biomedical Engineering and Informatics*, vol. 1, pp. 576 – 579, Oct. 2011.

[9] K. Balasubramanian, "A flash ADC with reduced complexity", *IEEE Transactions on Industrial Electronics*, vol. 42, pp. 106 – 108, Aug. 2002.

[10] C.H. Chan, *et al*, "A voltage-controlled capacitance offset calibration technique for high resolution dynamic comparator", *International SoC Des. Conf*, pp. 392-395, Nov. 2009.

[11] Y. Xu, *et al*, "Offset-Corrected 5GHz CMOS dynamic comparator using bulk voltage trimming: Design and analysis", *IEEE New Circuits Syst. Conf*, pp. 277-280, June 2011.

[12] B.W. Chen, *et al*, "A 3-GHz, 22-ps/dec Dynamic Comparator using Negative Resistance Combined with Input Pair", *IEEE Asia-Pacific Conf. Circuits Syst.*, pp. 648-51, Dec. 2010.

[13] C.C. Liu, *et al*, "A 10-bit 50-MS/s SAR ADC With a Monotonic Capacitor Switching Procedure", *IEEE J. Solid-State Circuits*, vol. 45, pp. 731-740, March 2010.

[14] B. Wicht, T. Nirschl and D. Schmitt-Landsiedel, "Yield and Speed Optimization of a Latch-Type Voltage Sense Amplifier", *IEEE J. Solid-State Circuits*, vol. 39, pp. 1148–1158, June 2004.

[15] D. Schinkel, E. Mensink, E. Klumperink, E. van Tuijl and B. Nauta, "A Double-Tail Latch-Type Voltage Sense Amplifier with 18ps Setup+Hold Time", *IEEE ISSCC Digest of Technical Papers*, pp. 314-315, Feb. 2007.

[16] B. Goll and H. Zimmermann, "A 65nm CMOS Comparator with Modified Latch to Achieve 7GHz/1.3mW at 1.2V and 700MHz/47µW at 0.6V", *IEEE ISSCC Digest of Technical Papers*, pp.328–329, Feb. 2009.

25

Designing Efficient Logistic System of Fresh Agricultural Products for Small Farms

Parung J.[1], Santoso A.[1], Prayogo D. N.[1], Angelina M.[1], Tayibnapis A. Z.[2], Djoemadi F. R.[2]

[1]Industrial Engineering Department, University of Surabaya, Surabaya, Indonesia
[2]Economic Department, University of Surabaya, Surabaya, Indonesia

Email address:
jparung@staff.ubaya.ac.id (Parung J.), dnprayogo@staff.ubaya.ac.id (Prayogo D. N.)

Abstract: This research paper presents a framework of an efficient logistics system of fresh agricultural products for small farms. This paper illustrates efficient logistic system with a case study of apple products in East Java – Indonesia, in order to increase competitiveness against imported apples. The research focuses are small farms, farmer groups, middlemen, and distributors. Data are obtained from observations and interviews with the farmer's groups which are representing small farms and distributors which are treated using process activity mapping (PAM), value stream mapping (VSM), and activity-based costing (ABC). Data analysis has identified value added and non-value added activities. Based on the analysis, this paper proposes an alternative efficient logistics system. However, this research has not been fully integrated due to insufficient data for designing holistic efficient logistics system for apples product by taking into account country and regional conditions.

Keywords: Logistic System, Small Farms, Value Added and Non-value Added Activities

1. Introduction

Logistics is part of a supply chain associated with the movement and storage of goods and at the same time with regard to the flow of money and information. Based on this understanding, it is known that the logistics is related to the ease of supply of goods in the region. It also indicates that, logistics plays a key role in the growth of the industry and the economy of a region. Logistic is becoming a key role because logistics activity is a major cost component for businesses and intertwined with many other economic activities.

The success of a holistic logistics management becomes an important requirement which directly or indirectly have an impact on the economic growth of the region. Indicators of success in logistics management can be viewed from the aspect of availability, flexibility and cost efficiency. The indicators related to the availability of goods and services at the right time and place while having the flexibility of the amount and delivery time in an area with a logical price.

But these three indicators of the success of the logistics often becomes irrelevant for fresh agricultural products due to the following factors: a). Availability of fresh agricultural products are influenced by the short time-gap between the time the of supply to the time of consumption; b). Total supply is inconsistent due to the influence of the season; c). Prices fluctuate according to season and d). The local government policy towards supply of fresh agricultural products varies among country and region.

Fresh agricultural products are products that should be consumed immediately to prevent damage without further preservation process. Therefore, the products must be delivered as soon as possible at a reasonable cost using the right logistics system. But before designing efficient logistics system; the farming activities, distribution, and transportation should be taken into account.

East Java province is one of the largest fresh agricultural producers in the country. The fruit products come from different plantation locations, such as Batu, Pasuruan, Probolinggo, and Blitar; while vegetables come from plantation locations in Batu, Lumajang, and Bondowoso.

Most of the consumers of agricultural products that were produced in East Java are living in the entire province of East Java, Jakarta and even reach out to almost all provinces in Kalimantan and Sulawesi. However, this fact remains controversial for producers. On the one hand, agricultural products from East Java province reach the wider consumer, but on the other hand, the initial benefits received by producers, especially yeoman and group of farmers not significantly increased. If conditions are unfavorably untreated, it can decrease the number of producers. A decrease in the number of producers will have an impact on the availability of products that are not continuous; the price is not stable so it will give a negative effect on efforts to improve the region's economy. That means handling of fresh agricultural products need special attention.

2. Literature Review

The main issue addressed in this study is designing a framework of the efficient logistics system of fresh agricultural product that is able to improve the welfare of society and the economy of the region.

2.1. The Concept of Logistics and Logistics Management System

According to Hutchinson [1], logistics is a process of getting things right (the right item), in the right amount (in the right quantity), at the right time, at the right place for the right price. Logistics management terminology according to the Council of Supply Chain Management Professionals (CSCMP), is part of supply chain management which is to plan, implement and control the level of efficiency and effectiveness of the flow and storage of goods/services, money and information from upstream to downstream and vice versa from the point the origin of the goods up to the point where the goods are used or consumed in order to meet customer demand [2] [3].

2.2. Logistics Cost

A key component of a logistics system is a logistics cost. Logistics costs are a logical consequence of the use of resources (resources) on each of logistics activity [4]. These logistics costs occur in each chain logistics system, thus forming a total logistics cost charged to the product being sold. Logistics costs have become one of the important components in the cost of the product and the selling price of products. Should logistics cost is naturally high, then the price of agricultural products is high [5]. One way to control logistics cost is to keep supply chain activities controllable. Logistics costs are important to note because, according to the World Bank survey, the logistics cost in Indonesia is one of the highest in the world. According to the survey by the World Bank, Logistics Performance Index (LPI) of Indonesia is less competitive among neighbors. As an example; logistics costs along the 55 kilometer (km) in Indonesia is about US $ 550, while in Malaysia is only US $ 300 [6]. There are six indicators measured in the Logistics Performance Index (LPI). The six indicators are:

a. The efficiency of the process of clearance customs and excise office services.
b. The quality of transport infrastructure and trade.
c. Ease in setting the price shipment

d. Competence and quality of logistics services by logistics service providers (3PL)
e. The ability to perform track and trace shipments.
f. Timeliness of distribution.

2.3. Fresh Agricultural Products

The price of fresh agricultural products is depending on logistics cost and their freshness upon arriving at the point of sales location. According to Pingxia et al [5], most of the research on logistics cost of fresh agricultural products, concentrated in the simple economical analysis. Therefore, research on how to maintain freshness of the product is becoming an important area in an agricultural supply chain network. Supply Chain of fresh agricultural products is different from other product supply chains. The important difference between agricultural supply chains and other supply chains is the significant change in the value of products throughout the entire supply chain until the points of final consumption [7]. Longer channel of supply chain would increase transportation distance of product due to the geographic separation of each channel, starting from a farm to the point of consumptions.

According to statistic analysis conducted by Roeger and Liebtag [8], transportation costs are a significant component of the final prices for fresh products. Therefore, one of the problems to be solved in an agricultural product is how should members of agricultural supply chain take into account the transportation cost and time, to appropriately make the most decisions. Study of Ferro et al [9] found that 3 features that characterize the fresh fruit supply chain are long supply lead times, uncertainty of supply and demand, and small margins. These features cause a need for efficiency in logistics management.

3. Data Collection and Methodology

This research is using constructive research paradigm. To construct a framework of an efficient logistics system, this research uses holistic approach by combining quantitative and qualitative strategy. This research mainly focused on quantitative approaches to streamline logistics costs. However, a qualitative approach is also used to increase costing comprehension.

For a case study of fresh apples, data collection is conducted through questionnaires, direct observation and interviews with the origin farmers, group of farmers, middlemen and distributors. Data of demand and logistics cost is obtained from the origin of apple farmers, middlemen and retailers. Data of transportation cost is obtained from 3PL, middlemen and distributor. Data related to the logistic activities including: operation process charts with time consumed of each process, human resources activities, and number of demand are collected from direct observation combined with interviews to the relevant functional person. The data then processed using the method of process activity mapping, value stream mapping, and activity based costing.

4. Designing Framework of Efficient Fresh Agricultural Products

A framework of efficient logistics system for fresh agricultural product starting from identification supply chain network, until making out logistics activities of each stage of supply chain. After identifying process, the next step is collecting relevant data.

A related data are human resources activities, driver cost, and processing time of each manual activity. Data then processed using process activity mapping (PAM), value stream mapping (VSM), and activity based costing (ABC) methods. Process activity mapping is used to determine the allocation of resources and time required of each activity. Value stream mapping is used for mapping processing time and lead time of activities based on visual observation related to the current conditions. Value stream mapping is also used to distinguish value added and non value added activity. Non-value added activities consist of necessary and unnecessary activities. Unnecessary activities are categorized as wastes for the logistics activities.

In order to prioritize types of waste to be handled, Pareto chart will be used. Activity-based costing is used to identify cost drivers of each activity and categorized them into SC drivers. Unnecessary non value added is using time as a unit then convert into financial unit (cost). Ishikawa diagram is used as a method for finding cause and effect relationships of each significant and important waste. While processing time is determined by people who work on, then it is possible to merge some activities with the aim of reduce processing time. A proposed framework of efficient logistics system is shown in Figure 1.

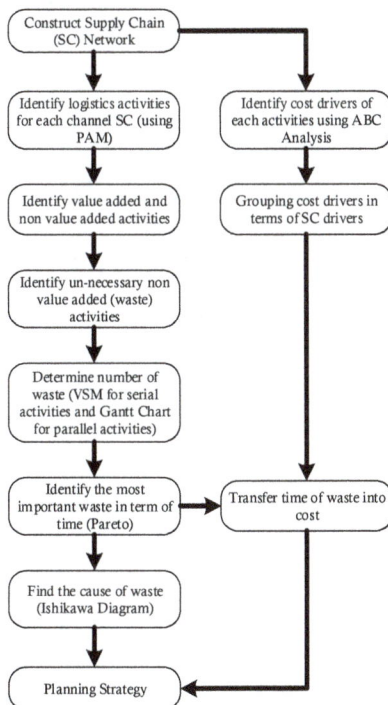

Figure 1. Framework of efficient logistic system of fresh agricultural products.

5. Case Study

5.1. Apple's Supply Chain in East Java-Indonesia

Supply Chain of fresh apples produce by small farm (yeoman) and group of farmer in East Java-Indonesia is shown in Figure 2.

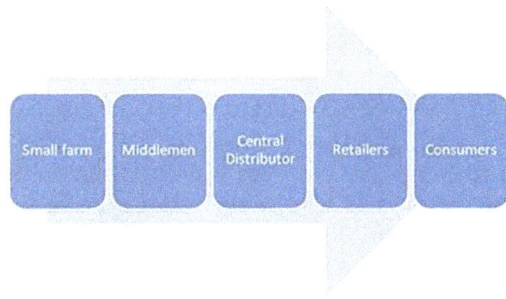

Figure 2. Apple Supply Chain in east Java-Indonesia.

For a case study of fresh apples, data collection is conducted through direct observation and interviews with the group of farmers and distributors. Data related to the activities undertaken are human resources activities, costs, and processing time. Data then processed using the method of process activity mapping (PAM) [8], value stream mapping (VSM) [9], and activity based costing (ABC) [10]. At the last activity-based costing is used to calculate the total cost per kg of apples for each different grade. Data processing results are then analyzed to do improvement of non-value added activity

5.2. Discussion

Supply chain in distributing apples consists of 5 entities, namely farmers, middlemen, the central distributor, retailer, and consumer. Farmers plant and nurture the apples to be harvested. Harvested apples will be sold to middlemen. Central distributor collects and stores apples from middlemen then distributing apples to various retailers who will sell directly to consumers.

Data which are collected from farmers and middlemen is processed using process mapping activity in order to know the percentage of time for value added activity and non-value added activity. The following Table. 1 and Table. 2 presenting the allocation of time for value added activity (VAA) and non-value added activity (NVAA) based on logistics processes in the group of farmers namely KTMA and central distributor namely UD. Buah Segar.

Table 1. Allocation of process activity mapping at KTMA.

Process	No. Activities	Time (mins)	%	VAA (mins)	% VAA	NVAA (mins)	%
Operation	9	13,655	91.0%	13,655	91%	0	0.0%
Transportation	1	30	0.2%	0	0%	30	0.2%
Inspection	0	-	0.0%	0	0%	0	0.0%
Storage	1	1,320	8.8%	0	0%	1.320	8.8%
TOTAL	11	15,005	100.0%	13,655	91%	1.350	9.0%

Table 2. Allocation of process activity mapping at UD. Buah Segar.

Process	No. Activities	Time (mins)	%	VAA (mins)	% VAA	NVAA (mins)	%
Operation	3	615	23.56%	615	23.56%	-	0.00%
Transportation	1	75	2.87%	0	0.00%	75	2.87%
Inspection	1	180	6.90%	0	0.00%	180	6.90%
Storage	1	1,740	66.67%	0	0.00%	1,740	66.67%
TOTAL	6	2,610	100%	615	23.56%	1,995	76.44%

Outputs of process activity mapping were processed using value stream mapping to obtain lead time, value added and non-value added activity. Through process activity mapping it is known that required lead time for KTMA is 250 hours with a value added activity = 225 hours. Hence, the required lead time for UD. Buah Segar = 43.5 hours with value added activity 10.25 hours with remains time are non-value added activities.

The total cost per kg of apples at KTMA was calculated using conventional methods; this is due to the same activities performed for each grade apples. Here is the calculation of the total cost per kg of apples at KTMA: The total cost per kg of apples in KTMA = total cost of the activity KTMA/output = IDR 62,373,650/8,000 kg = IDR 7,797 / kg

Based on direct observation, the selling price of apples in the supermarket is IDR 35,990 / kg, thus the total cost per kg of apples from KTMA constitute 21.66% of the total selling price in the supermarket. The total cost per kg of apples at UD. Buah segar for grade AA and AB grade is calculated using the conventional method for the same activity and activity based costing methods for different activities.

Through observation using value stream mapping, the biggest cost to the farmer or middleman is transportation costs. Another problem at the farmer level is long picking time and high cost for picking apples. Other issue at the middlemen is a long time to collect apples.

Identification of in-efficiencies at the level KTMA find in-efficiencies in the activity of cleaning, planting, harvesting, sorting, and storing apples. While the in-efficiency at the distributor level, mostly in apple storage. Furthermore, to improve efficiency of KTMA, some of the activities should be merged; namely merging cleaning activity with planting, harvesting with sorting, and eliminating the storage of apples. This proposal will generate lead time = 225 hours with value added activity = 225 hours. Besides a shorter time, these improvements will lower the total cost per kg of apples at KTMA. At the distributor level; is proposed to immediately send apples that had finished packaged on the same day. With this improvement proposal, the storage time decreases from the original 1,740 minutes to 840 minutes.

6. Conclusion

Efficient logistics systems for fresh agricultural product is important for supply chain involving small farms in its supply chain channel, because of the use of labor instead of mechanization. Consequently, logistics process time for small farm is influenced by weather conditions and facilities out of reach of small farms.

Integration of logistic activities among different channel of a supply chain for a fresh agricultural product has become a critical problem in increasing competitiveness due to the deterioration, freshness, and availability constraints. Efficiency discussed in this case is the efficiency in terms of time and cost simultaneously.

Application of the framework has merged some manual activities and decrease lead-time that can be converted into cost.

In order to increase competitiveness of local apples compared to the import products, the efficiency and at the same time margin of small farmers should be increased by government support. Types of support possibly by increasing the road construction investment for reducing the transportation time, Secondly, increasing the freshness of apples by supporting farmers with specific warehouse, and transportation vehicles and sharing results of agricultural research.

References

[1] Hutchinson, N.E., "An Integrated Approach to Logistics Management". 1987, Prentice-Hall.

[2] Simchi-Levi, David, Kaminsky, P., and Simchi-levi, E., "Designing and Managing The Supply Chain: Concepts, Strategy, and Case Studies", 2007, 3rd Edition, McGraw-Hill.

[3] Chopra, S. and Meindl, P. "Supply Chain Management Strategy, Planning, and Operation. 5th edition", 2013, Pearson Education Limited, England.

[4] Christopher, M., "Logistics and Supply Chain Management", 2011, 4th edition. Pearson.

[5] Pingxia Shang, Yuanze Xu; Jingmei Xue, "The Analysis based on the Logistics Cost Controling of the Fresh Agricultural Products", 2014, International Conference on Logistics Engineering, Management and Computer Science.

[6] The World Bank, http://lpi.worldbank.org/about access 15th October 2015.

[7] Van der Vorst J.G.A.J; Beulens, A.J.M and Van Beek, P "Modeling and simulating multi-echelon food systems", 2000, European Journal of Operational Research, Volume 122, Issue 2.

[8] Roeger E, Leibtag E. "Produce sourcing and transportation cost effects on wholesale fresh fruit and vegetable prices" In: Poster presented at the agricultural & applied economics association 2011 AAEA, CAES, & WAEA joint annual meeting, Pittsburgh.

[9] Esteban Ferro, Tsunchiro Otsuki, John S. Wilson, "The effect of product standards on agricultural exports", 2015, Food Policy, Vol.50.

[10] Pude, G. C., Naik, G. R., Naik, P. G., "Application of Process Activity Mapping for Waste Reduction a Case Study in Foundry Industry", 2012, International Journal of Modern Engineering Research.

[11] Tapping, D., Luyster, T., Shuker, T."Value Stream Management", 2002, Productivity Press, New York.

[12] Douglas T. Hicks, "Activity Based Costing: Make it Work for small and Mid-Size", 2002, 2nd edition, John Wiley and Sons, New York.

26

A New Method for Ground Moving Targets Tracking Using Radar Based on Compressed Sensing

Wang Xue-Jun

School of Electronic Information Engineering, Beihang University, Beijing, China

Email address:

sarahwangxj@sina.com

Abstract: In this paper, we propose a Compressed Sensing (CS) based method under the unknown sparse degree to track ground moving targets using Pulse-Doppler (PD) radar. We use the sparsity of delay-Doppler plane in the process of disposing PD radar echo to set up a sparse signal model in each pulse interval. At the state prediction stage, we can get the predicted values of target states by dynamic equations, with which we can build a delay-Doppler grid that is used to form orthogonal dictionary. At the state update stage, we can get the target state estimation through reconstruction algorithm, so as to realize precise tracking of targets. The problem of target tracking by PD radar will be transformed into the reconstruction of the sparse signal, which is accomplished by getting the location of targets in the grid, as a result of achieving ground target tracking based on Orthogonal Matching Pursuit (OMP) [1]. Then, aiming at the sparsity problem in the method of target tracking based on Orthogonal Matching Pursuit, we propose a new target tracking method based on Sparsity Adaptive Matching Pursuit (SAMP) algorithm [2]. Numerical simulations show that our tracking method can not only provide the equivalent computational time, but also get better tracking performance than the KF-based tracking.

Keywords: Compressed Sensing (CS), Pulse-Doppler (PD) Radar, Target Tracking, Orthogonal Matching Pursuit (OMP), Sparsity Adaptive Matching Pursuit (SAMP) Algorithm

1. Introduction

Target tracking by radar is the process that people use all kinds of observation and computing methods to realize the procedure of target states modeling and tracking using radar. This technology is widely used in the military and civil fields, such as airborne early warning, multi-target attack, ballistic missile defense, Marine monitoring, battlefield surveillance in military field and traffic control system, intelligent traffic control system in civil field.

In target tracking of radar, the environment generally includes air targets, ground targets and sea targets. Among them, the air and sea targets are relatively easy to track due to the little effect of environment and their research is also gradually mature from the perspective of international literature. However, the ground target tracking has some certain practical difficulties, such as the limitations of terrain shade, minimum detectable rate, the seriousness of ground clutter, the restriction of road network, the density and flexible of targets, etc.

CS is provided to alleviate the pressure of huge amount of information demand caused by the signal sampling, transmission and storage [3]. Compared with the traditional Nyquist sampling theorem, the difference of CS is that we can use a observation matrix reversing the basic shift matrix to make the transformation of high-dimensional signals cast onto a low-dimensional space, and reconstruct the original signal with a high probability by solving a nonlinear problem [4]. AS the CS theory can reduce the cost of sampling and computing, it has important influences and practical significance on many disciplines and fields. In recent years, some experts and scholars applied the CS theory to the radar signal processing and obtained many important achievements [5] [6] [7] [8].

In this paper, we prove that the targets are sparse in delay-Doppler plane through the establishment of delay-Doppler grid and achieve accurate reconstruction of signals after realizing the sparse representation of echo signal by setting up orthogonal dictionary. In addition, we use the particularity of the grid to search for location of the grid corresponding to the largest coefficient while rebuild signals. Finally, we get the accurate approximation of target measurements.

We mainly discuss the ground target tracking method based on CS using PD radar, analyze the features of receiving echo [9] and apply the CS theory to the tracking procedure [10], which can reduce the computational time and improve the tracking accuracy. Therefore, the research of this dissertation has very vital significance.

2. Tracking Model

2.1. Target Dynamic Model

Suppose that there are m targets moving in a straight line in our interesting area and each target's motion state can be showed by its position and velocity, so we can get the mth target's dynamics at the kth pulse interval as

$$X_K^m = \left[x_k^m, y_k^m, z_k^m, \dot{x}_k^m, \dot{y}_k^m, \dot{z}_k^m \right]^T$$

However, the PD radar track the targets by the delay and Doppler instead of position and velocity. In the kth pulse interval, the delay and Doppler of mth target could be described as: $\tau_k^m = \frac{2*R^m}{c}$ $f_k^m = \frac{2*V_k^m}{\lambda}$

where R^m denotes the distance between the mth target and radar and V^m denotes the radial velocity of mth target. If the initial position of radar is (x_k^0, y_k^0, z_k^0), then

$$R^m = \sqrt{(x_k^m - x_k^0)^2 + (y_k^m - y_k^0)^2 + (z_k^m - z_k^0)^2}$$

$$V_k^m = \frac{x_k^m \dot{x}_k^m + y_k^m \dot{y}_k^m + z_k^m \dot{z}_k^m}{R_k^m}$$

Hence, we can form a state dynamic equation in kth pulse interval, as

$$\bar{X}_k^m = F\bar{X}_{k-1}^m + w_k^m$$
$$\bar{X}_k^m = [\tau_k^m, f_{dk}^m]$$

$$F = \begin{bmatrix} 1 & 0 & T & 0 \\ 0 & 1 & 0 & T \\ 0 & 0 & 1 & 0 \\ 0 & 0 & 0 & 1 \end{bmatrix}$$

F denotes state transform matrix, w_k^m denotes the state model error, and T denotes the state sampling interval, which is equal to the pulse repetition interval (PRI).

2.2. PD Measurement Model

We use the linear frequency modulation (LFM) pulse signal as the transmitting signal, and suppose its pulse width as T_p, carrier frequency as f_c, bandwidth as B, so the emission signal can be represented as

$$s(t) = Arect(\tfrac{t}{Tp})e^{j\pi Kt^2 + j2\pi f_c t} \tag{1}$$

Where K=B/T$_p$ denotes the modulation slope. Suppose there are m targets moving in our interesting area, then the echo signal in the kth pulse interval can be expressed as:

$$S(t) = \sum_{m=1}^{M} A_m rect(\tfrac{t-\tau_m}{Tp})e^{j\pi K(t-\tau_m)^2 + j2\pi f_c(t-\tau_m) + j2\pi f_{dm}(t-\tau_m)} + w(t) \tag{2}$$

We can get the sampling signal after sampling of S(t) with sampling interval ts=1/fs=1/2B, as

$$s(nt_s) = \sum_{m=1}^{M} A_m rect(\tfrac{nt_s-\tau_m}{Tp})e^{j\pi K(nt_s-\tau_m)^2 + j2\pi f_c(nt_s-\tau_m) + j2\pi f_d(nt_s-\tau_m)} + w(nt_s) \tag{3}$$

Where n=1, 2, 3...N denotes the number of sampling, and let Am=1. If

$$\phi(n,\tau_k^m, f_{dk}^m) = rect(\tfrac{nt_s-\tau_k^m}{Tp})e^{j\pi K(nt_s-\tau_k^m)^2 + j2\pi f_c(nt_s-\tau_k^m) + j2\pi f_{dk}^m(nt_s-\tau_k^m)} + w(nt_s) \tag{4}$$

Then

$$s(nt_s) = \sum_{m=1}^{M} \phi(n,\tau_k^m, f_{dk}^m) + w(nt_s) \tag{5}$$

Where $w(nt_s)$ denotes Gaussian noise and obeys the normal distribution of N(0,Q),where Q is noise covariance.

$$\phi(n,\tau_k^m, f_{dk}^m) = \left[\phi(t_1,\tau_k^m, f_{dk}^m), \phi(t_2,\tau_k^m, f_{dk}^m)...\phi(t_{N-1},\tau_k^m, f_{dk}^m) \right]_{N*1}^T$$

$$w(nt_s) = [w(t_0), w(t_1)...w(t_{N-1})]^T$$

2.3. Sparse Measurement Model

In order to apply the sparsity of delay-Doppler to target tracking procedure, we study an equivalent sparse representation method of echo signal model in PD radar. In the kth pulse interval, we discrete the delay-Doppler plane into N1×N2 grids, and if the number of targets M<<N1×N2, then the targets are sparse in delay-Doppler plane [11]. After converting the measurement model of echo signal into the corresponding sparse measurement model, the position of nonzero elements in the sparse vector correspond to the position of targets in grid. The size of resolution for delay and Doppler in grid is determined by the signal bandwidth and pulse duration, as 1/2B and 1/T [12]. The size of resolution limits the size of grid, so the grid partition can't be infinite. Under the same target scene and tracking conditions, we generally use the same grid resolution [13]. As is show in Figure 1, It's the grid of N1×N2.

The value that is corresponding to the grid point is show in Figure 2.

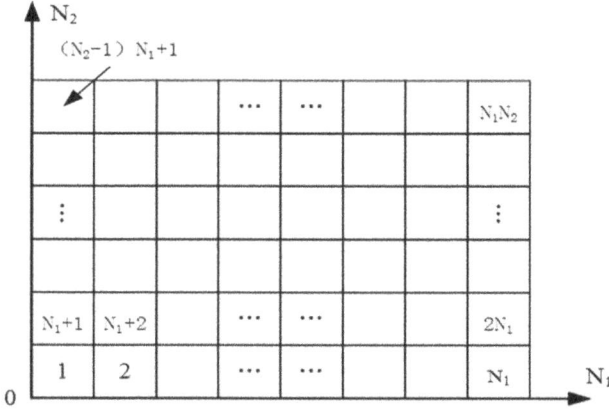

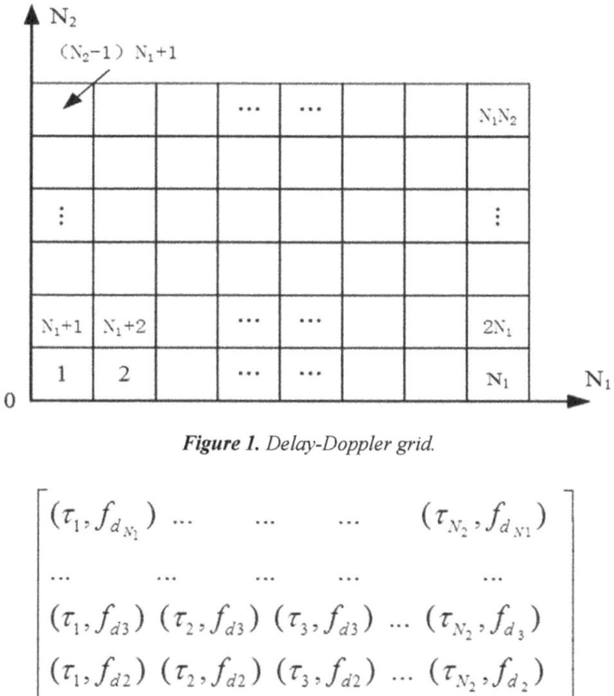

Figure 1. Delay-Doppler grid.

Figure 2. Values grid points.

Then we can get the sparse expression of echo signal as:

$$S(nt_s) = \sum_{i=1}^{N1} \sum_{j=1}^{N2} \Psi(\tau_k^i, f_d^j)\varphi + w(nt_s) = \Psi_{N \times Nc}\varphi_{Nc \times 1} + w \quad (6)$$

Of which $N_c = N_1 \times N_2$ denotes the sum of grids, $\Psi_{N \times Nc}$ denotes the sparse dictionary and $\varphi_{Nc \times 1}$ denotes sparse coefficient matrix.

$$\varphi = [\varphi_{11}, \varphi_{12}, \dots \varphi_{1N}, \varphi_{21}, \dots \varphi_{2N}, \dots \varphi_{ij}, \dots \varphi_{N1N2}]^T_{Nc \times 1}$$

$$\Psi(\tau_k^i, f_d^j) = rect(\tfrac{nt_s - \tau_k^i}{Tp})e^{j\pi K(nt_s - \tau_k^i)^2 + j2\pi f_c(nt_s - \tau_k^i) + f_{dk}^i(nt_s - \tau_k^i)}$$

$$= \left[\phi(t_0, \tau_k^i, f_{dk}^j), \ \phi(t_1, \tau_k^i, f_{dk}^j) \dots \phi(t_{N-1}, \tau_k^i, f_{dk}^j)\right]^T_{N*1}$$

2.4. Observation of Sparse Signal

According to the principle of CS, sparse signal must remain the geometric properties under the influence of observation matrix, so if we want to get a perfect reconstruction, we must make sure that the observation matrix won't let the two different k-sparse signal map into the same sample collection, and this requests the matrix consisted of m column vectors from observation matrix is singular. Some relevant scholars proved that the observation of signals won't destroy the geometry nature of signals when the observation matrix and sparse matrix are irrelevant [14]. The random Gaussian matrix has a special nature: For a M×N random Gaussian matrix, when meets M>=CKlogN/K (K denotes the sparsity of signal), it can guarantee high precision to restore the original signal [15]. So we choose the random Gaussian matrix as the observation matrix and we can get the observation vector containing original signal information after projecting the sparse signal to measurement matrix.

3. Tacking Filter

Typical target tracking procedure is generally consisted of two stages, namely the predict stage and update stage. In the prediction stage, The predicted state at kth pulse interval can be obtained by the target's estimated state at k-1th pulse interval as $\bar{X}_k^m = F\bar{X}_{k-1}^m + w_k^m$.

Then, in the update stage, we can get the target's estimation state at the kth pulse interval using new measurement values to modify the predicted state. And the OMP algorithm can be used to reconstruct the signal in the update stage so that we can get the best sparse approximation of objective measurement values.

3.1. OMP-Based Ground Target Tracking Using PD Radar

OMP algorithm belongs to greed iterative algorithm avoiding the optimization problem of solving minimum L0 norm. It uses the prior information of signal sparsity to choose the most matching column vector from measurement matrix to construct the sparse approximation of original signal. In the procedure of target tracking by PD radar, we firstly make the sparsity decomposition of receipt signal and then we will get the sparse vector containing target's measurement information. Here is the process of OMP-based ground target tracking using PD radar.

1) Calculate the target's prediction state in the kth pulse interval by the target dynamic equation and set up the delay-Doppler grids by the target's prediction state;

2) Build the sparse dictionary according to the grid values, and make sparse decomposition of original echo signal [16];

3) Initialize the allowance as original echo signal and

compute the correlation coefficient of allowance and sparse dictionary, which is to make inner-products of allowance and each column in sparse dictionary;

4) Save the column corresponding to the largest absolute value to the index set and put this column in a support set;

5) Use the least square method to make sparse approximation of signal;

6) Update the allowance;

7) Continue the iteration until the allowance is less than a certain minimum value;

8) The column serial number in index set corresponding to the location of grid is the target's location.

3.2. SAMP-Based Target Tracking Using PD Radar

In the reconstruction process, the OMP algorithm is relatively simple and fast. But the OMP algorithm itself needs to attain the accurate reconstruction under the condition of known sparse degree [17]. In fact, the sparse degree of signal is usually unknown, for that the number of target in delay-Doppler plane is unknown before the sparsity estimation. Aiming at this problem, we put forward a method of SAMP-based target tracking using PD radar that can tracking target in the condition of unknown sparsity. The tracking procedure is as follows.

1) Calculate the target's prediction state in the kth pulse interval by the target dynamic equation and set up the delay-Doppler grids by the target's prediction state;

2) Build the sparse dictionary according to the grid values, and make sparse decomposition of original echo signal [16], the sparsity of signal is unknown;

3) Initialize the allowance as original echo signal and compute the correlation coefficient of allowance and sparse dictionary, which is to make inner-products of allowance and each column in sparse dictionary;

4) Save the columns corresponding to the first L absolute values to the candidate set and use the least square method to make original sparse approximation of signal;

5) Save the columns of candidate corresponding to the first L absolute values to support set and use the least square method to make final sparse approximation of signal.

6) Update the allowance, if it's bigger than original value, then change to the next stage and increase L;

7) Otherwise, continue the iteration until the allowance is less than a certain minimum value;

8) The column serial number in index set corresponding to the location of grid is the target's location.

4. Numerical Simulations

In this section, we will show the performance of CS-based target tracking using PD radar through the numerical simulations. We will describe and discuss the simulation results. As the target's prediction state provides a good approximation of estimation state in each pulse interval, so we can only choose a part of delay-Doppler plane which contains all of the delay-Doppler values in our interesting area to set up the grid according to the target's prediction state in kth pulse

interval [18].

Our coordinate system is based on inertial coordinate system, and the ground projection point of radar's original position is the origin of coordinates with the north direction for OX axis, the west direction for OY axis, and the line which is vertical to OX-OY plane for the OZ axis. The PD radar at an altitude of 1km is moving with a speed of 30m/s in a straight line. It's original position is as (0, 0, 1000) m, carrier frequency is as f_c=1Ghz, bandwidth is as B=100MHz, pulse width is as T_p=0.05us, pulse repetition interval is as PRI=10ms. There is one target moving in our interesting area, and it's original position is (3000,150,0)m, the component speed of OX axis is 10m/s and 15m/s is for OY axis. Assume that the target is making uniform linear motion and the noise is Gaussian noise with SNR=10db. Keep the delay resolution of 3.33ns (corresponding to the range resolution of 0.5m), and keep Doppler resolution of 6.7Hz (the speed resolution of 1m/s).

4.1. Numerical Simulations of OMP-Based Ground Target Tracking Using PD Radar

Firstly, we use OMP-based method to track the round target and compare its performance with KF filter. The results of numerical simulations are as follows.

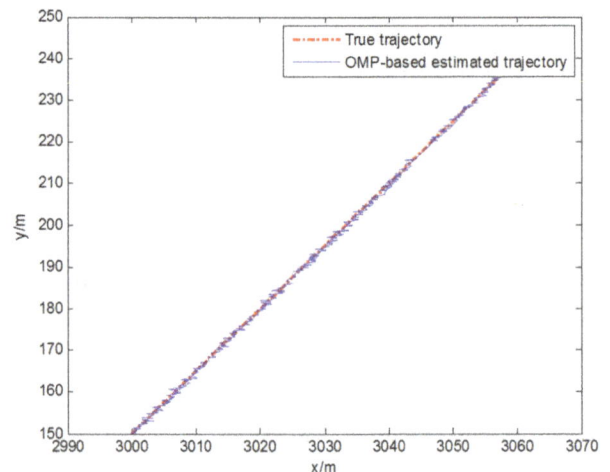

Figure 3. *Motion curve.*

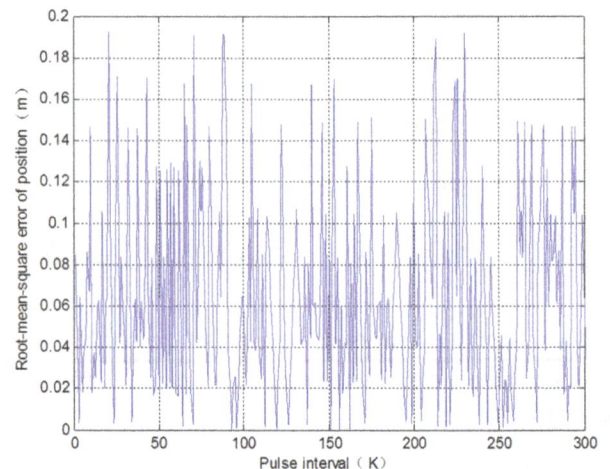

Figure 4. *Root-mean-square error of position.*

The OMP-based approach took 3.24s to estimate the target state of 300 pulse intervals on average. We compare this tracking result with KF filter of which the simulation results are as follows.

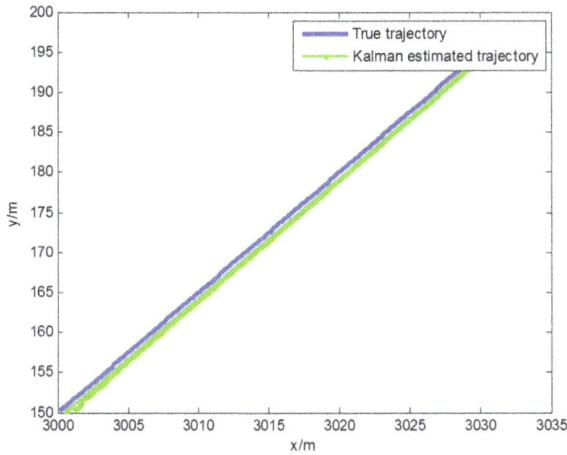

Figure 5. Motion curve.

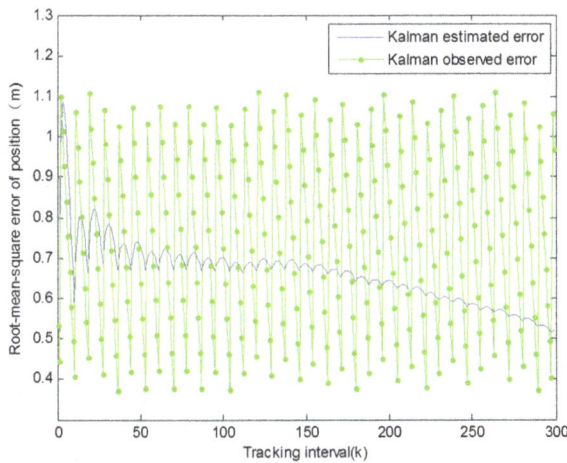

Figure 6. Root-mean-square error of position.

The KF filter method took 2.84s to estimate the target state of 300 pulse intervals on average. And their contrast is as follows.

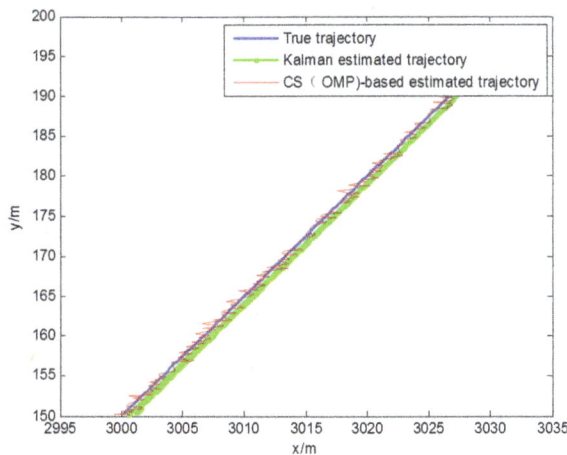

Figure 7. Motion curve.

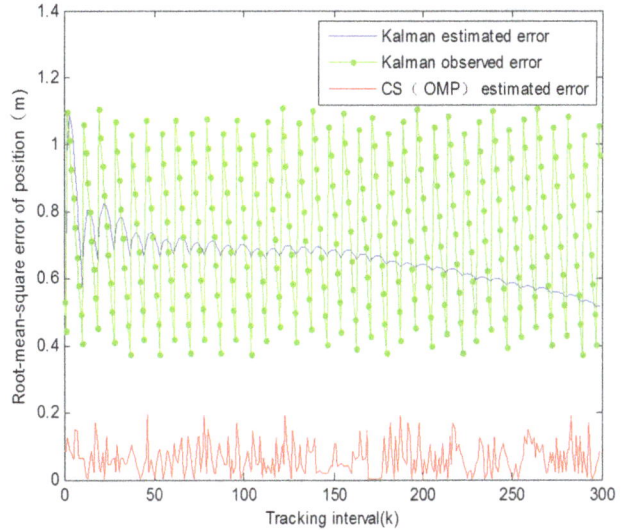

Figure 8. Root-mean-square error of position.

From Figure 6 and Figure 7, we found that the OMP tracking approach could provide better precision with corresponding time compared with the traditional KF filter.

4.2. Numerical Simulations of SAMP-Based Ground Target Tracking Using PD Radar

In view of the fact that the OMP-based method needs the known sparsity of signal in advance, we develop a new method of SAMP-based ground target tracking method using PD radar. The data of simulation is the same as OMP-based method.

In SAMP-based method, the option of L is an important point. According to the literature [19], the range of L is as $1 \leqq L \leqq K$, where k is the sparsity of signal. As this paper is about single target tracking, so we let L=1.

We use SAMP-based to track ground target and compare its performance with OMP-based method. The simulation results are as follows.

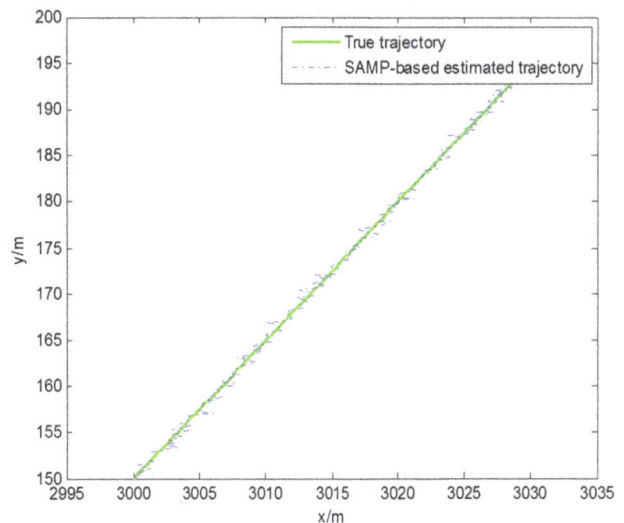

Figure 9. Motion curve.

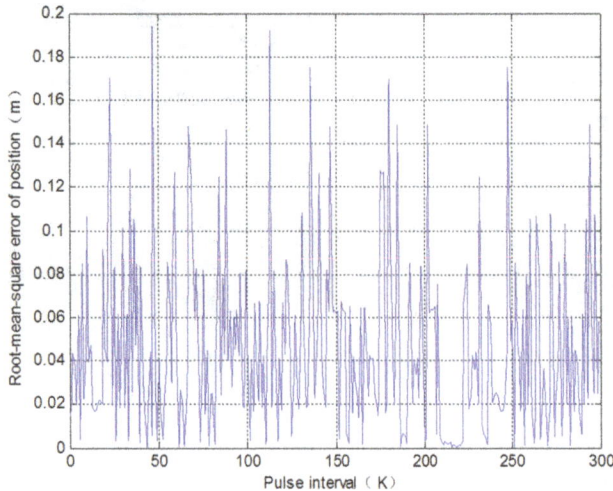

Figure 10. *Root-mean-square error of position.*

The SAMP-based method costs 3.85s to estimate the target state of 300 pulse intervals on average. We compare this tracking result with OMP-based method and the simulation results are as follows.

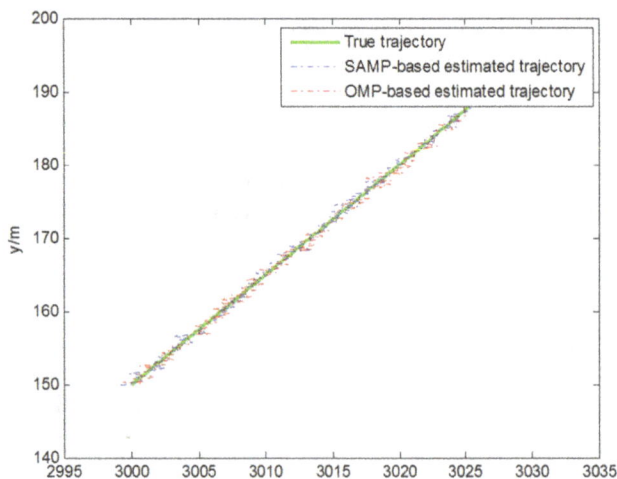

Figure 11. *Motion curve.*

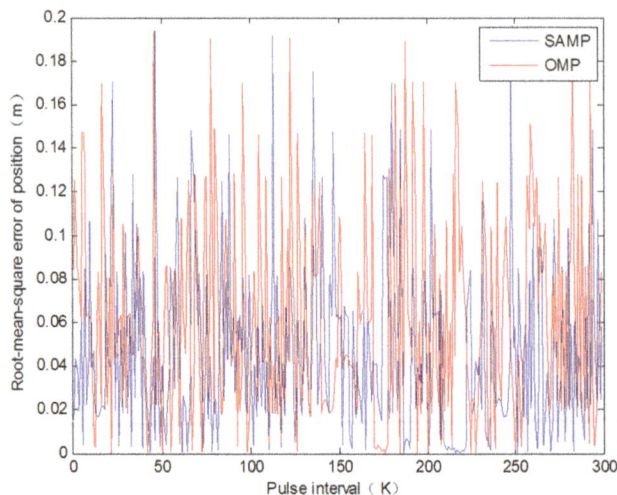

Figure 12. *Root-mean-square error of position.*

From Figure 10 and Figure 11, we can see that SAMP-based method not only solve the problem of unknown sparsity, but also provide the considerable tracking precision.

In conclusion, compared with the traditional KF filter, our CS-based ground target tracking method using PD radar can provide better precision while costs the quite computational time.

5. Conclusion

In this paper, we proposed a Compressed Sensing (CS) based method under the unknown sparse degree to track ground moving targets using Pulse-Doppler (PD) radar. Basing on the analysis of PD radar echo, we transformed the target tracking problem into the reconstruction of sparse signal in CS theory through setting up the orthogonal dictionary by dispersing the delay-Doppler plane.

The simulation results showed that our OMP-based ground target tracking method could provide better precision than traditional KF filter under the considerable computational time. In addition, the SAMP-based ground target tracking method could not only resolve the problem of target tracking with unknown sparsity, but also provide a better tracking performance than KF filter. However, there were some shortcomings in this paper, such as without considering the effect of ground clutter [20]. In the future, we will extend the CS-based ground target tracking method to multi-targets tracking and let the targets scene conform to the realistic conditions.

References

[1] Joel A. Tropp, Anna C. Gilbert. Signal Recovery From Random Measurements Via Orthogonal Matching Pursuit [J]. IEEE Transactions on Information Theory, 2007. Vol. 53, No. 12. 4655-4666.

[2] Thong T. Do, Lu Gan, et al. Sparsity Adaptive Matching Pursuit Algorithm for Practical Compressed Sensing [C]. Asilomar Conference on Signals, Systems and Computers, Pacific Grove, California, 2008, 581-587.

[3] Dai Qionghai, Fu changjun, et al. Compressed Sensing Research [J]. Chinese Journal of Computers, 2001. Vol. 34, No. 3. 428-434.

[4] Emmanuel J. Candes, Terence Tao. Decoding by Linear Programming. IEEE Transactions on Information Theory, 2005, Vol. 51, No. 12. 4203-4215.

[5] Ye Linmei. Compressed Sensing-Based Radar Signal Processing Applications [D]. Xiamen: Xiamen University, 2014.

[6] J Liu, DQ Han, CZ Han, et al. Adaptive compressed sensing based joint detection and tracking algorithm for airborne radars with high resolution. In: International Conference on Information Fusion, 2014, 1-8.

[7] M Miller, J Hinze, M Saquib, et al. Adjustable Transmitter Spacing for MIMO Radar Imaging With Compressed Sensing [J]. IEEE Sensors Journal, 2015. Vol. 15, No. 11. 6671-6677.

[8]Z Liu, X Wei, X Li. Aliasing-Free Moving Target Detection in Random Pulse Repetition Interval Radar Based on Compressed Sensing [J]. IEEE Sensors Journal, 2013. Vol. 13, No. 7. 2523-2534.

[9]Gou Yonggang, Pi Yiming, et al. Target Tracking System Simulation of Airborne Pulse Doppler Radar [J]. Journal of Chinese Electronics Science Institute, 2009, Vol. 4, No. 1. 82-85.

[10]Han Fang. Signal Application Research of Pulse Doppler Radar [D]. Haerbin: Harbin Engineering University, 2007.

[11]Spatyabrata Sen, Arye Nehorai. Sparsity-based Multi-Target Tracking Using OFDM Radar [J]. IEEE Transactions on signal processing, 2011. Vol. 59, No. 4. 1902-1906.

[12]Phani Chavali, A Low-Complexity Sparsity-Based Multi-Target Tracking Algorithm for Urban Environments [A]. IEEE Radar Conference, 2011, 309-314.

[13]Yang Jun, Zhang Qun, et al. Compressed sensing-based Multi-Target Tracking Using Cognitive Radar [J]. Journal of Radar, 2014.

[14]Baraniuk R G. Compressive sensing [lecture notes]. IEEE Signal Processing Magazine, 2007. Vol. 24, No. 4. 118-121.

[15]Jacques L, Hammond DK, et al. Dequantizing compressed sensing: When oversampling and non-gaussian constraints combine. IEEE Transactions on Information Theory, 2011. Vol. 57, No. 1. 559-571.

[16]Shi Guangming, Liu Danhua, et al. Compressed Sensing Theory and It's Theory Development [J]. Electronics Journals, 2009. Vol. 37, No. 5. 1072-1080.

[17]Li Fanghua. Compressed Sensing-based Target Location Algorithm by Radar [D]. Changsha: Hunan University, 2012.

[18]Satyabrata Sen. Adaptive OFDM Radar for Target Detection and Tracking [D]. St. Louis: Washington University, 2010.

[19]Tong T Do, Lu Gan, et al. Sparsity adaptive matching pursuit algorithm for practical compressed sensing. In: Proc of the 42nd Asilomar Conference on Signals, Systems and Computers. Pacific Grove, 2008, 581-587.

[20]PB Tuuk, SL Marple. Compressed sensing radar amid noise and clutter using interference covariance information [J]. IEEE transactions on Aerospace & Electronic Systems, 2014. Vol. 50, No. 50. 887-897.

27

Enhanced EHD and Electrostatic Propulsion Devices Based on Polarization Effect Using Asymmetrical Metal Structure

Taku Saiki

Department of Electrical and Electronic Engineering, Faculty of Engineering Science, Kansai University, Suita, Japan

Email address:

tsaiki@kansai-u.ac.jp

Abstract: Electro hydro dynamic (EHD) and electrostatic propulsion devices were developed in the 1920s by Thomas Townsend Brown. One such device, called a "lifter", has no moving parts and, in the air, operates on electrical energy. It is a fashionable device and has a very simple structure, basically consisting of a narrow wire electrode and a large, flat one. However, it has a low ratio of propulsion force to unit electrical input power. According to theory, the propulsion force it generates depends on the interaction between the ion density of the ionized air and the charges on the surface of the large electrode. EHD and electrostatic propulsion models using the polarization effect are proposed to improve the ratio of the propulsion force to unit electrical input power. The propulsion device generates propulsion force through the use of an asymmetrical metal structure with charges generated by the polarization effect. The propulsion force the new devices generated for the same electric energy was 5.7 times higher than that of a basic type lifter owing to additional propulsion force being generated by the maximum polarization effect in the experiments. It was found that combining other effects with this polarization effect results in the ratio of generated propulsion force to electric power being close to 100N/kW when the electric power is high. This value is as high as that of a helicopter. We also performed numerical analysis was also performed for capacitances and charges for various kinds of EHD and electrostatic propulsion devices. An optimized system was developed and is discussed in this paper.

Keywords: EHD, Electrostatic Propulsion, Polarization, Electric Field, Charge, Electron

1. Introduction

A lifter that is made light in weight by using a high voltage supply floats in the air. For such lifters, only electrical energy is needed to obtain the propulsion force in the air. Normal propulsion devices such as rockets need materials to be propelled. Lifter-type propulsion devices have no moving parts as helicopters do, and can be expected to be used as Unmanned Aerial Vehicles (URVs) in the future [1].

The Biefeld-Brown effect was discovered by Paul Alfred Biefeld and Thomas Townsend Brown in the 1920s [2]. Brown also proposed and was awarded patents for a number of electro hydro dynamic (EHD) propulsion devices, such as lifters [1,3]. The principle, which is called ion craft, should be the same as that of the lifter. Many movies of lifters floating in the air can be seen on the Internet, and a number of papers on the theory of lifters to generate propulsion forces have been published[4-12]. After the Biefeld-Brown effect was published, many discussions on it were held. Mainly, the principles behind the effect were discussed and many theories

to explain the generated propulsion forces were proposed. One view, that the effect can be explained by "generation of momentum by ion wind", was shown by the USA's National Renewable Energy Laboratory. However, recent rigid research indicates that the principle of the propulsion can be explained as the electrical forces between charges in ionic wind and the electrons on the large flat electrode[6-8]. Thus, it has been shown that the lifter propulsion is based on a principle that differs from conventional principals. Rockets work on the basis of the equation of motion.

The objectives in developing the propulsion devices are achieving 1) low propulsion force to electrical input energy, 2) low propulsion force per unit volume, and 3) low propulsion force per unit weight. Some papers [12,13] have previously reported experimental results for improving the ratio of generated propulsion force to electric energy. Thus, at present, heavy, high voltage sources cannot be mounted on propulsion devices.

In this paper, I propose an improved model for the structure of the rounded electrode used to improve the charge density of ion wind. It is a model with a simplified small negative electrode to make the ion source small. In addition, propulsion devices using the polarization effect, such as a cascading model, were fabricated to improve the ratio of propulsion force to electrical input energy.

2. Principle of EHD and Electrostatic Propulsion Devices

Simply put, a lifter consists of a tin wire electrode and a large, flat electrode [1]. The principle for producing the propulsion force of a lifter and EHD propulsion devices was proposed in recent research activities shown in the literature [8,11]. An equation to calculate the generated propulsion force of a lifter is shown next in references [8] and [11]

$$F = \int_v \rho E \cdot dv . \qquad (1)$$

In the equation, E is the electric field generated from the charges on the large flat electrode and v is the effective volume area in which ionized air flows near the large electrode. The equation for calculating ρ, the ion density near the large electrode in the air, is

$$\rho = \frac{Id}{\mu VA} . \qquad (2)$$

Here, I is the current, d is the gap between the tin wire electrode and the large flat electrode, μ is the ion mobility in the atmosphere, V is the voltage between the electrodes, and A is the profile area of ionized air floating near the large electrode. It shows that all the generated ions do not interact with the large electrode. Here, if we set E=V/d, eq. (1) is simplified and we obtain

$$F = \frac{Id}{\mu} . \qquad (3)$$

Here, I describe the meaning of eq. (1). The charges on the large electrode attract the ion wind by the electric field between the generated ions and the large flat electrode. As a result, propulsion force is produced. The observed current at the power supply results from the fact that ions exchange the electrons on the large electrode (interaction between the ion wind and the large flat electrode). Detailed numerical estimations of EHD fluids, space charge, and potential in the ionized air flow are shown in reference [5-9].

The ratio of generated power to electrical input power is given by

$$\theta = \frac{F}{P} = \frac{I}{\mu E} . \qquad (4)$$

Below, I simply describe the calculation method for the capacitance between the electrodes and the charge on them. The method is shown in eq. (1), according to the principle for producing propulsion force. Evaluating the quantity of the charge on the electrode when a high voltage is applied results

in improved propulsion force. However, the electrical charge is not always proportional to the propulsion forces the EHD devices generate.

In this work, I used a rounded electrode for the structure of the EHD device in order to 1) use the model as a unit module, 2) achieve uniform divergence of the ionized air in the radial direction, and 3) efficiently protect the ion density in the rounded electrode from degradation.

I used metal structure for the device to enhance 1) the charges on the electrodes owing to the polarization effect, 2) the emission of ions from the negative needle electrode, and 3) the propulsion force produced by the electrical field between the ion wind and the large electrode or between the ion wind and the polarization electrodes.

For the idea for polarization effect in EHD devices, one [1] of Brown's patents proposed that an EHD device in which a dielectric material is sandwiched between the electrodes produces high propulsion force. The details for this experiment differed from those given in Brown's patent in that propulsion forces are generated by metal electrodes for which an asymmetrical structure is used to produce the polarization effect.

Here, the needle electrode was connected to the negative and the large electrode to the positive voltage. The structure for generating ions was simplified.

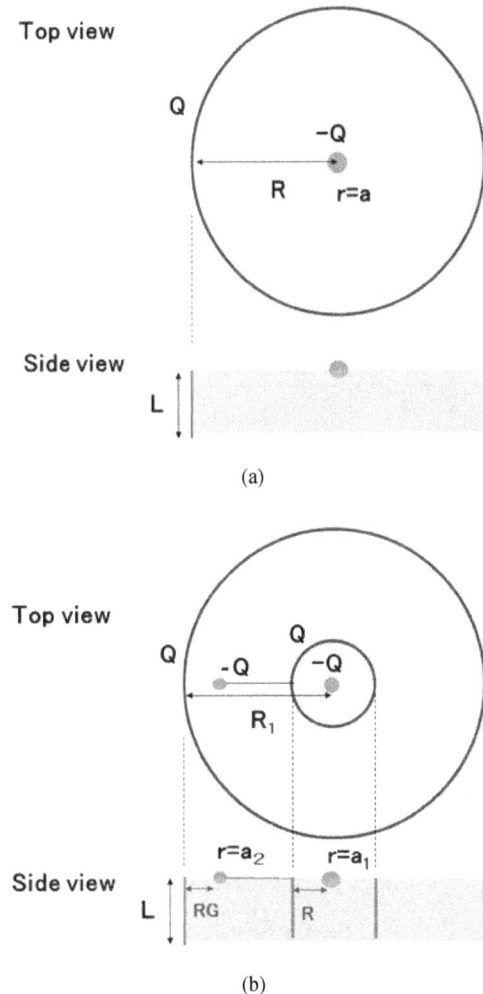

(a)

(b)

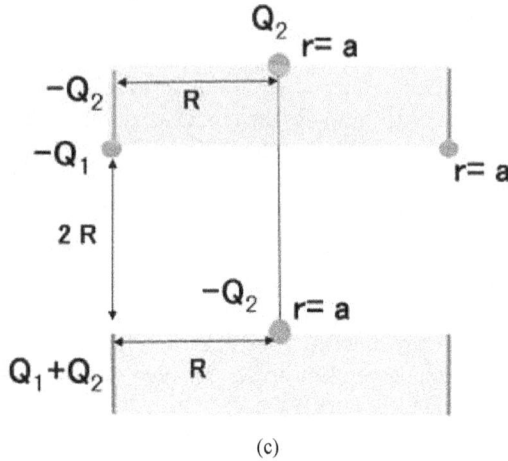

(c)

Fig. 1. Calculation models. (a) normal model 1a, (b) double ring model 1b, (c) polarization pole model 1c

For the model shown in Fig. 1(a) I used a rounded electrode, in contrast to the conventional lifter that uses flat plate electrodes. Here, I neglected ion flow because its calculation is very complicated. The equation to calculate the electrical potential on the radial axis r for model 1a is

$$V(r) = \frac{-Q}{4\pi\varepsilon_0} \cdot \left[\frac{1}{r} - \frac{\ln\left[\frac{L+\left(L^2+(R-r)^2\right)^{1/2}}{R-r}\right]}{L} \right]. \quad (5)$$

Here, L is the width of the large electrode and R is the radius of the large electrode. Thus, the equation for the electrical potential difference V_{ab} between both electrodes is

$$V_{ab} = \frac{Q}{4\pi\varepsilon_0} \cdot \left[\frac{1}{a} - \frac{1}{b} + \ln\left[\frac{R-a}{R-b} \cdot \frac{L+\left(L^2+(R-b)^2\right)^{1/2}}{L+\left(L^2+(R-a)^2\right)^{1/2}} \right] \right]. \quad (6)$$

Here, $0 < a, b < R$, and it is assumed that b is very close to R. The equation for the capacitance between the electrodes for model 1a is

$$C_{ab} = \frac{4\pi\varepsilon_0}{\left[\frac{1}{a} - \frac{1}{b} + \ln\left[\frac{R-a}{R-b} \cdot \frac{L+\left(L^2+(R-b)^2\right)^{1/2}}{L+\left(L^2+(R-a)^2\right)^{1/2}} \right] \right]}. \quad (7)$$

The third term in eq. (6) can be neglected because it is very small compared to the first term. If L>> R, eq. (7) is more simplified and changed to

$$C_{ab} = 4\pi\varepsilon_0 \frac{b \cdot a}{b-a}. \quad (8)$$

For calculating model 1b, there is one metal rod electrode to produce polarization connected directly to inner electrode. RG is the gap length as shown in Fig. 1(b). RG is very much shorter than L. Here, C_{a2b2} is generated capacitance governed by eq. (7) to set the space gap and C_{a1b1} is the capacitance between the negative needle electrode and the inner radial electrode governed by eq. (8). The total capacitance for model 1b is determined by calculating the series connection of C_{a1b1} and C_{a2b2}. The equation for calculating the capacitance for model 1b is

$$C_2 = \frac{C_{a1b1} \cdot C_{a2b2}}{C_{a1b1} + C_{a2b2}}. \quad (9)$$

If the number of metal rods needed to generate polarization is n, the capacitance is rewritten as

$$C_2' = nC_2. \quad (10)$$

In the case of model 1c, the capacitance between the upper and lower ring electrodes is

$$C_0 = 2\pi\varepsilon_0 \frac{2\pi R}{\ln(2R/a)} \quad (11)$$

The capacitance generated by using a center electrode to generate polarization is the same as that for model 1b. The capacitance C_2 is derived by calculating the series connection of two C_{ab} shown in eq. (7). The total capacitance C_3 for model 1c, derived by calculating the parallel connection of C_2 and C_0, is

$$C_3 = C_0 + C_2. \quad (12)$$

It is clear that eq. (12) and eq. (9) indicate increased capacitances in each of the propulsion models.

Here I return to the discussion about the propulsion force. For model 1b, the generated propulsion force obtained by reconsidering eq. (1) is

$$F_2 = \int_v (\rho_1 E_1 + \rho_2 E_2) \cdot dv. \quad (13)$$

Here, E_1 is the electric field producing the charges on the inner radial ring to generate polarization, E_2 is the electric field producing the charges on the large outer radial electrode, ρ_1 is the ion density near the inner radial ring to generate polarization, and ρ_2 is the ion density near the large outer radial electrode.

For model 1c, the generated propulsion force obtained by reconsidering eq. (1) is

$$F_3 = \int_v [\rho_3 E_3 + \rho_4 (E_2 + E_4)] \cdot dv. \quad (14)$$

Here, E_3 is the electric field producing the charge $-Q_2$ on the upper ring, E_4 is the electric field producing the charges Q_2 on the lower ring electrode, ρ_3 is the ion density near the upper radial electrode (the ions are produced at the edge of the polarization rod at the center of the module), and ρ_4 is the ion density near the lower ring electrode (the ions are produced at the edge of the upper ring electrode). Here, the polarity of ρ_3 is positive and that of ρ_4 is negative. From eq. (13) and eq. (14), comparing them to eq. (1), it is concluded that the propulsion forces that both model 1b and model 1c generate should be modified. Here, it should be noted that it is not always necessary to exchange the electrons between the electrodes when the forces are produced. In other words, the current is not needed to generate the propulsion forces from the polarization electrodes.

Finally, the ratios of generated power to electrical input power for model 1b and 1c are given by $\theta_2=F_2/P$, and $\theta_3=F_3/P$, respectively.

3. Experimental Setup

3.1. Propulsion Model

Three types of model A units, which were actually fabricated in this work, are shown in Fig. 2. These models, which produce propulsion force, can be set in series or parallel. I also fabricated two other types of models. In one, the ion wind flows downward in the ring electrode. In the other, the ion wind expands in the radial direction and flows outside the ring electrode. In all the A model units, the ion wind flows inside the ring electrode (Fig. 2). The distance between the negative needle electrode and the positive large ring electrode in the vertical direction was set to be 4 cm in each case.

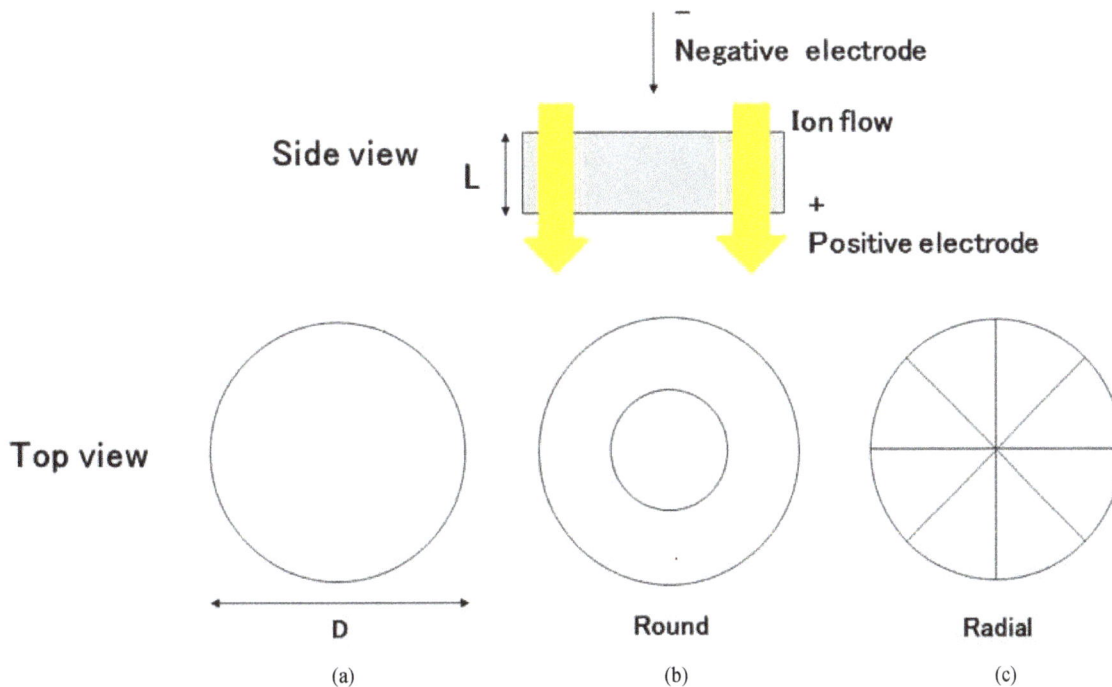

Fig. 2. Enhanced EHD and electrostatic propulsion devices. Model A units: (a) normal Aa, (b) with added inner ring electrode Ab, (c) with added radial electrode Ac.

Table 1. Model A parameters.

Model	Diameter D (cm)	Height L(cm)	Diameter of Inner Ring D_1(cm)	Height of Inner Ring L_1(cm)
Aa1	8	3	—	—
Aa2	16	5	—	—
Ab	16	5	8	5

The aluminum foils used as large electrodes were 10 μm in thickness. To improve the capacitance of the devices, various kinds of electrodes were added inside the large ring electrodes. The Aa model has a single ring electrode, the Ab model has another ring electrode inside the outer ring electrode, and the Ac model has a star electrode inside the outer ring electrode. These models were used to conduct an experiment to produce high propulsion forces. The size parameters are shown in Table 1. The weights of the model Aa1, Aa2 and Ab were 0.6g,

1.5g, and 1.8g, respectively.

The devices were set in series to enhance the propulsion force for the same electrical input power. In Fig. 3 depicting model B1, the cascading of two devices is shown at the left side and the cascading of n devices is shown at the right. Model at the left side in Fig. 3 was used in this experiment. For the model at the left side, the distance between the upper ring electrode and the lower ring electrode was set to be 3 cm. A wire was connected on the upper ring electrode and the end was closed to the lower ring electrode with a gap of 1 cm. It was normally difficult to connect the devices in series because in some cases the propulsion forces disappeared due to ion flow or the positive and negative electrode settings. The upper ring electrode generates polarization. Negative charges are collected at the edge of the wire connected on the upper ring electrode. The weights of the model B1 was1.1g.

○Cascading
Two module

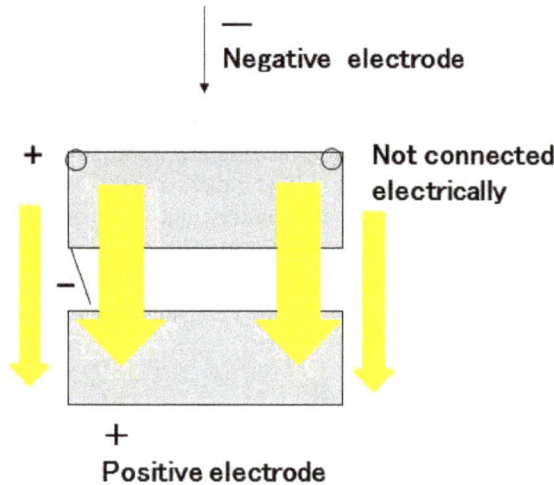

○Cascading
N module

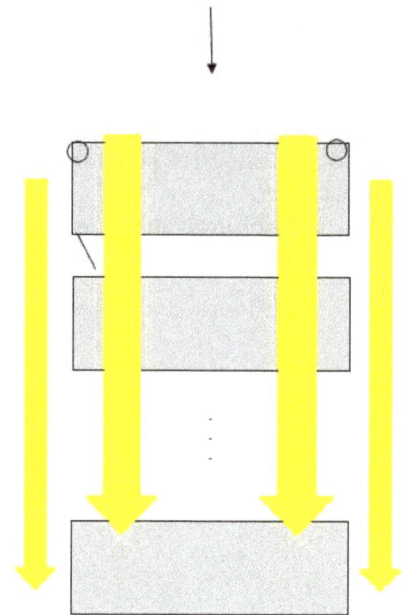

— Negative electrode

+   Not connected
      electrically

−

+
Positive electrode

Fig. 3. *Enhanced EHD and electrostatic propulsion devices. Model B1.*

Model B2 (Fig. 4) has a metal wire at the center of the module to generate polarization. Positive ions are produced from the upper edge of the wire, generating a propulsion force to the upper ring electrode by the positive ion flow. Thus, two propulsion forces, which work on both the upper ring electrode and the lower ring electrode, are generated. The weights of the model B2 was 3.2g. The model is adapted to Fig. 1(c).

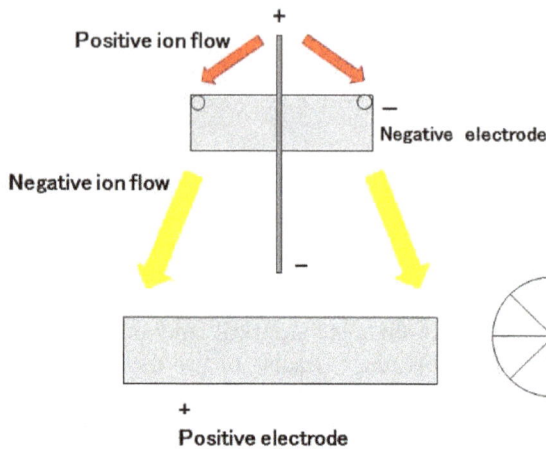

+
Positive ion flow

Negative electrode

Negative ion flow

−

+
Positive electrode

(a)   (b)

Fig. 4. *Enhanced EHD and Electrostatic propulsion devices Model B2. (a) Structure, (b) photo image.*

In model B3 (Fig. 5), the ion wind flows in the radial direction. Enhanced propulsion force is produced by the polarization effect. The model is adapted to Fig. 1(b). Four small plate electrodes are set in the radial direction inside the outer large ring electrode. Negative ions generated from the upper negative needle electrode are trapped the outer large ring electrode and finally the current is supplied from a high voltage source. The structure is proposed because an EHD generator should be set in the center of the device. In the future such generators will use a weakly ionized plasma (such as a flame) to generate an ultra-high voltage [15].The weights of the model B3 was 3.8g.

Side View

Top View

(a)

(b)

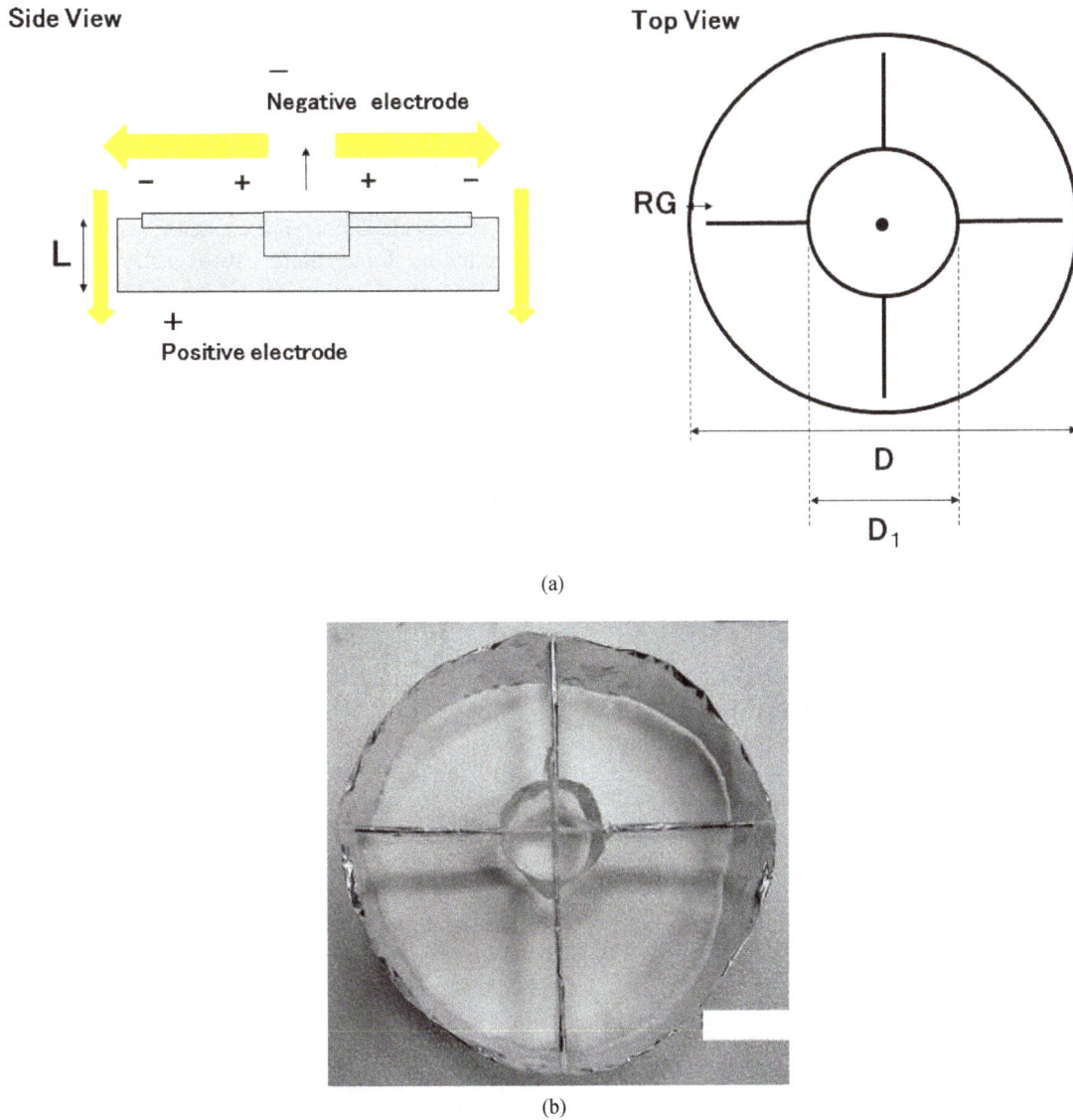

Fig. 5. Enhanced EHD and Electrostatic propulsion devices Model B3. (a) Structure, (b) photo image.

The structure parameters of model B are shown in Table 2.

Table 2. Model B parameters.

Model	Diameter of Ring D(cm)	Height L (cm)	Diameter of Inner (Upper) Ring D_1(cm)	Height of Inner (Upper) Ring L_1(cm)	Gap (cm)	Distance between Top and Bottom Rings(cm)
B1	8	3	8	3	1	3
B2	16	5	8	3	Upper 2 Lower 3	9
B3	32	5	8	3	1	—

3.2. Measurement of Propulsion Force

The instrument used to measure the propulsion forces of the EHD devices is shown in Fig. 6.

The propulsion devices were connected to a DC high voltage generator. A six stage Cockcroft Walton circuit was used to generate the voltage. A sinewave generator was connected to the Cockcroft Walton circuit. In this experiment, a DC power supply (PR36-3A, TEXIO, Japan) was used to generate a sinewave with a high output voltage and 25 KHz frequency. The output voltage was evaluated by using a voltage tester (CD731a, SANWA, Japan) and a high voltage probe (HV-60SANWA, SANWA, Japan). The maximum measured output voltage of the Cockcroft Walton circuit was 48 kV. The current was evaluated by connecting 22k ohm resistors in series to the propulsion devices and measuring the voltage. A dielectric plastic rod with a diameter of 1.0 cm was set on a gravimeter (ACS-20, ASONE, Japan). The propulsion force was measured by reading the decrease in the weight.

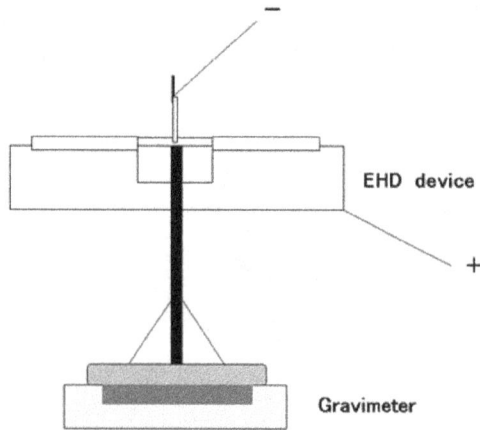

Fig. 6. Instrument for measuring propulsion force of the enhanced EHD propulsion devices.

4. Results

4.1. Calculated Capacitance and Charge on Electrodes

The calculated results on the capacitance of model 1a are shown as a function of L in Fig. 7. Here, R is set to be 4 cm and is fixed at that length. It was assumed that the calculating space is in the vacuum.

The results clarified that the capacitance for model 1a, shown in Fig. 1(a), the capacitance increased gradually and saturated as L increased.

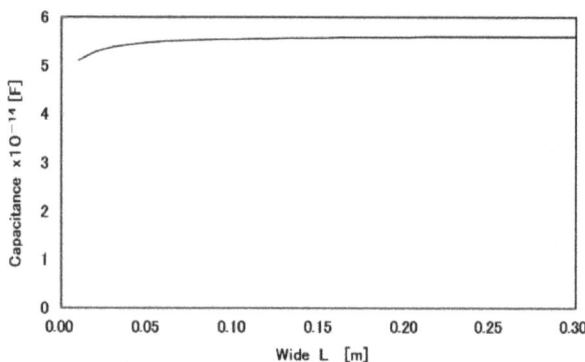

Fig. 7. Calculated capacitance for model 1a as a function of L.

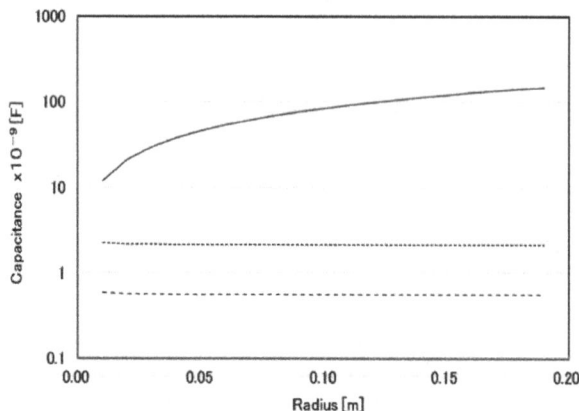

Fig. 8. Calculated capacitance results for each propulsion model.

The calculated results of the capacitance for each model are shown in Fig. 8. The dashed line shows the results for model 1a, the dotted line shows the results for model 1b, and the solid line shows the results for model 1c. a and a1 are all set to be 0.005 [m]. b was set to be R-0.005[m]. To simplify the calculation, L was set to 0.03 [m], assuming that R1 is adequately longer than R. n was set to be 4 and RG was set to be 1 cm in the calculation. The capacitances for model 1a and 1b degraded as R increased. However, the capacitances for model 1c increased as R increased due to the structure.

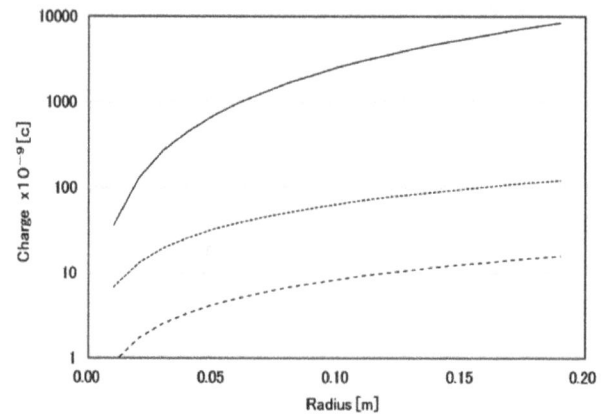

Fig. 9. Calculated charge results.

The calculated results for the total charges on the large electrode are shown in Fig. 9. The dashed line shows the results for model 1a. The dotted line shows the results for model 1b, and the solid line shows the results for model 1c. For the input voltage to the propulsion devices, I consider the discharge property between the needle and flat electrodes, i.e., the limit of 15 [kV] per cm for the discharge determined [14]. Thus, the maximum input voltage was given by an equation of $15 \times R$ [kV]. For model 1c, the calculated charge was obtained by adding the charge on the upper ring electrode to that on the lower ring electrode. For model 1b, it was obtained by adding the charge on the inner ring electrode to that on the outer large ring electrode. The charges for model 1a, 1b, and 1c increased as R increased. The charges for model 1c were the largest for all the models.

4.2. Measured Propulsion Force

The results obtained for the measured propulsion force for model Aa1, Aa2, and Ab are shown in Fig. 10. The black circle dots show the results for model Aa2. The white circle dots shows the results for model Aa1. The white square dots shows the results for model Ab. The data shown by the dots in Fig. 10(a) is consistent with that shown by the dots in Fig. 10(b). The calculated electrical input power in Fig. 10(b) was obtained by multiplying the measured current data. When the input voltages were 22 and 26 kV, the measured current levels were 0.15 and 0.26 mA, respectively.

When the electrical input power was 8 W, it can be seen that the force that model Aa2 generated was 2.5 times higher than that of model Aa1. The maximum generated force was 1.8 g at an input voltage of 43 kV. However, the generated force was gradually saturated as the input voltage and the electrical input power increased. In another experiment, the force measured for model Aa2 for L= 10 cm was 2.0 g at an

input voltage of 43 kV. In spite of using a long L, the generated force was only slightly improved.

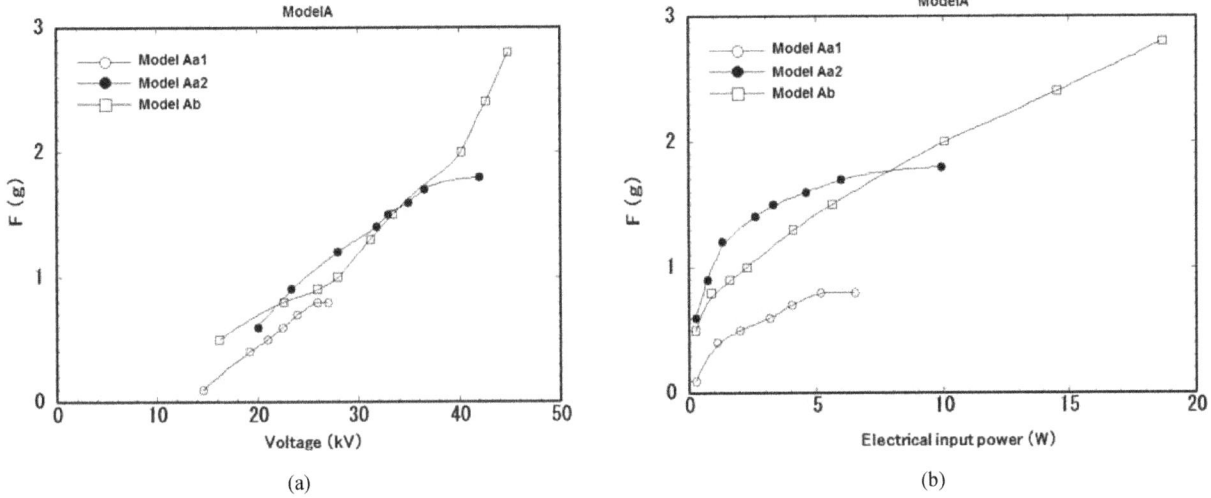

Fig. 10. Measured propulsion force results for model Aa1, Aa2, and Ab. (a) Voltage-force property, (b) Power-force property.

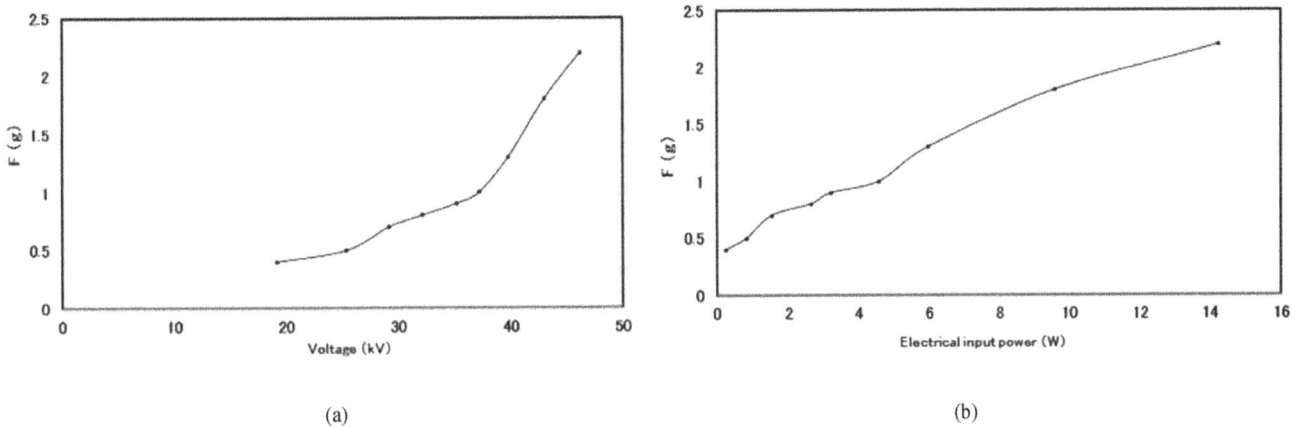

Fig. 11. Measured propulsion force results for model B1. (a) Voltage-force property, (b) Power-force property.

When the electrical input voltage was 44 kV, it can be seen that the propulsion force that model Ab generated was 1.6 times higher than that of model Aa1. The maximum generated force was 2.8 g at an input voltage of 44 kV. However, the generated force was not saturated as the input voltage and the electrical input power increased. In another experiment, in which a star electrode and two inner multi-rings were added to model Aa2, the maximum generated forces for them were 2.3 and 3.2 g, respectively.

The measured propulsion force results obtained for model B1 are shown in Fig. 11. If the generated power values shown in Fig. 10(a) are compared with those shown in Fig. 11(a), when the electrical input voltage was 46 kV, it can be seen that the force that model B1 generated was 3 times higher than that of model Aa1. The maximum generated force was 2.3 g at an input voltage of 44 kV. The generated force markedly increased at an input voltage of 38 kV as shown in Fig. 11(a).

The measured propulsion force results obtained for model B2 are shown in Fig. 12. If the generated power values shown in Fig. 10(a) are compared with those shown in Fig. 12(a), when the electrical input voltage was 43 kV, it can be seen that the force that model B2 generated was 2 times higher than that of model Ac. The maximum generated force was 3.6 g at an input voltage of 43 kV. The generated force increased as a function of a value close to the square of the input voltage as shown in Fig. 12(a).

Measured propulsion force results obtained for model B3 are shown in Fig. 13. The maximum generated force was 5.1 g at an input voltage of 48 kV. The generated force increased as a function of a value close to the square of the input voltage as shown in Fig. 13(a). If the generated power values shown in Fig. 10(a) are compared with those shown in Fig. 13(a), it can be seen that the maximum generated force that model B3 generated was 6.4 times higher than that of model Aa1.

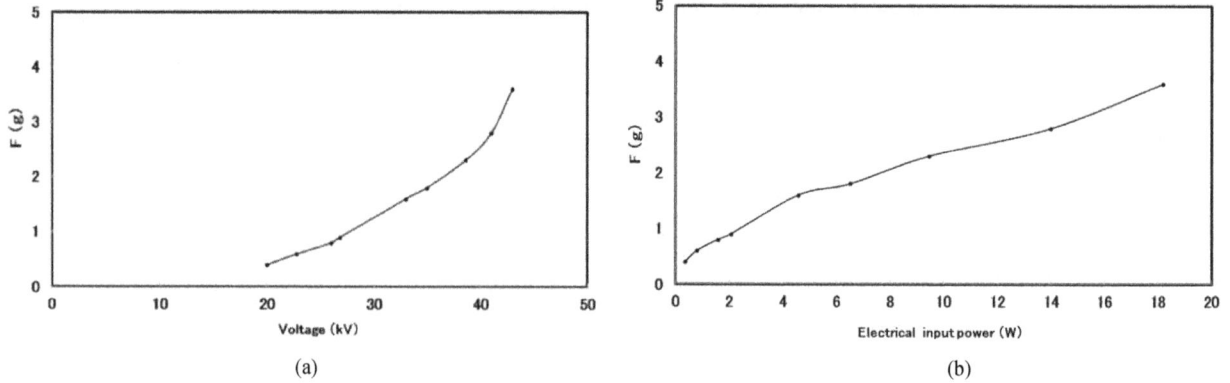

Fig. 12. Measured propulsion force results for model B2. (a) Voltage-force property, (b) Power-force property.

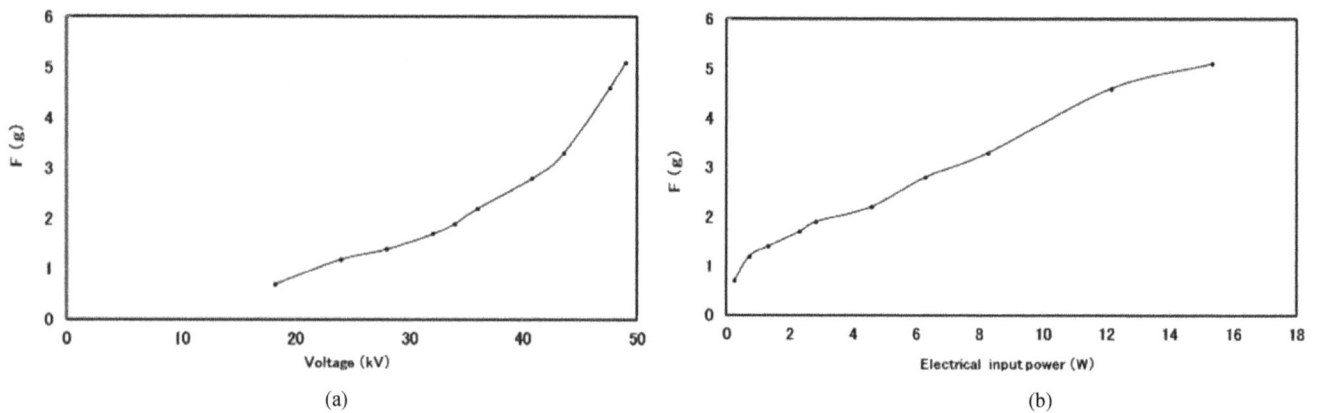

Fig. 13. Measured propulsion force results for model B3. (a) Voltage-force property, (b) Power-force property.

5. Discussion

I proposed modified EHD and electrostatic devices to improve the ratio of the propulsion force to electrical input power. In the numerical analysis, the results of the calculated total charge on the surface of the large electrode were enlarged due to the polarization effect. From the propulsion force results obtained for the devices governed by eq. (1) and eqs. (13) (14), because the forces are proportional to the charges on the surface of the large flat electrodes, it can be concluded that the polarization effect improves the forces.

The calculated results lead one to expect that the force that model B3 generates will be 7 times higher than that of model Aa1 owing to eq. (1) and eq. (13). If the devices are operated at a high voltage, the gap used will be longer, the stored charges on the large and flat electrodes will be larger, and the quantity of the charges will be enlarged proportional to the gap length approximately.

Table 3. Comparison of measured and calculated propulsion force.

Model	Voltage (kV)	Current (mA)	Power (W)	Fexp(g)	d (m)	Fcal(g) Eq.(3)	α Fexp/ Fcal	θ (N/kW)
Aa2	42	0.15	9.9	1.8	0.09	1.1	1.6	1.8
Ab	45	0.39	19	2.8	0.09	1.7	1.6	1.4
B1	46	0.31	14	2.2	0.066	0.94	2.4	1.6
B2	43	0.42	18	3.8	0.09	1.8	2.1	2.1
B3	48	0.26	12	4.6	0.066	0.81	5.7	3.8

In the experiment on the forces that model A and B generated, the models showed high performance with respect to the ratio of force to electrical input power.

Especially for model B3, A large propulsion force are generated whenever ion flows in the horizontal direction. The ion wind cannot be contribute to the enhanced propulsion force. This should prove that the lifter work by the electrostatic propulsion as mentioned in the reference[6,7].

The electrical input power, measured generated power, calculated generated force, and ratio are shown in Table 3.

The gap length d is the value for which 1 cm was added to the gap length from the edge of the radial electrode to the negative needle electrode in Table 3. I used eq. (3) to calculate the theoretical values of the propulsion force the devices generated. The ion mobility was set to be 2.15×10^{-4} (m^2/Vs).

The ratio of generated force to electrical input power of a basic lifter with a single narrow electrode and a single flat aluminum electrode, as shown in reference [1,2], was estimated to be 1.3 N/kW.

It has been clarified from this experiment that the total

amount of ions generated from the small electrode in the conventional lifter model is large. For a large, flat electrode, however, the ion density is low.

The 0.4 mA current for model Aa2 with 16-cm outer diameter is the same as that of the basic lifter at the input voltage of 46 kV. However, the generated force for model Aa2 is 1.6 times higher than that of the basic lifter as shown in Table 3. The model A structure with downstream ion flows was able to confine ions generated from the negative electrode and prevent the ion density from being low close to the large radial electrode.

The forces that the models B1, B2, and B3 generated are rigidly proportional to the square of the input voltage because the electric field intensity increases when the number of ions generated increases at the same time.

The forces that model Aa generated were rigidly generated in proportion to the input voltage because the electric field intensity is constant when the number of generated ions increases.

The forces that models Aa1 and Aa2 generated saturated when the input voltage was 43 kV. This is because the generated ions were saturated in the air, the ion density consequently reached the space charge limit, and the ions were repelled near the large electrode.

For model Ab, the inner ring electrode was set in the center of the large ring electrode and the measured generated force was more than twice that measured for model Aa2. The reason is that the surface area of the positive large electrode increased. However, a model having too many ring electrodes will not result in an increase in the generated force.

For model B1 using the polarization effect, the ring electrodes were cascaded and the force measured for it was 3 times as large as that measured for model Aa1.

For model B2 using the polarization effect, there was a pole to generate polarization. The force measured for model B2 was close to 2 times larger than that measured for model Ac.

The force measured for model B3 was 4.6 g at an input voltage of 46 kV and electrical input energy of 12 W. The ratio of the generated force to electrical power was 4N/kW. The value was maximum in all the proposed models. The current of model B3 was 50% lower than that of model Aa1 for the same input voltage of 24 kV. The ratio of the generated force to electrical input power was 15N/kW at 23 kV input voltage. This is one order higher than that of the basic lifter, for which the ratio was 8.3N/kW at 28 kV input voltage. The generated force for input voltage was not saturated. The property is ideal and well consistent with the theory. The polarization plate electrodes should be set to be tilted for the large rounded electrode. The maximum angle for generating force should exist. In these experiments, I was able to modify the ratio of the generated force to the electrical input power.

However, the ratio of the generated force to unit weight has not been modified yet. Using a high voltage generator with higher output voltage and longer gap length will enable the ratio of generated force to unit weight to be improved. Specially, the ratio of the models B1, B2, and B3 should be substantially modified. The aluminum foil thickness used in the experiment was 10 μm; this thickness should have an optimized value.

If the aluminum foil is very thin, the potential that confines the charge will degrade and thus charges on the metal plate cannot be confined efficiently. Basically, the structure of the large electrode must be maintained to store a lot of charges. In the future, the optimized thickness should be studied.

Modifications should also be made to obtain a higher ratio of the generated force to electrical input power. Using multiple large, flat electrodes as positive electrodes should make it possible to improve the ratio, since it should enhance the electrical field intensity. The experimental data in reference [12] shows the effect to improve the ratio of generated force to electrical input power clearly. However, the effects of enhancing the electric field intensity due to the edge and flat surfaces of electrodes are mixed. Thus, we cannot divide them. In reference [12], a 6.5 N/kW ratio was obtained when the small electrode was negative s in a high voltage and high power regime. The models I showed may be optimized, in which case it should be able to obtain a higher ratio than this. It is hard to set B2 modules in cascade but it is possible to set them in parallel. Models B1 and B3 can be set in series. Setting devices in series, as shown at right in Fig. 3, will result in an increased ratio of generated force to electrical input power. Using a multi-electrode structure comprising a large flat electrode, multi-stage polarization electrodes, and using long gap between electrodes and more high input voltage [13] should make it possible to raise the ratio to over 100N/kW at a low electrical input power and less 100N/kW at a high electrical input power. Using long gap between electrodes results in improving the ratio of generated force to electrical input power as shown in eq. (4) because the electric field is weak for the same input voltage [13]. The ratio at a high electrical input power would be higher than that of a conventional helicopter using a motor, for which the ratio is around 70N/kW.

Combining model B3 with model B1 will result in generating a force of over 10 g. Moreover, using weakly ionized plasma to produce ultra-high voltage, as in EHD generators [15], we should use the optimized system to reduce the weight of the high voltage power supply and improve the ratio of generated force to electrical input power.

6. Conclusion

This paper discussed various kinds of proposed EHD and electrostatic propulsion devices using asymmetrical polarization electrodes that I propose as means to improve the ratio of generated propulsion force to electrical input power. The levels of force that the devices generated were measured in experiments. Electrical charges on the surfaces of electrodes were estimated numerically with an aim to improving the generated force. The models substantially improved the generated force; for the same electric energy, the force they generated was 5.7 times higher than that of a basic type lifter. This was due to additional propulsion force being generated by the polarization effect.

Future subjects in this research will include using 1) a high voltage supply with reduced weight achieved by using a flame-jet or low-density-plasma EHD generator and 2) multiple polarization electrodes as means to improve the generated propulsion force.

References

[1]   T. T. Brown, "Electrokinetic Apparatus," U.S. Patent N°2949550, 1960.

[2]   http://www.jlnlab.com/

[3]   http://www.blazelabs.com/

[4]   M. Tajmar, "Biefeld–Brown Effect: Misinterpretation of Corona Wind Phenomena", AIAA Journal 42(2) (2004).

[5]   L. Zhao and K. Adamiak, "Numerical analysis of forces in an electrostatic levitation unit," J. of Electrostatics, **63**, pp. 729-734 (2005).

[6]   L. Zhao, K. Adamiak, "EHD gas flow in electrostatic levitation unit", J. of Electrostatics, **64**, pp. 639–645 (2006).

[7]   A. A. Martins, M. J. Pinheiro, "Modeling of an EHD corona flow in nitrogen gas using an asymmetric capacitor for propulsion", J. of Electrostatics, **69** (2), pp. 133–138 (2011).

[8]   R. Ianconescu, D. Sohar, and M. Mudrik, "An analysis of the Brown- Biefeld effect," J. of Electrostatics, **69**, pp. 512–521 (2011).

[9]   M. Chen, L. Rong-de, Y. Bang-jiao, "Surface aerodynamic model of the lifter," J. of Electrostatics, **71**(2), pp.134–139 (2013).

[10] F. X. Canning, C. Melcher, and E. Winet, "Asymmetrical Capacitors for Propulsion," NASA, NASA/CR—2004-213312 (2004).

[11] J. Wilson, H. D. Perkins, and W. K. Thompson, "An Investigation of Ionic Wind Propulsion," NASA, NASA/TM—2009-215822 (2009).

[12] M. Eiant and R. Kalderon, AIP advances, **4**, 077120-1-20 (2014).

[13] K. Masuyama and S. R. H. Barret, "On the performance of electro hydrodynamic propulsion", Proc. of Royal Soc. A, **469** (2154), 20120623-1-16 (2013).

[14] T. Kouno, "High Voltage Engineering", Asakura Publishing, Tokyo (1995) (Chapter 2, in Japanese).

[15] T. Saiki, "Study on High Voltage Generation Using Flame Column and DC Power Supply", J. of Electrostatics, **70**, pp.400-406 (2012).

28

Comprehensive Analysis on Characteristics of SiC Power Device

Shi Mingming[1], Lu Wenwei[2], Ge Le[2]

[1]Jiangsu Electric Power Research Institute, Nanjing, China
[2]Nanjing Institute of Technology, Nanjing, China

Email address:
simon8612@126.com (Shi Mingming), luwwnjit@126.com (Lu Wenwei), supertiger_bear@126.com (Ge Le)

Abstract: Analyzing the research status and development trend of SiC (Silicon Carbide) power device, this article describes the latest research results of switching characteristics and power loss characteristics of SiC power device. With detailed analysis on switching characteristics of Schottky Barrier Diode (SBD) and MOSFET, this paper emphasizes on the differences between them and the corresponding power devices. The comparison study between power loss characteristics of MPS and valve loss of silicon carbide thyristor for ultra-high voltage, also and the differences in power loss of switching power supply between SiC MOSFET and Si MOSFET provide scientific basis for the optimal selection and application of SiC power device.

Keywords: SiC, Power Device, Switching Characteristics, Power Loss Characteristics

1. Introduction

Power electronic devices are the vital basis of power electronic devices. The characteristics of the devices have vital effects on the technical index and performance of the device [1]. Nowadays, the outstanding advantages of SiC materials encourage researchers to research and develop SiC power electronic devices with high performance, actively promoting its commercialization process to earn a wider range of application advantage and potential. In the field of high frequency, high temperature and high power electronic applications, advantages and great application of power electronic devices SiC power electronic devices has the incomparable potential to Si semiconductor device [2].

SiC power devices commercialization greatly promoted the study of applied technology SiC power devices, particularly in the areas of aerospace, electric vehicles, exploration, power systems and new energy power generation, researchers system-level benefits assessment of the SiC power device, drive circuit design, multi-tube technology and the expansion of power converters to achieve high temperature and many other technical problems explored and studied.

Currently, SiC SBD, SiC JFET and SiC MOSFET have successfully commercialized. Due to the different characterisctics of semiconductor materials, there are some differences in the electrical characteristics between SiC power devices and Si power devices [2-6]. In order to ensure the correct use of SiC power devices and give full play to its advantages, which based on the converter system of SiC power devices can obtain better performance, we need thorough analysis and study on switching characteristics, loss characteristics and parameters of SiC power device, especially on switching characteristics and loss characteristics [5-7].

With rapid development of silicon carbide materials, SiC power semiconductor devices have been widely concerned in the field of switching power supply, being a potential alternative to the Si semiconductor devices. In order for better application of SiC power devices, this paper studies the switching characteristics of SiC SBD and MOSFET, the power loss characteristics of MPS, valve loss of Silicon Carbide thyristor for ultra-high voltage and the differences in power loss of switching power supply between SiC MOSFET and Si MOSFET.

2. Analysis on Switching Characteristics of SiC SBD

Switching characteristics of power diode includes forward

recovery characteristics and reverse recovery characteristics. The following analysis is respectively on SiC and SBD [8].

2.1. Forward Recovery Characteristics

The formation of power diode voltage overshoot is mainly related to two factors: conductivity modulation effect and internal parasitic inductance effect. Due to the absence of conductivity modulation effect, SBD SiC can only be affected by parasitic inductance, and the zero forward recovery voltage of SBD SiC can be basically achieved by process improvement.

2.2. Reverse Recovery Characteristics

Reverse recovery characteristics of power diode is an important index of diode selection, mainly affected by the conductivity modulation effect and parasitic capacitance effect. Si fast recovery diode has conductivity modulation effect, with long reverse recovery time and large spike reverse recovery current. It can induce large voltage spikes on the stray inductance in the line, which may increase the voltage stress of power device. SBD SiC has no conductivity modulation effect, and reverse recovery is mainly affected by the parasitic capacitance in the circuit. As a result, the reverse current spike is small, the switching speed is fast, and the switching loss is small too.

3. Analysis on Switching Characteristics of SiC MOSEFT

3.1. Switching Characteristics

The switching characteristics of SiC MOSFET are mainly concerned with the nonlinear parasitic capacitance of poles. Table 1 gives the similar power level (1200V/10A) of SiC MOSFET and input capacitance Ciss, output capacitance Coss and Miller capacitance Crss capacitance value of Si MOSFET [9, 10]. These parasitic capacitance parameters have obvious influence on the transient switching process of power MOSFET.

Table 1. Comparison of MOSFET Parasitic Capacitance.

Devices Types	Ciss(pF)	Coss(pF)	Crss(pF)
CMF10120D(SiC)	928	63	7.45
IXTH12N120(Si)	3400	280	105

From table 1, the parasitic capacitance values of all poles of MOSFET SiC are much less than that of Si MOSFET in the similar power levels. Based on the switching process of MOSFET, the smaller parasitic capacitance value is, the faster switching speed of MOSFET can be, which can shorten the switching process time. So as to reduce the transition cross region of leakage current and drain source voltage in the switching process, that is to say, it can reduce the switching loss of the MOSFET.

3.2. Driving Characteristics

The performance of the grid drive circuit plays a key role of on the switching process of MOSFET. In the fast switching application of SiC MOSFET, we especially need to seriously consider the existing problems in the driving circuit [8].

From the point of reducing the conduction resistance, it is beneficial to set the driving voltage of SiC MOSFET higher. But for the fast switching process of SiC MOSFET, the current rate of change of drain is large and the voltage rate of change of drain source is large, through the parasitic inductance and the miller capacitance in the circuit, it coupled to the grid and connected to the source electrode through the grid resistor, forming a loop eventually. Both ends of grid resistance will cause serious voltage spike. Due to the possible breakdown phenomenon for the grid in the switching process of SiC MOSFET, the value of positive driving voltage should be limited.

In the shutdown process, due to the lower open voltage of SiC MOSFET, the grid voltage spike may lead to mis-conducting. So we need to set up the negative bias voltage to enhance the anti interference ability of the grid. The most direct way to control the coupling is to reduce the parasitic parameter in the circuit and to reasonably select the driving resistance of external grid.

4. Analysis on Power Loss Characteristics of Silicon Carbide MPS

MPS diode is an integrated diode device on forming a net PN junction in the drift region of Schottky diode. The net structure design as shown in Figure 1 makes the depletion region of PN junction not be interlinked in the positive and zero situation. When it works positively in this way, with current flowing through multiple conductive channels under the Schottky barrier, the device is conductive. When it works negatively, PN junction and Schottky barrier is reverse biased, making the depletion region formed by a PN junction extend to the channel region. As soon as the reverse bias voltage is more than a certain value, the depletion layer of the Schottky barrier will overlap [11].

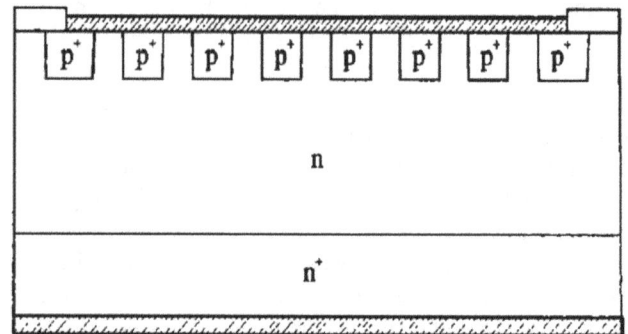

Figure 1. The structure of MPS diode.

When the depletion layer passes through, it will form a potential barrier in the channel region. So as to increase the reverse bias voltage from the depletion layer to the direction of the N+ substrate, it can support this barrier and shield the Schottky barrier from the voltage bias, suppressing the Schottky barrier reduction effect and eliminating the leakage current. As soon as the penetrating condition establishes, the other parameters stay constant in addition to the slight increase in the external leakage current caused by the space charge region.

Power consumption of device is determined by the I-V characteristics. For 4 H-SiC MPS devices, power consumption is composed of three parts: the first one is power consumption in forward conduction, decided by forward voltage dropping, forward conduction current and working cycle; the second one is power consumption in reverse blocking, depending on the reverse bias voltage, leakage current and working cycle; the third one is power consumption in switching process, due to the fast switching speed of MPS diode, the static power consumption is the main part.

4.1. Power Consumption in Forward Conduction

Assuming the forward conduction current is J_{FC}, the conduction time is t_{on}, the total working time is t_{total}. The power consumption P_{DF} can be calculated by the following formula:

$$P_{DF} = J_{FC} V_{FS} \frac{t_{on}}{t_{total}} \tag{1}$$

where V_{FS} is forward voltage drop of Schottky barrier and drift region.

4.2. Power Consumption in Reverse Blocking

Assuming the imposed bias voltage when reverse blocking is V_R, the conduction time is t_{on}, the total working time is t_{total}. The power consumption in reverse blocking P_{DR} is concluded from the following formula:

$$P_{DR} = V_R \left[q \frac{\bar{D}}{f} \frac{n_i}{N_D} + \frac{q n_i W}{f} \right] \frac{t_{total} - t_{on}}{t_{total}} \tag{2}$$

Combining the power consumption in forward conduction and reverse blocking of MPS, we can conclude the total power consumption P_D:

$$P_D = P_{DF} + P_{DR} \tag{3}$$

4.3. How Temperature Effect Power Consumption

When the device temperature rises, height of Schottky barrier, ionization rate of drift region impurity, electron mobility, diffusion coefficient, life and the intrinsic carrier concentration and so on, these factors will change[12], which also lead to the power consumption of MPS. Because SiC has a "frozen" effect that is not completely ionized at room temperature, and the ionization rate changes with the

change of temperature, we must take how temperature effect power consumption into consideration when calculating the relationship between temperature and power consumption.

In the temperature range of 300~900K, the intrinsic carrier of material SiC changes intensely with temperature. Although the diffusion coefficient and the lifetime of the electrons change with temperature, the change of the intrinsic carrier concentration is the main factor affecting the power consumption during reverse blocking.

5. Analysis of Valve Loss Characteristic of Ultra-High Voltage DC Silicon Carbide Thyristor

5.1. Circuit Model

Ultra-high voltage DC transmission system based on thyristor valve is a complex system composed of many electrical components. The situation is more complicated if the exchange system is considered for the coupling of the converter transformer. The circuit model is simplified without sacrificing the equivalence principle in this paper. We use ideal voltage source for AC system. Taking the need of commutation circuit into account, the converter transformer adopts the ideal transformer model with series transformer leakage inductance and think of existing stray capacitance of transformer lead wire end. Single pole converter circuit model is shown in Figure 2 [13-16].

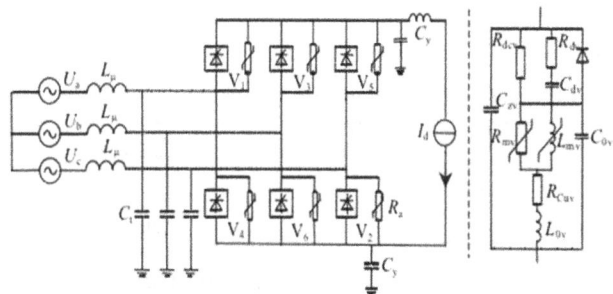

Figure 2. *Working Circuit of Converter Valve.*

L_μ—leakage inductance of converter transformer; C_r—stray capacitance from AC side; C_{ZV}—valve terminal capacitance; R_{dcv}—valve equivalent DC equalizing resistor; R_{dv}—valve damping resistance; C_{dv}—valve damping capacity; R_{mv}—equivalent resistance of valve reactor; L_{mv}—equivalent inductance of valve reactor; R_{CuV}—DC coil resistance of valve reactor; L_{0V}—stray inductance; C_{0v}—equivalent capacitance of valve reactor; C_y—stray capacitance from DC side; R_a—valve arrester.

5.2. Electrical Design Parameters of Converter Valve

The DC converter valve thyristor contains two design schemes in this paper: 1) use conventional commercial silicon devices, that is, Zhuzhou KPE5000-72 thyristor; 2) Silicon carbide KPD5000-400 thyristor is calculated by the characteristics of silicon carbide material, and the main parameters are compared as shown in Table 2 [17].

Table 2. Comparison of Main Parameters of Silicon Thyristor and Silicon Carbide Thyristor.

Parameter Name	Si Thyristor	SiC Thyristor
Repetitive Peak Off-State Voltage U_{DRM}/kV	7.0	40.0
Reverse Repetitive Peak Voltage U_{RRM}/kV	7.2	40.0
Repetitive Peak Off-State Current I_{DRM}/A	0.40	0.04
Reverse Repetitive Peak Current I_{RRM}/A	0.40	0.04
Off-State DC Voltage U_d/V	6450	32000
Off-State DC Current I_d/mA	50	10
Rated Junction Temperature /°C	120	250
State Threshold Voltage U_{T0}/V	1.25	4.40
On-State Slope Resistance r_t/mΩ	0.20	0.34

Due to the increase of the voltage resistance of silicon carbide single thyristor, a large number of thyristor series are reduced. The calculation of each single valve (pressure 200 kV) SIC thyristor series number is 12, but the silicon thyristor series number is 67.

When the converter valve is close to 90 degree, and when the load current is the maximum value of the silicon flow valve and silicon carbide flow valve, the operating parameters are shown in table 3. In the use of silicon carbide thyristor valve, when the valve is close to 90 degrees, we assume that the impact of the saturation reactor is unchanged. As the valve current passes zero, di/dt is constant. According to the reverse recovery charge relationship between silicon carbide thyristor and silicon thyristor, combining the reverse recovery charge characteristic curve, we can calculate the reverse recovery current peak value IRM=130A and the reverse recovery charge is Qn=760µC.

Table 3. Circuit Parameters of Silicon Valve and Silicon Carbide Valve.

Thyristor valve	U_{VN}/kV	I_d/A	α/°	L_μ/mH	di/dt/A/µs)	I_{RM}/A	Q_{rr}/µC
Si	177	5700	88	13.8	5.28	160	7600
SiC	177	5700	88	13.8	5.28	130	760

5.3. Method for Calculating Loss of Converter Valve Unit

The circuit theory points out that the instantaneous power consumption of a circuit element is equal to the product of voltage and current on both ends of the element (P=UI), and if the circuit element is a resistance element, the power consumption of the resistance element is converted into Joule heat. In a working cycle T, the average loss of the resistive element is:

$$P_{av} = \frac{1}{T}\int_0^T UI dt = \frac{1}{T}\int_0^T I^2 R dt \qquad (4)$$

In the study of the basic components of commutation valve's electrical model, the electrical model of the thyristor under different working conditions is a represent of series of resistance, voltage or voltage source and the resistor; R_m represents core loss of saturable reactor, R_{Cu} represents saturated reactor winding. The loss of each component can be calculated in accordance with the formula (4) as long as the voltage and current at the ends of these resistors are determined.

5.4. Loss of Converter Valve and Junction Temperature of Thyristor Under Different Working Conditions

Calculated by the loss of the converter valve components, the commutation components will inevitably generate losses at runtime. Loss associated with thyristor (loss of state, loss of opening, loss of turn off and loss of off state) will increase the junction temperature of thyristor. The blocking ability of thyristor is closely related to the junction temperature of thyristor. If the thyristor junction temperature exceeds the rated junction temperature, the thyristor may lose its ability to block, resulting in the loss of the normal operation of the converter valve [18]. In order to ensure the junction temperature of thyristor, the exsiting DC converter valve is equipped with water cooling system.

5.5. Economic Benefits of Silicon Carbide Thyristor

It can be seen that the SiC thyristor can reduce the loss of the DC converter valve under various working conditions. These savings can be provided to the user, bringing the dual benefits of energy saving, environmental protection and economic profit. Taking ±800 kV, 5 kA DC exchange commutation as an example, we estimate the direct economic benefits to use silicon carbide single valve instead of silicon single valve. Assume the DC converter valve is under the rated operating conditions in 365 days of each year [17]. Silicon carbide single valve power loss is 136kW, so the use of silicon carbide single valve annual power consumption is 397MW, and the use of a single valve annual power consumption is 726MW. In a DC project, if the use of SiC thyristor valve can save 397MW·h annually, according to the ordinary residents electricity fee 0.5 yuan/kW·h to calculate, the electricity loss annual saved can directly gain profit of 17.5 million yuan.

6. Comparative Analysis of Power Loss of SiC MOSFET and Si MOSFET in Switching Power Supply

6.1. Power Loss Analysis of Switching Power Supply

The power loss of MOSFET in switching power supply is mainly the conduction loss and switching loss [19]. Specific analysis are shown as follows:

6.1.1. Conduction Loss

Conduction loss of MOSFET P_Q in switching power supply is decided by the conduction resistance $R_{Q(on)}$, the calculation of conduction loss P_Q is shown in the formula (5):

$$P_Q = I_{Prms}^2 \cdot R_{Q(on)} \qquad (5)$$

where I_{Prms} is the current value flowing through the switching tube. Under the condition of keeping I_{Prms} the same, the PQ is proportional to the conduction resistance $R_{Q(on)}$. The size of $R_{Q(on)}$ is changing with the junction temperature T of MOSFET, Specific relationship is $R_{Q(on)} \infty T^{\gamma}$. where γ is a constant, γ of Si is 2.42, but γ of SiC is 1.3 [20]. In the high junction temperature of 135 C, the resistance of SiC MOSFET is only increased by 20% while the Si MOSFET is up 240%. Therefore, the SiC MOSFET device is suitable for working under high temperature environment, and comparing with Si MOSFET device, only smaller heat sink is required.

6.1.2. Switching Loss

Switching losses are generated due to the switching time of the MOSFET. In the process of MOSFET's communication and interruption, due to the effective voltage and current working at the same time, the switch stack of MOSFET has a long time to cause the loss of MOSFET. The calculation of switching loss P_{SW} is shown as the formula (6):

$$P_{SW} = P_{SW(on)} + P_{SW(off)} = \int_{t_{on}, t_{off}} u_{ds}(t) i_{d}(t) dt \qquad (6)$$

Switching loss P_{SW} of MOSFET Mainly includes opening loss $P_{SW(on)}$ and turn off loss $P_{SW(on)}$. The size of P_{SW} is not only related to drain source voltage of switching tube u_{ds} and drain current i_{d}, but also related to the opening time t_{on} of the switch tube and the turn off time of t_{off}. the smaller Miller Capacitance MOSFET between grid and drain is, the faster switching speed of MOSFET is, the smaller the switching loss is [21].

Under the condition of the same power bus voltage U_S and the same output power, the switching loss of the flyback switching power supply is mainly the turn off loss $P_{SW(off)}$. Due to the drain source voltage u_{ds} reduces to close to 0 from the power bus voltage U_S in the opening process, the drain current rises slowly. Therefore, relative turn off loss $P_{SW(off)}$ and opening loss $P_{SW(on)}$ is very small, almost negligible.

6.2. Comparative Test of the Switching Characteristics of SiC MOSFET and Si MOSFET

6.2.1. Switching Speed of SiC MOSFET and Si MOSFET

The switching speed of SiC MOSFET and Si MOSFET is compared with the test respectively, which shows change of grid source voltage of SiC MOSFET and Si MOSFET in the switching process. Experiment shows that SiC MOSFET is much faster than Si MOSFET. Therefore, the turn off loss $P_{SW(off)}$ of SiC MOSFET will be significantly less than the turn off loss of Si MOSFET.

6.2.2. Comparative Test of the Power Loss of MOSFET

The opening loss, turn off loss, conduction loss and total loss of SiC MOSFET and Si MOSFET in a switching period are calculated by using the test calculation software coming with TPS2024 isolation channel oscilloscope. Because the switch speed of SiC MOFET is significantly faster than Si MOSFET, so there is a significant difference in the turn off

loss. The turn off loss of Si MOSFET is 6.26W, and the turn off loss of SiC MOSFET is 61.0W. Due to the limitations in PWM chip driving voltage in the test, the grid drive voltage of MOSFET is 12V, the conduction resistance of SiC MOSFET is about 3 times more than that under the 20V grid voltage. So there are no obvious advantages of SiC MOSFET in conduction loss, both being 2.59W. Due to the switching loss of the flyback switching power supply is mainly the turn off loss, and the opening loss is very small, the opening loss of the two devices is 0 in the test.

If the grid drive voltage of SiC MOSFET is up to 20V, the conduction resistance will be reduced to 1/4 under the 12V grid voltage state.

6.3. Comparative Test of the Efficiency of Switching Power Supply

Compared with the high switching speed of Si MOSFET and SiC MOSFET, the power loss in flyback switching power supply can significantly reduce. In order to verify the effect of the power loss of SiC MOSFET in the switching power supply, we test the switching power supply input power of SiC MOSFET and Si MOSFET respectively [22].

It shows that the switching power supply input power of SiC MOSFET is 1.28 kW, and the switching power supply input power of Si MOSFET (effective) is 1.33 kW, with a difference of 0.05kW. It is a part that loss of Si MOSFET higher than that of SiC MOSFET, accounting for 3.8% of the total input power. It indicates that even if we directly use SiC MOSFET instead of Si MOSFET, it will also improve the efficiency of nearly 4%, and if the grid drive voltage is up to 20V, the increase of the efficiency will be more obvious.

7. Conclusion

We study the characteristics and parameters of SiC SBD and SiC MOSFET in this paper. The main contents are shown as follows:

1) To describe the research status of the application of SiC power devices, and to point out the problems of the application of SiC power devices.

2) To study the switching characteristics of SiC SBD and SiC MOSFET, and to compare and analyze the switching characteristics of SiC power device and Si power device.

Through the analysis of the power loss characteristics of silicon carbide MPS, the valve loss of high voltage DC silicon carbide thyristor and the power loss of Si MOSFET and SiC MOSFET in the switching power supply. It can be concluded that silicon carbide device has following advantages:

1) We can get the best power loss characteristics by choosing appropriate structure parameters. Therefore, the application of MPS is for high temperature, high power system.

2) If using silicon carbide thyristors to replace silicon thyristors, due to under the condition of the same trigger angle, total power loss is greatly reduced. It can significantly reduce the requirements of cooling equipment, more benefit to the normal operation in the DC system from the aspect of

operation, and can save large amounts of electricity per year, bringing economic benefits that can not be ignored.

3) SiC MOSFET is a high voltage switching device with excellent performance. It not only has a good blocking ability and a low turn-on voltage, it also has the extremely fast switching speed, which can be used for high voltage and high frequency switching power converter field.

References

[1] Qian Zhaoming, Sheng Kuang. Development and perspective of high power semiconductor device [J]. Converter Technology & Electric Traction, 2010, (1): 1-9.

[2] Chen Zhiming, Li Shouzhi. Wide band gap semiconductor power electronic devices and their applications [M]. Beijing: China Machine Press, 2009.

[3] Agarwal A K. An overview of SiC power devices [A]. ICPCES 2010 [C]. 2010, 1-4.

[4] Agarwal A, Callanan R, Das M, et al. Advanced HF SiC MOS devices [A]. Power Electronics and Applications, 13th European Conference 2009 [C]. 2009. 1, 8-10.

[5] Jiang Dung, Burgos R, Wang Fei, et al. Temperature-dependent characteristic of SiC devices: performance evaluation and loss calculation [J]. IEEE Transactions on Power Electronics, 2012, 27(2): 1013-1024.

[6] Glaser J S, Nasadoski J J, Losee P A, et al. Direct comparison of silicon and silicon carbide power transistors in high-frequency hard-switched applications [A]. APEC2011 [C]. 2011. 1049-1056.

[7] Alatise O, Parker-Allotey N A, Mawby P. The dynamic performance of SiC Schottky barrier diodes with parasitic inductances over a wide temperature range [A]. PEMD2012 [C]. 2012, 1-6.

[8] Zhao Bin, Qin Haihong, Ma Ceyu, et al. Exploration of switching characteristics of SiC-based power devices[J]. Advanced Technology of Electrical Engineering and Energy, 2014, 33(3): 18-22.

[9] http://www.cree.com/.

[10] http://www.ixys.com/.

[11] Niu Xinjun, Zhang Yuming, Zhang Yinmen, et al. Analysis of Power Dissioation Characteristics of MPS Based on SiC [J]. RESEARCH & PROGRESS OF SSE, 2003, 23(2): 193-198.

[12] Zhang Yuming, Study on silicon carbide materials and devices, doctoral dissertation, Xi'an Jiao Tong University, 1998. 4.

[13] IEC61803-1999, Determination of power losses in high-voltage direct current converter [S].

[14] Cepek M. Loss measurement in high voltage thyristor valves [J]. IEEE Trans on Power Delivery, 1994, 9(3): 1222-1236.

[15] Kimbark E W. Direct current transmission: Vol. I [M]. New York: John & Sons, Inc, 1971: 21-25.

[16] Uhlmann E. Power transmission by direct current [M]. Heidelberg, New York: Springer-Verlag Berlin, 1995: 37-41.

[17] Jin Rui, Lei Linxu, Wen Jialiang, et al. Discussion on Power Loss of HVDC Converter Valves Adopting Silicon Carbide Thyristors [J]. Power System Technology, 2011, 35(3): 8-13.

[18] Wen Jialiang, Zha Kunpeng, Gao Chong, et al. Research and development of whole-set operational test UHVDC thyristor valves [J]. Power System Technology, 2010, 34(8): 1-5.

[19] Sha Zhanyou, Wang Yanpeng, Ma Hongtao, etc, Optimal design of switching power supply [M]. Beijing: China Electric Power Press, 2009: 260-261.

[20] Hosseini Aghdam M G, Thiringer T. Comparison of SiC and Si power semiconductor devices to be used in 2.5 kW DC-DC converter[C]. International Conference on Power Electronics and Drive Systems, PEDS, 2009: 1035-1040.

[21] Nathabhat Pankong, Tsuyoshi Funaki, Takashi Hikihara. Characterization of the gate-voltage dependency of input capacitance in a SiC MOSFET [J]. IEICE Electronics Express, 2010, 7(7): 480-486.

[22] Cao Hongkui, Chen Zhibo, Meng Linan. Comparative Analysis of SiC MOSFET Power Losses in Switching Power Supply [J]. Journal of Liaoning University of Technology (Natural Science Edition), 2014, 34(2): 82-85.

29

Research on Selectivity of Co-processing Protection Methods for Ship Distribution Network Based on Communications

Huang Jing, He Huiying, Zhou Shiwan, Zhang Tao[*]

Electrical Engineering Department, School of Electrical Engineering, Naval University of Engineering, Wuhan, China

Email address:

836148867@qq.com (Zhang Tao)

[*]Corresponding author

Abstract: Relaying system is a distributed discrete structure, each protector and detection unit is distributed in logic and physics. In this text, the authors propose a selective protection method of ship distribution network based on the strategy of communication coordinate, designing the protector to intelligent structure which has the function of fault current detection, calculation of current rising rate, communication and coordination processing. For communication coordinated strategy in ship power distribution network, the application of selective protection research, to improve the effectiveness of the selective protection of ship power distribution network is of great significance.

Keywords: Power Distribution Network, Selective Protection, Collaborative Strategy, Communication

1. Introduction

With the increasing capacity of ship power system, the structure is increasingly complex, Higher request is made to the relay protection of ship power system [1-4]. Ship distribution series is increasing, upper and lower level protectors often trip at the same time when appearing large short circuit current [5-7]. For promising the safe operation of ship distribution network, for completing well selective protection, for improving the electric energy quality of ship power system and maintaining the continuity of power supply, it has great significance to carry out the research of selective protection of intelligent ship distribution network. [8-12]

In this text, we propose a selective protection method of ship distribution network based on the strategy of communication coordinate, designing the protector to intelligent structure which has the function of fault current detection, calculation of current rising rate, communication and coordination processing. Analysis principle of fault location based on communication and coordination, proposing the selective promotion strategy based on communication and coordination, giving the coordinate acting relationship of protector and general principle. Giving a specific coordination strategy of each level protector aimed at distribution network of three levels. Completing the design of the structure of ship distribution system protector based on communication cooperation strategy, completing setting up hardware platform. The experimental result tested the effectiveness of communication cooperation strategy.

2. Analysis of Short Circuit Fault Location Principle Based on Multi-computer Communication

To combine the relay protection system, in this paper, each of the protection system to protect device designed to the corresponding intelligent node, this intelligent node have detection function, fault protection setting function, communication function and protection function. Detection function and protection setting calculation is protector uses current sensor for detecting current flows through the protection of the installation points and detect the current through A/D converter to intelligent module, intelligent modules completed to calculate current rising rate and analysis and fault judgment; Use protector of communication function will determine the results and their own protector status is sent to the superior protector, at the same time receive lower judgment results and motion state, intelligent module of protective device integrated protector of higher and lower judgment results with their own judgment to make fault location and remove action, all protectors complete selective protection together, implementing the detection and selective protection of the distribution network fault.

The purpose of multi-computer communication is using serial communication function of SCM to send the determine results and control action of all levels protectors to the superior protector. Using current rising rate and current setting to realize fault current distributed determine and quickly setting, through the communication coordination to realize fault location and selective removal [12-13]. In typical radiant tertiary distribution network as an example, the protector as per level of protection from the machine directly, the whole distribution network form the main type of communication, as shown in Fig. 1.

In communication, assign a unique address to each protector first. And then define the communication information for three types of data: the first data status is value "2", expressing determination results for short circuit fault; the second data status is value "1", expressing the determination results for no short circuit fault; the third state of data is value "0", expressing not to receive data protection at a lower level. All communication data from down to sent to the superior device.

Completed defines communication data between protector, achieve short circuit fault location with communication data. In Fig. 1, third level protector is located in the end of the distribution network, as long as protector detected the short trouble, identified as the terminal short circuit protector. Secondary protector detected the fault and received triple protector data, fault occurs below the level 3 protector if the data is "2"; short circuit fault occur between secondary protector and tertiary protector if the data is "1"; If the data is "0", that means first protector has not accept from secondary protector, that also means secondary protector is damaged, so short fault appear below the first protector.

3. Selective Protection Based on Communication Coordination Strategy

All these researches are aimed at ship AC power distribution network, built on ship AC power distribution network topology system. As a result, analysis the ship AC power distribution network topology system is needed. Typical 3 level ship radiation power distribution network in Fig. 2 as an example, state in detail the distributed short fault decide and communication cooperate strategy ideology.

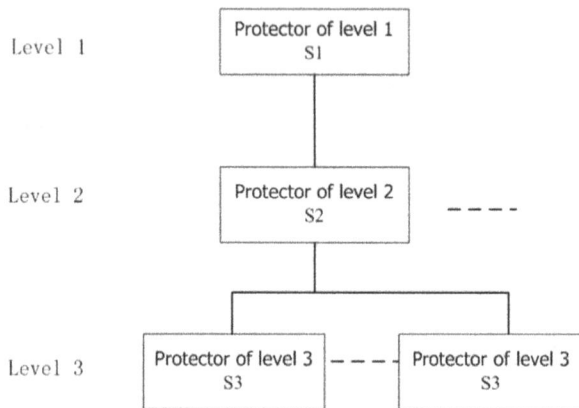

Fig. 1. Three level SCM communication model.

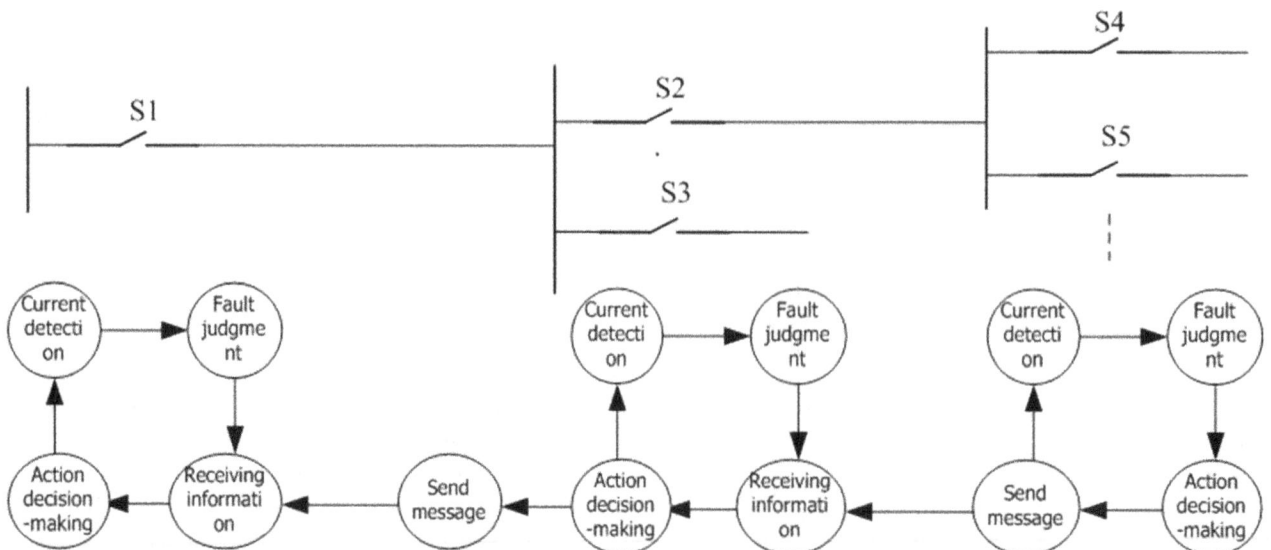

Fig. 2. Three level ship radiant power distribution network.

Shows as the figure, protector sends the judge results to superior protector through communication, accept judge results from subordinate protector at the same time, complete current detection, fault judgment and fault location. In that way, protector act or not next, accomplish selective protection or not are depend on communication cooperation strategy, that means for accomplishing selective protection, the important is the communication cooperation strategy of each level of protector.

Action dependence of each protector mainly include the two terms: The first is detection and judgment results of the protector itself; the second is data accepted in subordinate protector. The principle of protector: detect on the same level, calculation and short fault has judged, accept judgment result from subordinate protector, according to the previous section fault location principle, decide whether the protector to remove faults at the corresponding level. If the data received from lower level protector shows appearing short fault, protector at the corresponding level will not act; if the data received from lower level protector shows do not appear short fault, protector at the corresponding level will act to remove the fault.

Here is coordinated strategy of each level protector in three level distribution network.

Coordinated strategy of three level protector: For three level protector S4, S5, S6, S7 located at the protection position in the end of distribution network, when judged to be short fault according to current rising rate, protector will act to remove fault directly, and send judgment result and the action state of protector to superior protector.

Coordinated strategy of second level protector: Protector S2 detected short fault itself, also need to receive judgment result from S4 and S5, then completing fault location according to fault location theory. Making action decision combine detection and judgment result itself, and send the detection and judgment result and action decision to first level protection. Action strategy of second level protector as showed in table 1.

Table 1. *Protector S2 action decisions.*

Information of S4 and S5 / Determine results of s2	00, 01, 10, 11	12, 21, 22
Yes	Action	Inaction

Corresponding strategy of first level protection: First level protector S1 detected short fault, and receive judgment result from protector S2 and S3, making action strategy combine the information received, as shown in Table 2.

Table 2. *First level protector action decisions.*

Information of S2 and S3 / Determine results of s1	00, 01, 10, 11	12, 21, 22
Yes	Action	Inaction

4. Design of Protector Structural

Detection and protection setting calculation function is

protector detect the current flow through protection mounting point, the current detected flow through A/D converter will sent to intelligent module, intelligent module completing calculating and analysis the current rising rate received and making fault judgment; Using communication function, protector send the judgment result and action state to superior protector, and receive judgment result and action state from lower level protector at the same time, intelligent module of protector combines superior and subordinate judgment result and judgment result of itself, making fault location and removing action, disposing cooperate and the realize selective protection. Integral structural of protector is shown in Fig. 3.

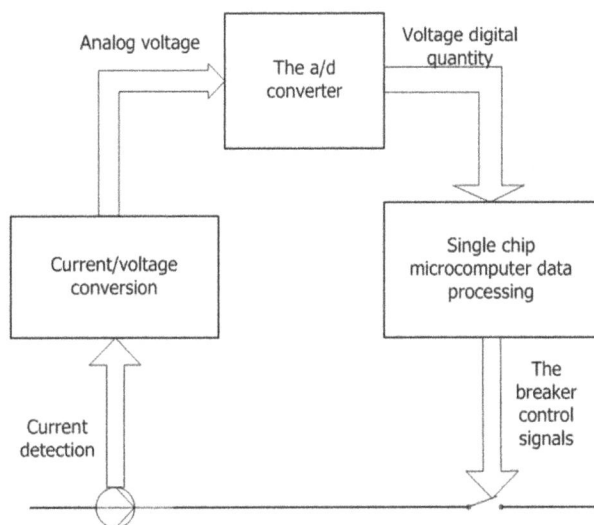

Fig. 3. *The whole structure of protector.*

Using hall current sensor to collect current flow through protector uninterrupted, A/D converter transform the current collected from sensor to digital signal and transport to SCM, SCM undertake all the function of intelligent module, completes data calculation, fault setting, communication processing, fault remove and output control. In the whole process, sampling frequency, conversion accuracy, conversion rate and coordinate strategy are main content to realize selective protection. Then choosing device aimed at the need of each module.

(1) Current Detection Module

Current detection module mainly converts the current flow through protector in ship distribution network to small-signal current that can be read, then convert the current to voltage signal. This text mainly aimed at detecting and protecting short fault, the current flow through protection is bigger when short-circuited, for meeting the require of detection and control, also do not affect the work of distribution network, we choosing hall current sensor to detect the current in distribution network. The voltage signal detected from hall current sensor is analog quantity, it cannot be disposed by SCM directly, converting the analog quantity to digital quantity that can be disposed by SCM. This text is mainly aimed at protecting the short fault in ship alternating current power system, the voltage signal detected is AC signal, and having a high requirement for converting accuracy.

(2) SCM Processing Module

For meeting the requirement of calculation setting, rapidly fault judgment and fast communication in devices, SCM processing module choose 32-bit processor ARM SCM LPC1768. ARM SCM integrate all kinds of modules, possess many distinctive characteristics: First, processing rapidly, powerful, bigger storage capacity and convenient for client upgrade and extension and transformation; Second, highly safe, the data is not easy to lose. Third, ARM SCM bring their own network communication port, it can combine peripheral devices and realize network communication. Fourthly, it can download and upgrade from network directly, maintain conveniently.

(3) Multi-computer Communication Based on CAN Bus

CAN bus support serial communication network of distributed control and real-time control, main feature: (1) CAN bus can use multi-communication, it is flexible and convenient; according to priority to bus access. (2) Using non-destructive bus verdict technology, high priority level node can send data preferentially, low priority level node stop sending data when sending. (3) It can meet the requirement of point-to-point and one-to-many sending and receiving data. (4) Far transmitting range, communication rate reach up to 1Mbps. (5) It uses data frame structure, possessing CRC detection, anti-interference capability of the data is high, low error rate. (6) It can remove connection automatically, make sure that other operation will not be disturbed when error seriously.

For realizing the judgment result of high level and low level protector can be communicated rapidly, meeting the need of protector one-to-many communication coordinate processing. This text chooses CAN bus that possessing multi-communication, higher anti-interference performance and higher real-time performance to complete multi-communication among protector.

5. Experiment Confirm of Selective Protection Method Based on Communication Coordinate Strategy

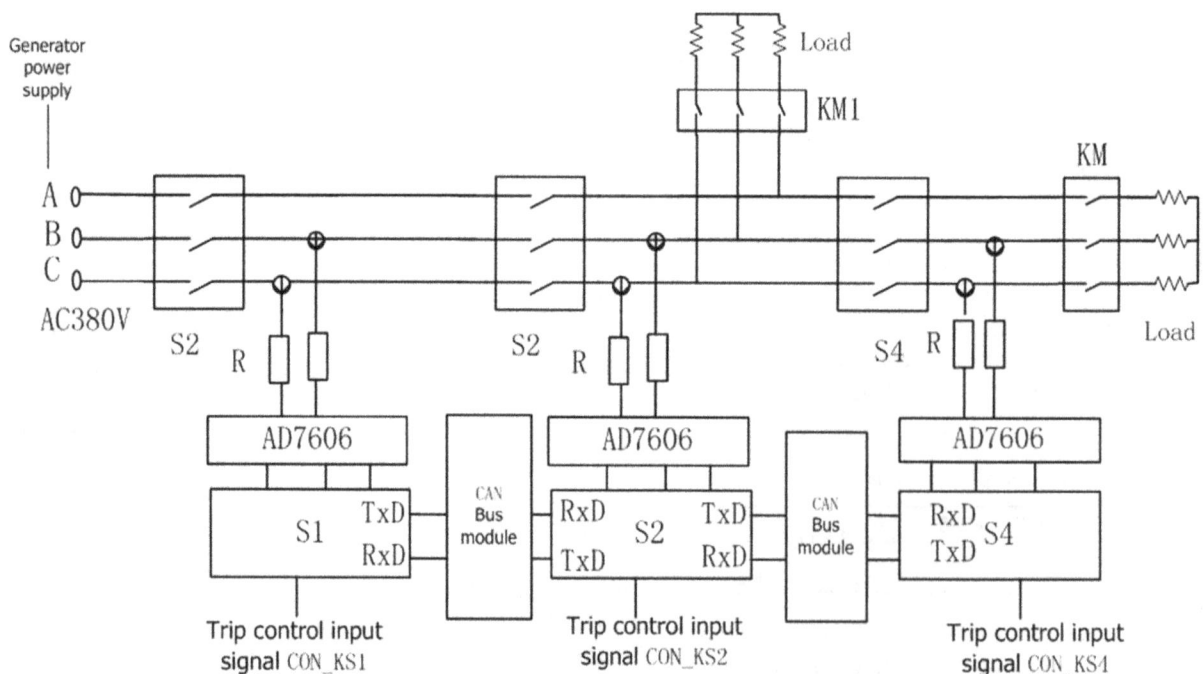

Fig. 4. Circuit of selective protection system.

For realizing selective protection of ship distribution network, simulating the method in this text of selective protection method based on communication coordinate strategy, in this chapter, we process the experiment confirm. Experiment purposes: Confirming whether the protection method based on communication coordinate strategy can realize selective protect or not, the rapidity of communication coordinate strategy among different protectors. Experiment is divided into three part: short current obtain, A/D converting and removing short fault selectively by communication method.

5.1. Experimental Circuit

Circuit of two and three level protector system is shown in fig. 4, in the Figure, S2 and S4 are the representations of two and three level protectors, KM is short circuit access control contactor which mainly controlling sudden load short fault. Protection system uses current sensor to collect current data, using four current sensor to collect the current of B phase and C phase, using the formula $i_a = -(i_b + i_c)$ to calculate current of A phase, the type of current sensor: LMK（BH）-0.66CT, transformation ratio is 900A/5V, A/C transfer resister is

10Ω.Generator set is single simulate synchronous generator, capacity: 5kVA, nominal voltage: 380V, nominal current: 12.55A. For limited experiment condition, synchronous generator cannot conduct short-circuit experiment repeatedly, in this experiment, we simulate short-circuit by sudden loading big load. Using the similarities between short current within a quarter periods to simulate short current, testing the function of selective protection system based on current rising rate and communication coordinate strategy.

System working state before short fault:S4 and S5 master contactor switch on, contactor KM switch off, the current flow through protector S2 and S4 is zero, current detected by four sensor are all zero.

Working state when in short-circuit: KM switching on, switch in short fault, short-circuit current flow through protector S2 and S4, all the four sensor detected short-circuit current. S2 and S4 remove the fault according to communication coordinate strategy. The physical map of selective protection is shown in Fig. 5.

Fig. 5. Physical map of selective protection system.

5.2. Experimental Data and Analysis

In the experiment, it is no-load before short-circuit, current flow through hall current sensor is 0; Current flow through hall current sensor is short-circuit current after short-circuit. For testing the selective protection method based on communication coordinate strategy in this text, next

simulating and analysis from different condition.

(1) Simulation: The selectivity and effectiveness of protection system when the end of transmission line appears short-circuit fault.

When appearing short fault in the end of transmission line, the waveform of short fault communication of two and three level protector short fault communication and S4 removing fault is shown in Fig. 6. Four waveform, 1) is current waveform of A phase, 2) is communication state of two and three level protector, 3) is release action control signal of three level protector S4. 4) is release action control signal of third level protector S2.

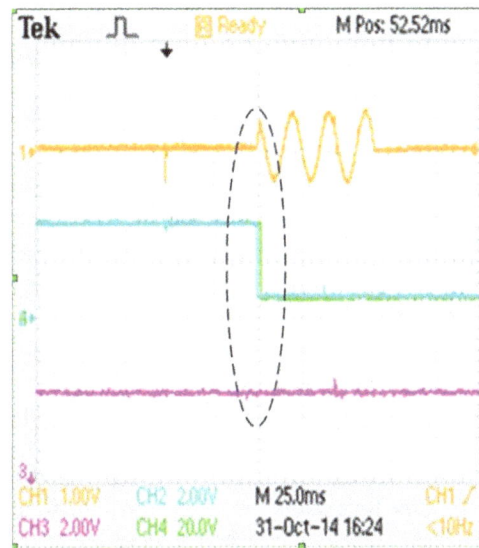

(a) Waveform of fault current excision

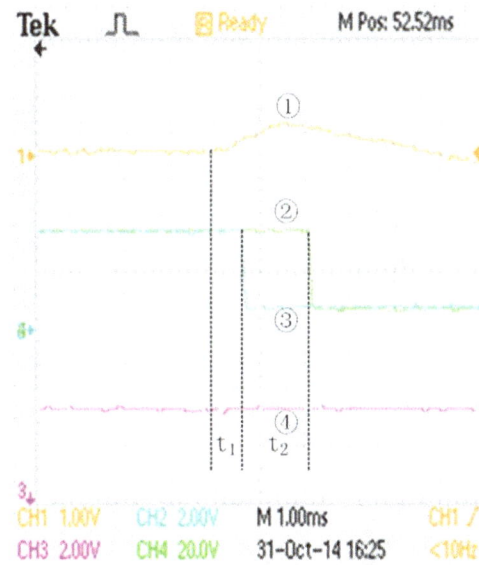

(b) Partial enlarged drawing

Fig. 6. Waveform of protector removing short fault.

It can be seen from "A" phase current waveform, the peak value of current is 20.42A, initial current rising rate up to 15.70A/ms.

In Fig. 6(b), t_1 is the interval time from initial moment to the time third level protector S4 output release act control signal, it is about 400µs. Within t1, the second and third level protector S2 and S4 completed current detection and fault judgment according to rising rate at the same time, the third level protector outputs release action control signal directly according to action strategy, touch off the breaker to switch off. t_2 is the interval time from S4 outputs release action and control signal to the second and third level protector S2 and S4 completed communicate, it is about 1ms. S2 reach judgment result of different action after communication according to cooperation strategy, thus the fourth waveform in Fig. which shows release do not output action and control signal do not appear lower jump.

This implies that protection system completed fault detection, calculation setting, communication cooperation judgment and release control the signal input, tested the accuracy and effectiveness of protection system communication cooperation when appear ending short-circuit. Because the chosen protector release executes some responses slowly in experiment, short current cannot be switch-off immediately after release input control signal, this lead to switch-off the short current after 75 ms, but thinking that we mainly research selective protection strategy, action speed of the switches will not affect the test of protection strategy.

(2) Test: The selective and effectiveness of protection system when second and third level protector appears short fault.

The waveform of second level protector remove short fault is shown in Fig. 7, the four waveform in Fig. top to bottom, the first is waveform of A phase, the second is release control and action signal output from second level protector S4, the third is release control and action signal output from second level protector S2, the fourth is waveform of second and third level protector communication state.

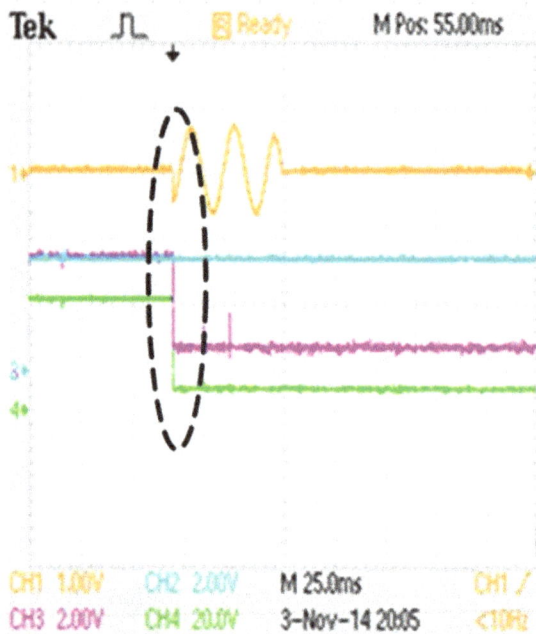

(a) Waveform of fault current excision

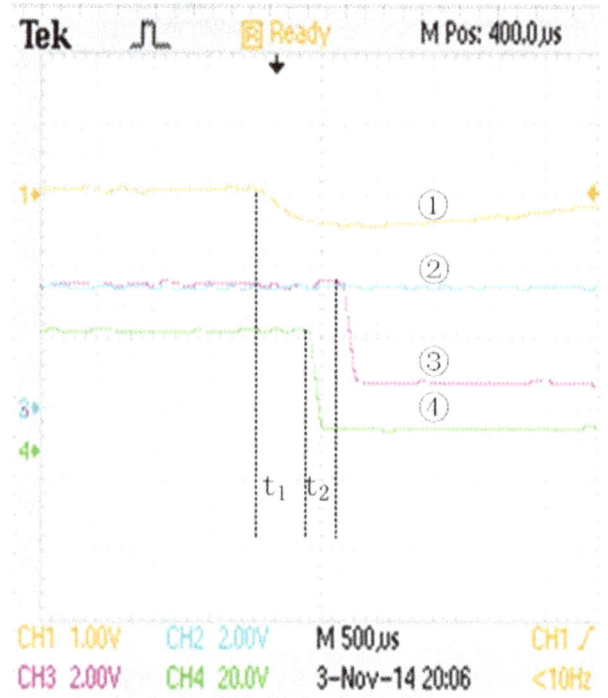

(b) Partial enlarged drawing

Fig. 7. *Waveform of protector S2 removing short fault.*

Seen from the waveform of "A" phase current, the peak of current is 19.82A, initial current rising rate up to 15.24A/ms. Rise rate judgment setting value of S4 and S2 is still 14.85A/ms.

In Fig. 7(b), t_1 is the interval time from initial time of short-circuit to the time second and third level protector completed communicating, it is about 400µS. Within t_1, the second and third level protector S2 and S4 completed current detection and fault judgment according to rising rate at the same time, third level protector do not tested short current and send the judgment result to the second level protector S2.second and third level protector completed communicating. t_2 is the interval time from the second and third level protector completed communicating to the time second level protector output the release action and control signal, it is about 200µs.Within t_2, the second level protector combines received judgment result from S4 with judgment result itself, reach to the needing action to remove fault according to communication cooperation strategy, outputting release action and control signal, thus the third waveform appears lower jump.

Experiment result tested the correctness and effectiveness of protection system according to communication cooperation strategy when the second and third level protector appears short fault. In a family way, because the chosen protector release executes some responses slowly in experiment, short current cannot be switch-off immediately after release input control signal, this lead to switch-off the short current after 75 ms, but thinking that we mainly research selective protection strategy, action speed of the switches will not affect the test of protection strategy.

6. Conclusions

Aimed at the question that it usually appears upper and lower levels protector switch-off at the same time when the short current is big, we present the ship distribution network selective protection method based on communication cooperation strategy. Giving a specific coordination strategy of each level protector aimed at distribution network of three levels. Completing the design of the structure of ship distribution system protector based on communication cooperation strategy, completing setting up hardware platform. The experimental result tested the effectiveness of communication cooperation strategy. Experiment test result reached that the ship distribution network selective protection method based on communication cooperation strategy is correct and effective. We carry out further research in the terms below: (1) In the context that ensure the correctness of test and judgment, we wish to increase the speed of communicating processing. (2) We wish to find a kind of breaker and contactor that response rapidly to increase the speed of fault switch-off.

Acknowledgements

This article is one of phased achievement of NSF "Coordination protection method of ship integrated power system based on multi-agent" (51207165).

Reference

[1] Liu Qiang. Technology discussion of smart power grid relaying [J]. Jiangsu electrical engineering, 2010(02): 82-84.

[2] Jiang Xiaoping, Zhang Peng. Design of smart relaying protector [J]. Association for science and technology BBS, 2011(11), 23-24.

[3] Ji Luming, Zhang Huailiang. Research of ship power distribution network structure [J]. Ship engineering, 2009(02): 35-38.

[4] Li lin, Shenbing, Zhuang Jinwu. Ship power system [M]. WuHan: The tides press. 2003.

[5] Yang Hu. Ship power system microcomputer protection and monitoring system research [D]. Wuhan: Huazhong university of science and technology, 2007, 06.

[6] Mihnko B D, Zoran M. Digital signal processing algorithm for arcing faults detection and fault distance calculation on transmission lines [J]. electrical power and energy system, 2001, 19(3): 165-170.

[7] Zhang Yongwen, Zhong Xiaoming. The selective protection of technology of low voltage power distribution system [J]. Electrical technology, 2007(09): 59-64.

[8] Amann N P. A Development of an adaptive protection scheme for shipboard power systems [D]. Mississippi State Univ, 2007.

[9] BO Z Q. A daptive non-communication protection for power lines BO scheme the delayed operation approach [J]. IEEE Transaction on Power Delivery, 2002, 17(1): 85-91.

[10] Ding Yixian, Zhao Ruifeng, Le ying. The selective protection of intelligent power distribution system [J]. Low voltage electrical appliances, 2013(08): 44-47.

[11] Feng Zongheng. Low voltage power distribution system selectivity and equipment selection of over-current protection [J]. Electrical engineering application, 2003(4): 17-23.

[12] Tu Jian, Zhang Xianhe, Liu Jinhua. [J]. Based on wavelet transform is the traveling wave ranging and protect the selective research of implementation Coal mine safety, 2009(11): 57-60.

[13] Yin Xianggen, Wang Yang, Zhang Zhe. Zone-division and tripping strategy for limited wide area protection adapting to smart grid [J]. Proceedings of the CSEE, 2010, 30(7): 1-7 (in Chinese).

[14] Shi Ji. The electric power communication and its application in the smart grid [J]. Digital Technology and Application. 2012(06): 50-51.

30

Distributed Generation Placement in Distribution Network Based on Power-Flow Relations

Hossein Karimianfard

Department of Electrical Engineering, Jahrom Branch, Islamic Azad University, Jahrom, Iran

Email address:
Karimianfard@Gmail.com

Abstract: Installing Distributed Generation (DG) influences distribution network stability and losses. The aim of the Distributed Generation placement (DGP) is to provide the best locations and sizes of DG units to optimize electrical distribution network operation and planning taking into account DG capacity constraints. In this paper a really formula based on Power flow for obtaining DGP and sizing to improve the performance of the distribution network is offered. The aim of this really formula is to eliminate complex and time-consuming algorithms in solving DGP problems. Using the proposed method, one can easily and without complex calculations obtains the size and location of DG units on the network. The proposed method has been tested on several networks and the test results show its effectiveness. The benefit of the proposed method is its low computational burden as well as its simplicity of implementation.

Keywords: Distributed Generation, Distribution Network, Power Flow, Power Loss, Voltage Profile

1. Introduction

The term distributed generation is defined as power generation technologies below 10 mw electrical output that can be sited at or near the load they serve. DG can be fitted with different strategies in the power network. For example, reducing power losses, cost reduction peak load, improving voltage profile or system reliability, all depend on the location and size of the DG [1]. Decision about DG placement is taken by their owners and investors, depending on site and primary fuel availability or climatic conditions. Although the installation and exploitation of DGs to solve network problems has been debated in distribution networks, the fact is that, in most cases, the distribution system operator has no control or influence about DG location and size below a certain limit [2]. In some works [3] the impacts of DG place on reconfiguration results have been studied. In [3], authors have dealt with network reconfiguration and DG placement simultaneously using Harmony Search Algorithm (HSA) based only on minimization of power losses. Ref [4] proposed a new method with transformation of variables to optimally allocate DG resources in a meshed network. They expressed the benefits of DG as a performance index, which was the minimization of network losses. Ref [5] placed DG units at the buses most sensitive to voltage collapse, and resulted in improvement in voltage profile, as well as decline in the power losses. Ref [6] proposed an analytical method to determine optimal location to place a DG in distribution system for power loss minimization. Ref [7] presented a multi-objective algorithm using evolutionary algorithm for sitting and sizing of DG in distribution system. Lots of methods have been applied to solve the optima DGP including heuristic and mathematical programming method [2].

In this paper a formula for solving distributed generation placement and sizing problem presented. The proposed formula has been validated on seven standard test systems. The rest of this paper is organized as follows: Section 2 gives the problem formulation and constraints of the problem, Section 3 the proposed formula and Section 4 the numerical example, Section 5 outlines conclusions.

2. Problem Formulation

2.1. Power Flow Equations

Distribution system power flow is calculated by the following set of recursive equations derived from the single line diagram shown in Fig. 1. Transfer admittance from node ith to jth and reverse, which are given as:

$$Y_{i,j} = \frac{1}{R_{i,j} + JX_{i,j}} \tag{1}$$

$$Y_{j,i} = \frac{1}{R_{j,i} + JX_{j,i}} \tag{2}$$

Line current I_i, which measured at the node of ith and the in positive direction of ith to jth be defined which is equal to:

$$I_i = V_i \sum_{j=0}^{n} Y_{i,j} - \sum_{j=1}^{n} Y_{i,j} V_j \quad j \neq i \tag{3}$$

Line current I_j, which measured at the node of jth and the in positive direction of jth to ith be defined which is equal to:

$$I_j = V_j \sum_{i=0}^{n} Y_{j,i} - \sum_{i=1}^{n} Y_{j,i} V_i \quad i \neq j \tag{4}$$

Active and reactive power flowing out of node ith and jth, which are given as:

$$P_i + JQ_i = V_i I_i^* \tag{5}$$

$$P_j + JQ_j = V_j I_j^* \tag{6}$$

Voltage magnitude at node ith and jth, which are given as:

$$V_i = \frac{P_i + JQ_i}{I_i^*} \tag{7}$$

$$V_j = \frac{P_j + JQ_j}{I_j^*} \tag{8}$$

Apparent powers S_i form node ith to jth, and S_j from node jth to ith, are equal to:

$$S_i = V_i I_i^* \tag{9}$$

$$S_j = V_j I_j^* \tag{10}$$

The power loss in the line section connecting nodes ith and jth may be computed as:

$$S_{Loss,i,j} = S_i + S_j \tag{11}$$

The total power loss of the system is determined by the summation of losses in all the line section, which is given as:

$$S_{Loss}^{Total} = \sum_{i=1}^{b} S_{Loss,i,j} \quad i \neq j \tag{12}$$

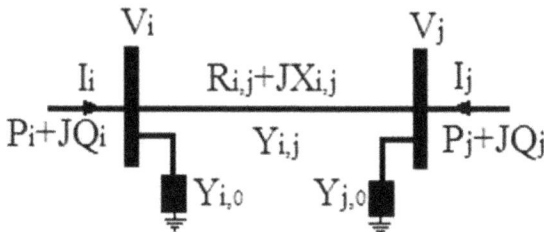

Fig. 1. *A representative branch of a Distribution System.*

2.2. Constraints of the Problem

Apparent Power flow through any distribution feeder must comply with the capacity of line, those are:

$$|S_i| \leq |S_{i,max}| \tag{13}$$

$$|S_j| \leq |S_{j,max}| \tag{14}$$

Node voltages and line currents limits:

$$V_{i,min} \leq |V_i| \leq V_{i,max} \tag{15}$$

$$|I_{i,j}| \leq |I_{i,j,max}| \tag{16}$$

Power conservation limits:

$$S_{Sub} + \sum_{i=1}^{nc} S_{DG,i} = \sum_{i=1}^{b} S_{Loss,i,j} + \sum_{i=2}^{n} S_{Load,i} \quad i \neq j \tag{17}$$

3. The Proposed Really Formula

In this section presented a formula for DG location and a formula for DG sizing. To do the following steps:

1) Convert radial network to the ring network with Close all tie switches in the network.
2) The DG location is equal to:

$$O_{Loc}^{DG} = n_{V,min,r} \tag{18}$$

Where $n_{V,min,r}$ is the node that has the lowest voltage profile in the ring network condition with close all tie switches.

3) Calculated Current in line section between nodes ith and jth in the ring network condition is equal to:

$$I_{i,j,r} = \frac{V_i - V_j}{Z_{i,j}} \quad \begin{array}{l} i \neq j \\ (i = 1,2,\ldots,b) \end{array} \tag{}$$

4) Calculated Apparent power flowing injected from nodes ith and jth in the ring network condition is equal to:

$$S_{i,j,r} = I_{i,j,r}^* V_n \tag{19}$$

Total Apparent power flowing in the ring network is equal to:

$$ST_{i,j,r} = I_{ref}^* V_{ref} + \sum_{i=1}^{b} S_{i,j,r} \tag{20}$$

5) The DG sizing is equal to:

$$O_{Siz}^{DG} = \frac{ST_{i,j,r}}{n} + S_{Load,(n_{V,min,r})} \tag{21}$$

where $S_{Load,(n_{V,min,r})}$ is the apparent power flowing at target node $(n_{V,min,r})$.

4. Numerical Example

4.1. 33-Node System

The line and load data of the 33-Node System are obtained from [8], and the total real and reactive power loads on the system are 3.715 MW and 2.3 MVAr, respectively. The initial power loss of this system is 0.2027 MW and 0.1351 MVAr. The lowest bus bar voltage is 0.9130 p.u., which occurs at node 18. Table 1 shown total apparent power injected from node ith to jth in base case and presence DG units in the ring network condition. The proposed method is able to model the DG as a negative constant P-Q load.

Table 1. Power Flow Results of 33- Node System, Ring Network Condition.

Nodes $i \rightarrow j$	Base case $ST_{i,j,r}$ (MVA)	Presence 1 DG $ST_{i,j,r}$(MVA)	Presence 2 DG $ST_{i,j,r}$(MVA)
1 2	3.6658 + 2.2672i	2.7273 + 1.7086i	2.1756 + 1.3326i
2 3	2.5932 + 1.5156i	1.8757 + 1.1333i	1.5343 + 0.9263i
3 4	1.2108 + 0.6457i	0.8874 + 0.4856i	0.6674 + 0.3613i
4 5	1.0941 + 0.5685i	0.7685 + 0.4063i	0.5478 + 0.2814i
5 6	1.0356 + 0.5395i	0.7089 + 0.3765i	0.4879 + 0.2514i
6 7	0.4338 + 0.1086i	0.3776 + 0.0922i	0.1675 - 0.0237i
7 8	0.2373 + 0.0104i	0.1779 - 0.0084i	-0.0327 - 0.1245i
8 9	0.3562 + 0.1845i	0.1872 + 0.0813i	-0.0855 - 0.0791i
9 10	0.0695 - 0.0095i	0.0644 + 0.0017i	-0.0218 - 0.0302i
10 11	0.0102 - 0.0293i	0.0043 - 0.0186i	-0.0818 - 0.0504i
11 12	-0.0342 - 0.0590i	-0.0408 - 0.0489i	-0.1268 - 0.0806i
12 13	0.2177 + 0.1731i	0.0868 + 0.0823i	-0.1499 - 0.0757i
13 14	0.1582 + 0.1384i	0.0267 + 0.0470i	-0.2097 - 0.1108i
14 15	0.0393 + 0.0588i	-0.0936 - 0.0338i	0.2299 + 0.1912i
15 16	0.2072 + 0.2230i	-0.0910 +0.0152i	0.0464 + 0.1119i
16 17	0.1476 + 0.2030i	-0.1512 - 0.0050i	-0.0136 +0.0918i
17 18	0.0877 + 0.1830i	-0.2113 - 0.0252i	-0.0736 +0.0717i
2 19	0.9767 + 0.6946i	0.7535 + 0.5164i	0.5425 + 0.3468i
19 20	0.8904 + 0.6565i	0.6651 + 0.4771i	0.4534 + 0.3070i
20 21	0.8029 + 0.6177i	0.5759 + 0.4372i	0.3638 + 0.2669i
21 22	0.3992 + 0.3061i	0.2773 + 0.2068i	0.1266 + 0.0804i
3 23	1.2953 + 0.8315i	0.8993 + 0.6081i	0.7773 + 0.5251i
23 24	1.2076 + 0.7832i	0.8100 + 0.5584i	0.6875 + 0.4751i
24 25	0.7943 + 0.5876i	0.3903 + 0.3582i	0.2659 + 0.2735i
6 26	0.5428 + 0.4113i	0.2714 + 0.2643i	0.2603 + 0.2549i
26 27	0.4837 + 0.3868i	0.2114 + 0.2392i	0.2001 + 0.2298i
27 28	0.4246 + 0.3624i	0.1514 + 0.2141i	0.1399 + 0.2046i
28 29	0.3652 + 0.3427i	0.0913 + 0.1940i	0.0796 + 0.1844i
29 30	0.6237 + 0.6639i	-0.0604 +0.2811i	-0.1983 +0.1854i
30 31	0.4224 + 0.0668i	-0.2620 - 0.3212i	-0.4000 - 0.4185i
31 32	0.2723 - 0.0031i	-0.4121 - 0.3912i	-0.5503 - 0.4886i
32 33	0.0622 - 0.1030i	0.3614 + 0.1053i	0.2236 + 0.0084i
21 8	0.3158 + 0.2726i	0.2091 + 0.1903i	0.1474 + 0.1463i
9 15	0.2275 + 0.1742i	0.0627 + 0.0593i	-0.1237 - 0.0691i
12 22	-0.3111 - 0.2668i	-0.1876 - 0.1665i	-0.0368 - 0.0401i
18 33	-0.0022 + 0.1430i	-0.3014 - 0.0653i	-0.1636 + 0.0316i
25 29	0.3778 + 0.3904i	-0.0311 + 0.1573i	-0.1570 + 0.0716i
$I_{ref}^* V_{ref}$	3.8383 + 2.3879i	2.7877 + 1.7428i	2.2055 + 1.3460i

$$ST_{i,j,r} = [3.8383 + 2.3879i\,] + [21.701 + 14.040i\,]$$

$$= 25.5393 + 16.4279i \text{ (MVA)}$$

$$O_{Loc}^{DG} = node\ 32$$

$$O_{Siz}^{DG} = \frac{25.5393 + 16.4279i}{33} + 0.21 + 0.1i$$

$$= 0.9839 + 0.5978i \text{ (MVA)}$$

and so on follow these steps to do next DG placement. Second DG placement in the presence first DG is equal to:

$$O_{Loc}^{2\,DG} = node\ 14$$

$$O_{Siz}^{2\,DG} = \frac{14.568 + 9.9557i}{33} + 0.12 + 0.08i$$

$$= 0.5615 + 0.3817i \text{ (MVA)}$$

Third DG placement in the presence Second DG is equal to:

$$O_{Loc}^{3\,DG} = node\ 25$$

$$O_{Siz}^{3\,DG} = \frac{9.9056 + 6.9659i}{33} + 0.42 + 0.2i$$

$$= 0.7202 + 0.4111i \text{ (MVA)}$$

The results of placement of single and multiple DG units with real and reactive power are presented in Table 2.

Table 2. Placement of Dg Units with Real and Reactive Power Capability for 33-Node System

	Size of DG in MVA	Node No.	Power Loss (MW)	Min Voltage p.u (Location)
Base Case	-	-	0.2027	0.91309 (18)
1 DG Unit	0.9839 + 0.5978i	32	0.0910	0.93431 (18)
2 DG Units	0.9839 + 0.5978i	32	0.0437	0.97771 (25)
	0.5615 + 0.3817i	14		
3 DG Units	0.9839 + 0.5978i	32	0.0243	0.98141 (18)
	0.5615 + 0.3817i	14		
	0.7202 + 0.4111i	25		
	Total size of 3 DG units		2.658 MVA	

4.2. 69-Node System

The line and load data of the 69-Node System are obtained from [9]; Total system loads are 3.802 MW and 2.694 MVAr. System line loss Without DG is 0.225 MW and 0.1022 MVAr. The lowest bus bar voltage is 0.90919 p.u, which occurs at node 65. The results of placement of single and multiple DG units with real and reactive power are presented in Table 3.

Table 3. Placement of Dg Units with Real and Reactive Power Capability for 69-Node System.

	Size of DG in MVA	Node No.	Power Loss (MW)	Min Voltage p.u (Location)
Base Case	-	-	0.225	0.90919 (65)
1 DG Unit	1.9209+ 1.3298i	61	0.0235	0.97314 (27)
2 DG Units	1.9209+ 1.3298i	61	0.0096	0.99239 (69)
	0.3658 + 0.2523i	21		
3 DG Units	1.9209+ 1.3298i	61	0.0073	0.99426 (50)
	0.3658 + 0.2523i	21		
	0.2063 + 0.1441i	69		
	Total size of 3 DG units		3.032 MVA	

Table 4. *Placement of Dg Units with Real and Reactive Power Capability for 118-Node System.*

	Size of DG in MVA	Node No.	Power Loss (MW)	Min Voltage p.u (Location)
Base Case	-	-	1.3020	0.8688 (77)
1 DG Unit	1.3790 + 1.0750i	113	1.0507	0.8688 (77)
2 DG Units	1.3790 + 1.0750i	113	0.7194	0.90949 (54)
	1.8084 + 1.1796i	71		
3 DG Units	1.3790 + 1.0750i	113	0.5498	0.94372 (77)
	1.8084 + 1.1796i	71		
	1.3327 + 1.0286i	51		
	Total size of 3 DG units		5.586 MVA	

4.3. 118-Node System

The test system is a hypothetical 11 kV with 118 node, and 15 tie lines. The system data is given in Ref [10]. The total power loads are 22.709 MW and 17.0411 MVAr. The results of placement of single and multiple DG units with real and reactive power are presented in Table 4.

4.4. 84-Node System

The line and load data of the 84-Node System are obtained from [11]; Total system loads are 28.35 MW and 20.7 MVAr. System line loss Without DG is 0.5320 MW 1.3743 MVAr. The lowest bus bar voltage is 0.92852 p.u, which occurs at node 10. The results of placement of single and multiple DG units with real and reactive power are presented in Table 5.

Table 5. *Placement of Dg Units with Real and Reactive Power Capability for 84-Node System.*

	Size of DG in MVA	Node No.	Power Loss (MW)	Min Voltage p.u (Location)
Base Case	-	-	0.5320	0.92852 (10)
1 DG Unit	1.6959 + 1.2540i	10	0.4332	0.94786 (84)
2 DG Units	1.6959 + 1.2540i	10	0.4053	0.94786 (84)
	1.2936 + 0.9165i	25		
3 DG Units	1.6959 + 1.2540i	10	0.3525	0.94881 (73)
	1.2936 + 0.9165i	25		
	1.6145 + 1.1823i	82		
	Total size of 3 DG units		5.694 MVA	

4.5. 136-Node System

The line and load data of the 136-Node System are obtained from [12]; Total system loads are 18.3138 MW and 7.9326 MVAr. System line loss Without DG is 0.3204 MW and 0.7029 MVAr. The lowest bus bar voltage is 0.93065 p.u, which occurs at node 117. The results of placement of single and multiple DG units with real and reactive power are presented in Table 6.

Table 6. *Placement of Dg Units with Real and Reactive Power Capability for 136-Node System.*

	Size of DG in MVA	Node No.	Power Loss (MW)	Min Voltage p.u (Location)
Base Case	-	-	0.3204	0.93065 (117)
1 DG Unit	1.1824 + 0.5201i	117	0.2545	0.96609 (113)
2 DG Units	1.1824 + 0.5201i	117	0.2367	0.96609 (113)
	0.9392 + 0.4148i	61		
3 DG Units	1.1824 + 0.5201i	117	0.2259	0.96609 (113)
	0.9392 + 0.4148i	61		
	1.1560 + 0.4951i	75		
	Total size of 3 DG units		3.575 MVA	

4.6. 70-Node System

The test system is a hypothetical 11 kV with 70 node, and 11 tie lines. The system data is given in Ref [13]. The total power loads are 4.468 MW and 3.059 MVAr. The results of placement of single and multiple DG units with real and reactive power are presented in Table 7.

Table 7. *Placement of Dg Units with Real and Reactive Power Capability for 70-Node System.*

	Size of DG in MVA	Node No.	Power Loss (MW)	Min Voltage p.u (Location)
Base Case	-	-	0.2275	0.90502 (67)
1 DG Unit	0.5132 + 0.3866i	65	0.1638	0.92791 (50)
2 DG Units	0.5132 + 0.3866i	65 49	0.1231	0.93269 (29)
	0.4317 + 0.2970i			
3 DG Units	0.5132 + 0.3866i	65 49 29	0.0892	0.9461 (15)
	0.4317 + 0.2970i			
	0.3751+0.25395i			
	Total size of DG units		1.619 MVA	

5. Conclusion

In this paper, a really formula based on power flow relations for obtaining the location and size of distributed generation has been presented. This formula is actually obtained from the power flow results. The proposed formula has been tested on several test networks and the results show its effectiveness. In fact, this formula is a good alternative to time-consuming and complex algorithms. The benefit of the proposed method is its low computational burden as well as its simplicity of implementation.

Nomenclature

n	Total number of nodes in the system.
b	Total number of branches in the system.
V_i, V_j	Voltage magnitude at node i zand j.
I_i, I_j	Equivalent current injected at node i and j.
$R_{i,j}$	Resistance of the branch between nodes i and j.
$X_{i,j}$	Reactance of the branch between nodes i and j.

$Y_{i,o}, Y_{j,o}$	Self-admittance at node i and j.
$Y_{i,j}, Y_{j,i}$	Transfer admittance from node i to j and reverse.
P_i, P_j	Active power flowing out of node i and j.
Q_i, Q_j	Reactive power flowing out of node i and j.
S_i, S_j	Apparent power flowing from node i to node j and reverse.
$Z_{i,j}$	Impedance of the branch between nodes i and j.
$I_{i,j,r}$	Current in line section between nodes i and j (in the ring network condition).
$S_{i,j,r}$	Apparent power injected from node i to j (in the ring network condition).
$n_{V,min,r}$	The node that has the lowest voltage profile in the ring network.
$S_{Load,(n_{V,min,r})}$	Apparent power flowing at target node $(n_{V,min,r})$.
O_{Loc}^{DG}	DG location.
O_{Siz}^{DG}	DG size.
$S_{DG,i}$	Apparent power supplied by DG unit at node i.
S_{Sub}	Apparent power supplied by the substation.
$S_{Load,i}$	Apparent power load at node i.
$P_{Load,i}$	Active power load at node i.
$Q_{Load,i}$	Reactive power load at node i.
$S_{i,max}, S_{j,max}$	Rower flow both at sending and receiving ends of particular line to be within the upper limit of the line.
$I_{i,j,max}$	Maximum current limit of line section between nodes i and j.
n_c	Total number of DG units.
V_n	Voltage magnitude at node n.

References

[1]   W El-Khattam, Salama M. M. A. Distributed Generation Technologies, Definitions and Benefits. Elect Power Syst Res 2004; 71: 119-128.

[2]   Pavlos S. Georgilakis, Nikos D. Hatziargyriou. Optimal Distributed Generation Placement in Power Distribution Networks: Models, Methods, and Future Research. IEEE Trans. Power Syst 2013; 28 (3): 3420–3428.

[3]   R. Srinivasa Rao, K. Ravindra, K. Satish, and S. V. L. Narasimham. Power Loss Minimization in Distribution System Using Network Reconfiguration in the Presence of Distributed Generation. IEEE Trans. Power Syst 2013; 28 (1): 317–325.

[4]   N. S. Rau and Y.-H. Wan. Optimum location of resources in distributed planning. IEEE Trans. Power Syst 1994; 9: 2014–2020.

[5]   H. Hedayati, S. A. Nabaviniaki, and A. Akbarimajd. A method for placement of DG units in distribution networks. IEEE Trans. Power Del 2008; 23: 1620–1628.

[6]   C. Wang and M. H. Nehrir. Analytical approaches for optimal placement of distributed generation sources in power systems. IEEE Trans. Power Syst 2004; 19 (4): 2068–2076.

[7]   G. Celli, E. Ghiani, S. Mocci, and F. Pilo. A multi-objective evolutionary algorithm for the sizing and the sitting of distributed generation. IEEE Trans. Power Syst 2005; 20 (2): 750–757.

[8]   M. E. Baran and F. Wu. Network reconfiguration in distribution system for loss reduction and load balancing. IEEE Trans. Power Del 1989; 4 (2): 1401–1407.

[9]   J. S. Savier and D. Das. Impact of network reconfiguration on loss allocation of radial distribution systems. IEEE Trans. Power Del 2007; 2(4): 2473–2480.

[10]   D. Zhang, Z. Fu, and L. Zhang, "An improved TS algorithm for loss- minimum reconfiguration in large-scale distribution systems. Elect. Power Syst. Res 2007; 77 (5-6): 685–694.

[11]   C. Su and C. Lee. Network reconfiguration of distribution systems using improved mixed-integer hybrid differential evolution. IEEE Trans. Power Del 2003; 18 (3): 1022–1027.

[12]   J. R. S. Mantovani, F. Casari, and R. A. Romero. Reconfiguração de sistemas de distribuição radiais utilizando o critério de queda de tensão. *SBA Controle Automação* 2000; 11 (3): 150–15.

[13]   Debapriya Das, H. Yin, S.S.H. Lee. A Fuzzy Multiobjective Approach for Network Reconfiguration of Distribution Systems. IEEE Trans. Power Del 2006; 21 (1): 202-209.

31

Detection of Cu(II) Ion in Water Using a Quartz Crystal Microbalance

Chi-Yen Shen, Yu-Min Lin, Rey-Chue Hwang[*]

Department of Electrical Engineering, I-Shou University, Kaohsiung, Taiwan

Email address:

cyshen@isu.edu.tw (Chi-Yen Shen), wasdxghj@gmail.com (Yu-Min Lin), rchwang@isu.edu.tw (Rey-Chue Hwang)

[*]Corresponding author

Abstract: Drinking water from a tap is a source of potential exposure to environmental contaminants. This requires that public water supplies should be regularly monitored for heavy metals. Many of heavy metal ions are retained and accumulated in water strongly. Consequently it has entered the food chain to threaten human health. A quartz crystal microbalance (QCM) based on a phosphate-modified dendrimer film was investigated for direct detection of Cu(II) metal ion in water. This QCM sensor exhibited the high sensitivity and the short response time to Cu(II) metal ion.

Keywords: Crystal Microbalance, Dendrimer, Metal Ion, Sensitivity

1. Introduction

The chemical industry is highly diverse and a large number of chemical products are used and produced in manufacturing processes. It has resulted in the formation of wastewater contaminated with organic and inorganic substances of very different composition and volume [1-2]. Hence, heavy metals released by industrial activities are increasingly being found in freshwater to affect the quality of water bodies. Pollution caused by heavy metal ions is a long-term and irreversible process. Consequently it is a significant and emerging threat to public health. Among heavy metal ions, copper has become a widely distributed pollutant in natural water as a result of the dumping of electronic trash and mining residues [3]. On the other hand, Cu(II) is the third most abundant transition-metal ion in the human body and is essential for the biochemical and physiological functions [4]. Its deficiency and excess lead to malfunction of the liver, heart diseases, neurological disorders, and deterioration of connective and bone tissues [1].

Most heavy metal ions in aqueous solution were measured using analytical instrumentation in laboratories such as atomic absorption spectroscopy [5], high performance liquid chromatography [6], and inductively coupled plasma-mass spectrometer [7]. These instrumentations are expensive and a well-trained operator is required. The advantages of a quartz crystal microbalance (QCM) are generally real-time, sensitive response and relatively inexpensive cost. A QCM oscillates at a specific frequency when an AC voltage is applied over its electrodes. The mass change per unit area at the QCM electrode surface relates to the observed change in oscillation frequency of the QCM, as shown by the Sauerbrey equation [8].

$$\Delta f = \frac{-2f_o^2 \Delta m}{A(\rho_q \mu_q)^{1/2}} \tag{1}$$

where Δf is the frequency change, f_o is the resonant frequency, Δm is the mass change, ρ_q is the density of quartz, μ_q is the shear modulus and A is Au electrode area. Frequency shifts of a QCM provide information on absorbed mass changes at QCM surface [9]. Therefore, the QCM was utilized in various fields such as environmental protection [10-12], lead acid batteries [13-16] and biomedicine [17-19].

For constructing functional and efficient chemical sensors, the modification of sensing films on the QCM electrode was been studied to greatly improve their properties [20]. Dendrimers are branched polymers with well-defined sizes and geometry [21-22]. Ability of dendrimers to coordinate metal ions in their interior branches or in the exterior units is developed the sensors with the selectivity and sensitivity of the analysis. These structural features endow them with the ability to encapsulate small guest molecules and act as nanocontainers for ions.

In this paper, a new design of sensor which combines high sensitivity of dendrimer with QCM detection has been developed for the detection of Cu(II) metal ion in water. The detection responses of sensor on Cu(II) metal ion sensing performance were studied quantitatively.

2. Experimental Methods

2.1. Preparation of Dendrimer Sensing Membrane

PAMAM dendrimer (ethylenediamine core, G5), whose chemical structure is shown in Figure 1, was dispersed in phosphorous acid solution with ultrasonic bath for 7 h and then rinsed by deionized water. In order to enhance the mechanical properties, the obtained phosphate-modified PAMAM dendrimer was mixed with 2.1wt% PVC in DMF solution.

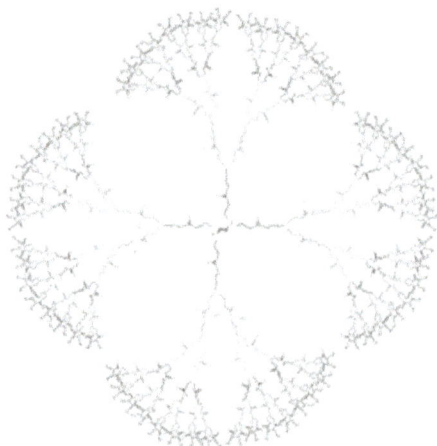

Figure 1. Chemical structure of ethylenediamine core G5 PAMAM dendrimer.

2.2. Fabrication of the QCM Sensor

The QCM device (Taitien Co., Ltd, Taiwan) with an operation frequency of 10 MHz was used in this work. The sensing membrane solution was dropped onto the electrode surface at room temperature. Then the samples were dried in the oven at 80°C. Finally, the samples were rinsed again with deionized water and dried with N_2 gas.

2.3. The QCM Experimental System

A QCM device based on the phosphate-modified dendrimer was coupled inside the flow cell (Figure 2), which was designed with a temperature controller shown in Figure 3 was maintained at 20°C, illustrated in Figure 4.

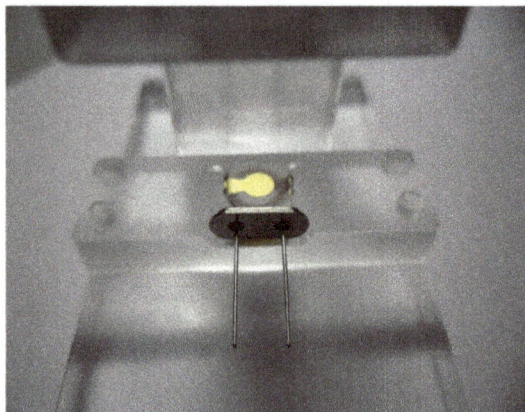

Figure 2. A QCM device coupled inside the flow cell in this work.

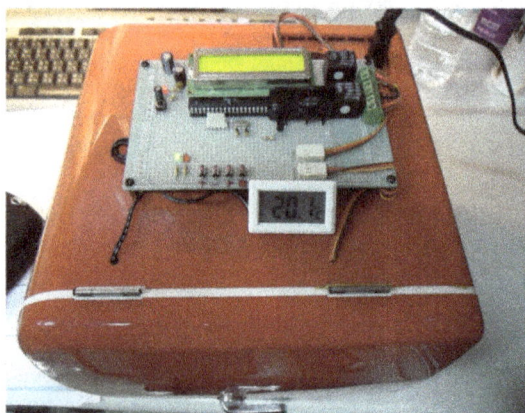

Figure 3. The temperature controller used in this work.

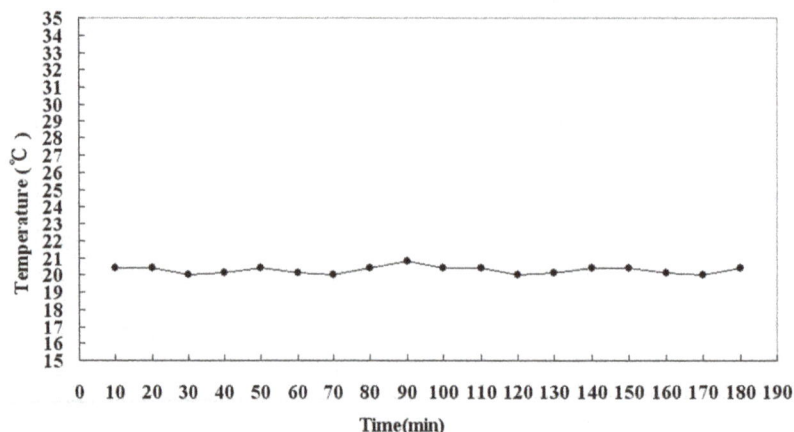

Figure 4. Output of the temperature controller used in this work.

The design of a crystal-controlled oscillator used as a QCM sensor in liquid is a difficult task because of the wide dynamic values of the resonator resistance that they should support during their operations [23]. The QCM experiences a large reduction in its quality factor that caused by the liquid. The Pierce oscillator design plotted in Figure 5 provides a great stability in frequency and a low phase noise, so the Pierce oscillator shown in Figure 6 was applied in this work to maintain the necessary loop gain and phase for the oscillation in a wide margin of values of the loss resistances of the QCM. Output spectrum of the QCM oscillator shown in Figure 7 indicates QCM generated stable oscillation occurred at 10 MHz.

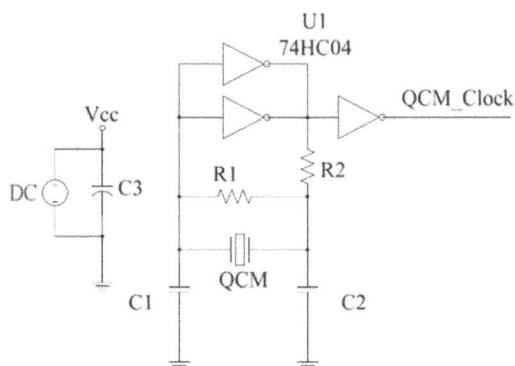

Figure 5. The circuit diagram of the Pierce oscillator.

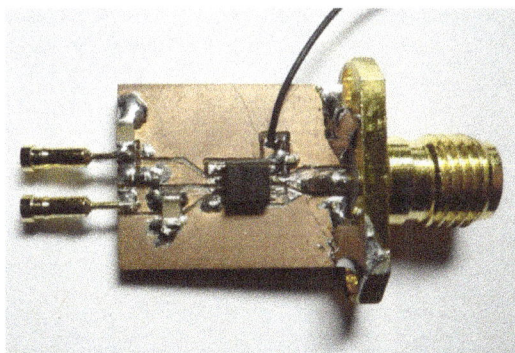

Figure 6. A photograph of the Pierce oscillator used in this work.

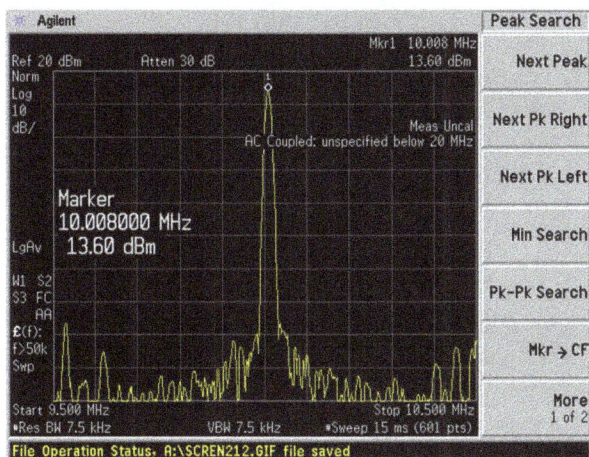

Figure 7. Output spectrum of the QCM oscillator in this work.

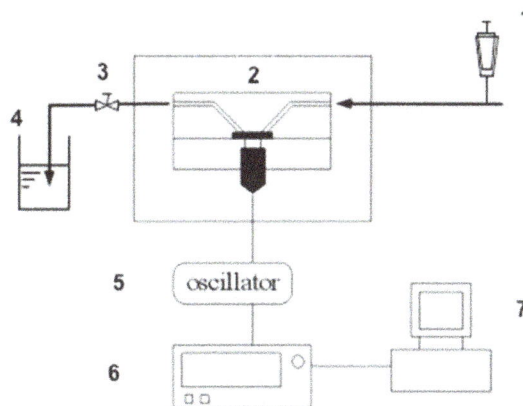

Figure 8. QCM measurement system. 1: micro-injector; 2: flow cell; 3: valve; 4: waste; 5: oscillator circuit; 6: frequency counter; 7: computer.

In this study, a counter (53131A, Agilent, USA) measured oscillation frequency of the QCM device. A computer through a GPIB interface controlled the counter, and the software for taking measurements was designed by LabVIEW 8.6 (National Instruments, USA). The QCM measurement system is illustrated in Figure 8.

2.4. Procedures for Metal Ion Detection

Various concentrations of metal ion solutions were prepared by mixing in various ratios with the deionized water. During the detection, 300 μl diluted metal ion solution were injected into the cell carefully with a micro-injector (KDS 200, KD Scientific, USA) at 0.01ml/min. After the frequency signal stabilized for several minutes, the QCM sensor generated a frequency-decrease after metal ions bounded with phosphorus groups in the surface of dendrimer. The frequency shifts in all experiments were calculated based on the average responses of the reactions with corresponding standard deviations of triplicate measurements.

3. Results and Discussion

3.1. Surface Morphology of the QCM Sensor

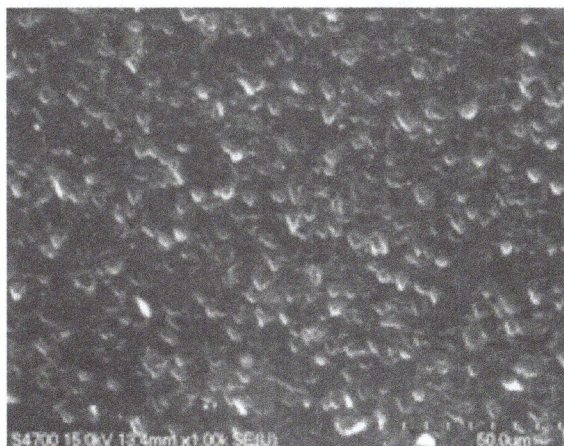

Figure 9. SEM images of the phosphate-modified dendrimer modified crystal electrode.

SEM was used to monitor and characterize the surface morphology of the QCM sensor after the modification process. Figure 9 shows that the surface morphology of the phosphorus-modified dendrimer modified Au crystal electrode has some small cell-like protrusions.

3.2. Effects of Fabrication Conditions of the QCM Sensor

The QCM sensor was fabricated with a drying step after the phosphate-modified dendrimer was deposited. The effect of drying time between 3 h and 5 h on metal ion detecting of the QCM sensor based on the phosphate-modified dendrimer with 2.1wt% PVC was shown in Figure 10. The frequency shift of the QCM sensor result from detecting 1 µM Cu(II) ions gradually increased with drying time from 3 h to 5 h. There are some small dark spots and lines were observed on the surface of these films at drying time of 5 h. Therefore, 4 h was selected as the optimal drying time in this study.

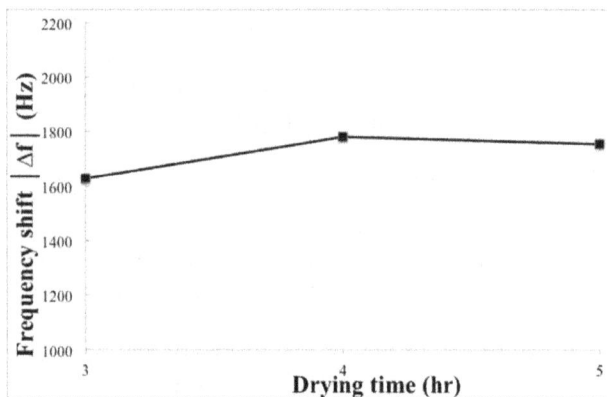

Figure 10. Effect of drying time on frequency shift of the QCM sensor based on the phosphate-modified dendrimer with 2.1wt% PVC, measured at 0.01 µM Cu(II) ions in pH 5.7.

3.3. Effect of pH Condition

It is known that the pH of a solution is important for metal trace detection. Figure 11 plots the effect of solution pH between 5 and 8 on the response of a QCM sensor based on the phosphate-modified dendrimer in 1 µM Cu(II) ions. The oscillation frequency of the QCM sensor increased with pH from 5 to 5.7, and decreased as the pH was increased further. Therefore, the solution with pH 5.7 was used in this study.

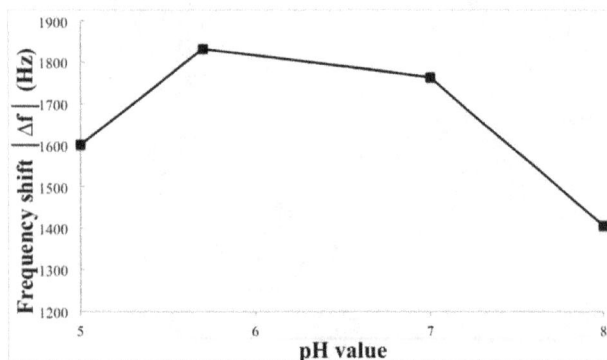

Figure 11. Effect of pH on frequency shift of the QCM sensor based on the phosphate-modified dendrimer, measured at 1 µM Cu(II) ions.

3.4. Characteristics of Metal Ion Detection

The results of detection curve of the QCM sensors based on the phosphate-modified dendrimer to various concentration of Cu(II) solution ranging from 0.0001 µM to 1 µM are plotted in Figure 12. An linear increased frequency shift was observed on the QCM sensor when increasing the Cu(II) ions concentration, revealing the absorption ability of the Cu(II) ions on the QCM sensor. Figure 13 shows the corresponding response time. The response time of the sensor was calculated as the time required to reach 90% of the saturation value upon exposure to Cu(II) ions. These QCM sensors had short response time that was less than 40 seconds.

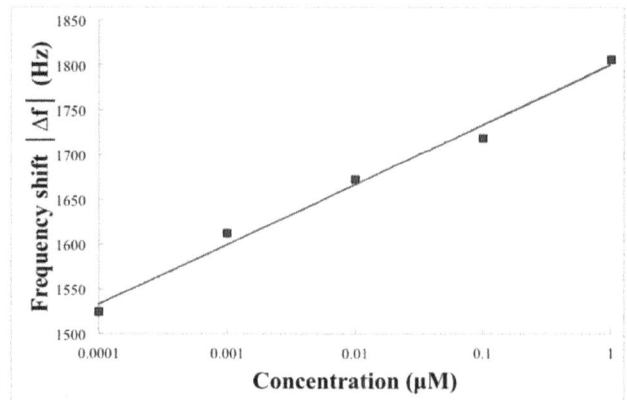

Figure 12. The frequency shift of the QCM sensors based on the phosphate-modified dendrimer to 10^{-4} µM – 1 µM Cu(II) ions in solution.

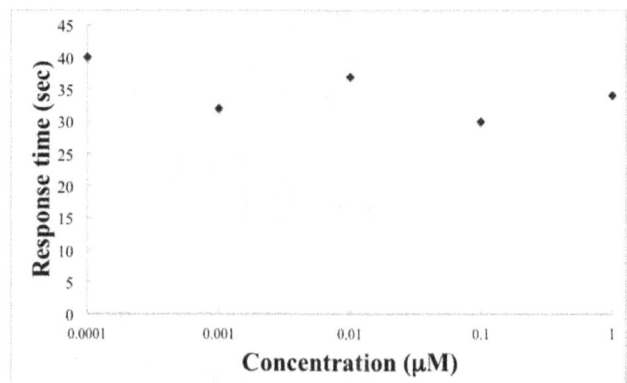

Figure 13. Response time of the QCM sensor based on the phosphate-modified dendrimer at different Cu(II) ions in solution.

4. Conclusion

In this study, a phosphate-modified dendrimer sensing membrane has been synthesized and characterized. It has been applied for Cu(II) metal ion sensing by QCM sensor at room temperature. The linear sensing range of Cu(II) ion was from 0.0001 µM to 1 µM. This developed QCM showed short response time that was less than 40 seconds. The experimental results indicated that the phosphate-modified dendrimer QCM sensor could be used for direct detection of Cu(II) metal ion with high sensitivity and fast response. We will develop this QCM sensor for detecting different metal cations, such as Ca(II), Mg(II), Zn(II), Co(II), Ni(II), Ag(I), and Fe(III), at room temperature in the future.

Acknowledgements

The authors thank the Ministry of Science and Technology, Taiwan, for partially supporting this research under Contract No. NSC 102-2221-E-214-003-MY3 and MOST 104-2622-E-214-005 -CC3.

References

[1] O. Zagurskaya-Sharaevskaya and I. Povar, "Determination of Cu (II) ions using sodium salt of 4-phenylsemicarbazone 1, 2 - naphthoquinone-4-sulfonic acid in natural and industrial environments," Ecological Processes, vol. 4, 2015, pp. 1-5.

[2] M. Qadir, M. S. Javier, and J. Blanca, "Environmental risks and cost-effective risk management in wastewater use systems," In: Wastewater, Springer, Netherlands, 2015, pp. 55–72.

[3] Z. Zhang, Z. Chen, C. Qu, and L. Chen, "Highly sensitive visual detection of copper ions based on the shape-dependent LSPR spectroscopy of gold nanorods," Langmuir, vol. 30, 2014, pp. 3625–3630.

[4] M. Ghaedi, F. Ahmadi, and A. Shokrollahi, "Simultaneous preconcentration and determination of copper, nickel, cobalt and lead ions content by flame atomic absorption spectrometry," Journal of Hazardous Materials, vol. 142, 2007, pp. 272–278.

[5] S. Pande, "Analytical applications of room-temperature ionic liquids: A review of recent efforts," Analytica Chimica Acta, vol. 556, 2006, pp. 38–45.

[6] J. S. Becker, M. Zoriy, A. Matusch, B. Wu, D. Salber, C. Palm, and J. S. Becker, "Bioimaging of metals by laser ablation inductively coupled plasma mass spectrometry," Mass Spectrometry Reviews, vol. 29, 2010, pp. 156-175.

[7] V. Chandrasekhar, S. Das, R. Yadav, S. Hossain, R. Parihar, G. Subramaniam, P. Sen, "Novel chemosensor for the visual detection of copper (II) in aqueous solution at the ppm level," Inorg Chem, vol. 51, 2012, pp. 8664–8666.

[8] G. Sauerbrey, "Verwendung von Schwingquarzen zur wägung dünner schichten und zur mikrowägung," Zeitschrift für Physik, vol. 155, 1959, pp. 206–222.

[9] X. Guo, Y. Yun, V. N. Shanov, H.B. Halsall, and W.R. Heineman, "Determination of trace metals by anodic stripping voltammetry using a carbon nanotube tower electrode," Electroanalysis, vol. 23, 2011, pp. 1052-1259.

[10] Q. Ji, S. B. Yoon, J. P. Hill, A. Vinu, J. S. Yu, and K. Ariga, "Layer-by-layer films of dual-pore carbon capsules with designable selectivity of gas adsorption," J. Am. Chem. Soc., vol. 131, 2009, pp. 4220-4221.

[11] K. Ariga, S. Ishihara, H. Abe, M. Li, and J. P. Hill, "Materials nanoarchitectonics for environmental remediation and sensing," J. Mater. Chem., vol. 22, 2012, pp. 2369-2377.

[12] L. Sartore, M. Barbaglio, L. Borgese, and E. Bontempi, "Polymer-grafted QCM chemical sensor and application to heavy metalions real time detection," Sens Actuators B Chem., vol. 155, 2011, pp. 539–544.

[13] D. D. Erbahar, I. Gürol, G. Gümüş, E. Musluoğlu, Z. Z. Öztürk, V. Ahsen, and M. Harbeck, "Pesticide sensing in water with phthalocyanine based QCM sensors," Sens. Actuators B, vol. 173, 2012, pp. 562-568.

[14] A. M. Cao-Paz, L. Rodríguez-Pardo, and J. Fariña, "Application of the QCM in lead acid batteries electrolyte measurements," Procedia Engineering, vol. 5, 2010, pp. 1260–1263.

[15] A. M. Cao-Paz, L. Rodriguez-Pardo, and J. Farina, "Density and viscosity measurements in lead acid batteries by QCM sensor," Proc. of 2011 IEEE International Symposium on Industrial Electronics, 2011, pp. 1290–1294.

[16] A. M. Cao-Paz, L. Rodriguez-Pardo, J. Farina, and J. Marcos-Acevedo, "Resolution in QCM sensors for the viscosity and density of liquids: application to lead acid batteries," Sensors, vol. 12, 2012, pp. 10604-10620.

[17] M. Tominagaa, A. Ohirab, Y. Yamaguchic, and M. Kunitakeb, "Electrochemical, AFM and QCM studies on ferritin immobilized onto a self-assembled monolayer-modified gold electrode," Journal of Electroanalytical Chemistry, vol. 566, 2004, pp. 323–329.

[18] K. N. Huang, C. Y. Shen, S. H. Wang, and C. H. Hung, "Development of quartz crystal microbalance-based immunosensor for detecting alpha-fetoprotein," Instrumentation Science & Technology, vol. 44, 2013, pp. 311-324.

[19] S. H. Wang, C. Y. Shen, T. C. Weng, P. H. Lin, J. J. Yang, I. F. Chen, S. M. Kuo, S. J. Chang, Y. K. Tu, Y. H. Kao, and C. H. Hung, "Detection of cartilage oligomeric matrix protein using a quartz crystal microbalance," Sensors, vol. 10, 2010, pp. 11633-11643.

[20] E. Biemmi, A. Darga, N. Stock, and T. Bein, "Direct growth of $Cu_3(BTC)_2(H_2O)_3 \cdot xH_2O$ thin films on modified QCM-gold electrodes – Water sorption isotherms," Microporous and Mesoporous Materials, vol. 114, 2008, pp. 380–386.

[21] M. W. P. L. Baars and E. W. Meijer, "Host-guest chemistry of dendritic molecules," Topics in Current Chemistry, vol. 210, Springer, New York, 2000, pp. 132-178.

[22] G. R. Newkome and C. D. Shreiner, "Poly(amidoamine), polypropylenimine, and related dendrimers and dendrons possessing different $1 \rightarrow 2$ branching motifs: an overview of the divergent procedures," Polymer, vol. 49, 2008, pp. 1-173.

[23] L. Rodriguez-Pardo, J. Fariña, C. Gabrielli, H. Perrot, and R. Brendel "Resolution in quartz crystal oscillator circuits for high sensitivity microbalance sensors in damping media," Sens. Actuators B, vol. 103, 2004, pp. 318–324.

32

An Optimized DAC Timing Strategy in SAR ADC with Considering the Overshoot Effect

M. Dashtbayazi[1], M. Sabaghi[2, *], S. Marjani[1]

[1]Department of Electrical Engineering, Ferdowsi University of Mashhad, Mashhad, Iran
[2]Laser and Optics Research School, Nuclear Science and Technology Research Institute (NSTRI), Tehran, Iran

Email address:
msabaghi@aeoi.org.ir (M. Sabaghi)

Abstract: In this paper, we report an ultra-low power successive-approximation-register (SAR) analog-to digital converter (ADC) by using a DAC timing strategy with considering overshoot effect to increase the sampling rate. This ADC is simulated for power supplies voltage of 0.6 V and 1.2 V in a 130-nm CMOS technology. The results indicate an ENOB greater than 9.3 bits for its full sampling-rate range (4 to 32 MS/s) with an FOM=5.3 to 9.3 fJ/conv-step.

Keywords: Data Converter, Overshoot Effect, Asynchronous Process, Power Efficiency, DAC Timing Strategy, Low Power Designs

1. Introduction

The SAR converters (ADCs) with a medium resolution (8 to 10 bits) and medium frequencies (a few tens to hundreds of MS/s) is one of the applications used in wireless networks and digital TV where small area, low power consumption and high sampling rate in single channel SAR are necessary [1, 2]. Thanks to technology scaling, the problem of long conversion time has been alleviated in these structures. The sampling rate is decreased with increasing the circuit resolution due to serially produce of the output bits. In order to increase the speed of SAR ADCs, different techniques have been recently reported such as multi-bit/step [3], time interleaving [4, 5] and the non-binary search algorithms [6]. Another appropriate method for speed up the sampling rate is the use of asynchronous process [7, 8]. S.W.M. Chen and et al. have proved that the maximum asynchronous conversion time is half of that in its synchronous counterpart [7]. Furthermore, the asynchronous process saves some area and power because it removes the need for some internal clock generator circuit.

In this work, to improve the speed of the converter with high power efficiency, a new approach to asynchronous processing by considering the overshooting effect is proposed which compared to the conventional method. A 10-bit 4-32 MS/s SAR ADC has been applied for investigation of the proposed asynchronous process on its speed.

The paper is organized as follows: section 2 briefly describes the conventional asynchronous process of the converter and its challenges; section 3 provides the details of the proposed asynchronous process and its circuit level implementation. The effectiveness of the target asynchronous process method compared to its conventional counterpart is also discussed in this section. Section 4 shows the effectiveness of the proposed asynchronous process technique by using extensive simulation results in 130nm digital CMOS process. Finally, in section 5, we conclude.

2. Conventional Asynchronous SAR ADC

Fig. 1 shows the block diagram of the SAR ADC based on the asynchronous process and the corresponding clock phases. As can be seen from path 2 in Fig. 1(a), a control signal (Valid) is produced by the comparator to confirm that the next clock cycle may start when the comparator's output is ready. On the other hand, the next clock cycle cannot start because the SAR logic and the DAC not generate the next comparison voltage at the input of the comparator as seen path 1 in Fig. 1(a). Consequently, this valid signal of path 2 needs a fixed delay before going to the comparator. For the total conversion phase using the asynchronous process, in average, less time are required as mentioned in [7]. In a conventional asynchronous process, the SAR logic delay and worst-case delay of the DAC settling time determine this fixed delay.

The ADC clock frequency strongly depends on the DAC settling time since the pulse width dedicated to each DAC capacitor as explained in [8]. Consequently, two design options for the DAC switching can be used in order to reduce the total DAC settling time as follows: 1) utilizing the same size switches with low equivalent resistance for all bits in the DAC in order to have a fast settling response; and 2) equalizing the settling time from MSB to LSB with scaled switches [8]. The DAC settling time decreases from MSBs to LSBs when the settling accuracy is designed to be the same for all bits. Since the step voltage is small, the pulse width dedicated for settling of LSB capacitors can be reduced to achieve higher speed.

Fig. 2 shows the schematic of proposed SAR structure in asynchronous conversion phase and its asynchronous process approach, where, τ_i and k_i are the time constant and accuracy of DAC settling for the DAC capacitor C_i, respectively. The architecture of Fig. 1(a) can be changed to Fig. 3 according to what was mentioned above. In this figure, two types of delay in the clock generation path considered as follows: 1) the fixed delay in order to the new decision requirements to be prepared by the control logic circuitry; 2) variable delay for the D/A array capacitors settling.

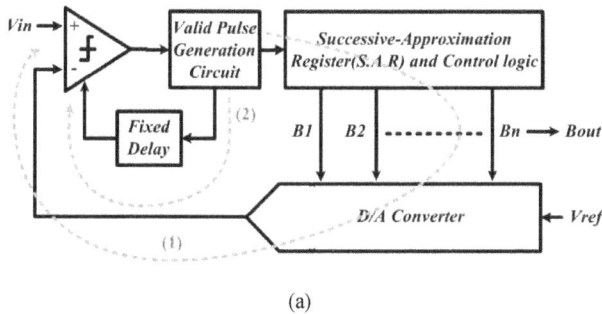

(a)

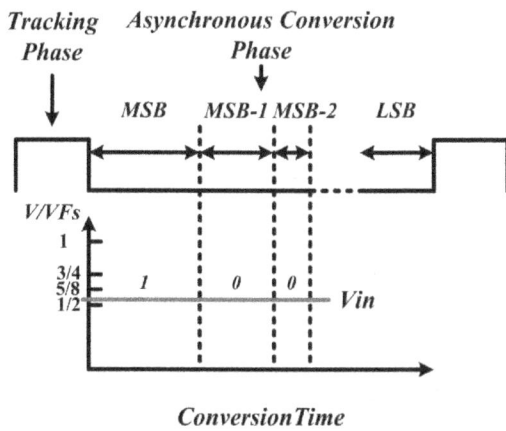

(b)

Fig. 1. *(a)Architecture of the conventional asynchronous SAR ADC (b) Conventional asynchronous process timings.*

3. The Proposed Asynchronous SAR ADC

3.1. Architecture

The *fixed delay* is determined based on the worst-case delay required for the DAC settling time in a conventional asynchronous SAR ADC as mentioned in the previous section. Since the settling times for each of the D/A array capacitors are different, the settling time can be reduced with switching of the DAC array. Therefore, the delay in path 2 of Fig. 1(a) reduces with moving from MSB to LSB capacitors. Consequently, the total A/D conversion time reduces. Thus, the delay in path 2 of Fig. 1(a) can be reduced when moving from MSB to LSB capacitors, and, hence, the total A/D conversion time can be reduced.

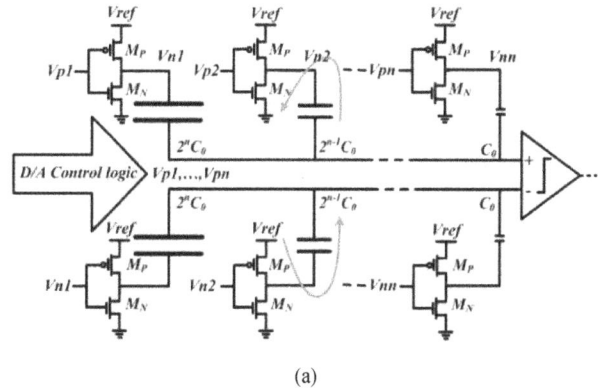

(a)

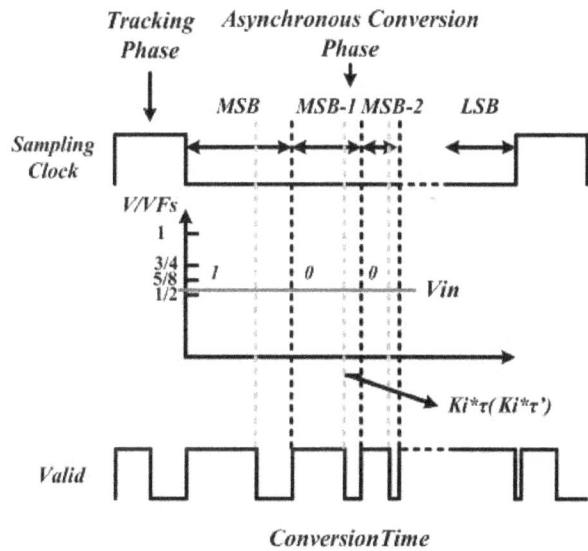

(b)

Fig. 2. *(a)Schematics of the SAR ADC in asynchronous conversion phase (the cycle associated with generating the (MSB-1) bit) (b) Proposed asynchronous process approach.*

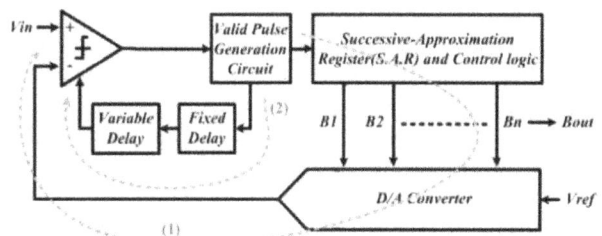

Fig. 3. *Architecture of the proposed asynchronous SAR ADC.*

3.2. Circuit Implementation

The circuit implementation and the timing diagram of the proposed asynchronous clock generator are presented in Fig. 4. The signal delay associated with the OR gate and the internal delay associated with the buffer chain causes fixed and variable delays, respectively. Due to utilizing the unary delay line, the buffer chain has a good accuracy.

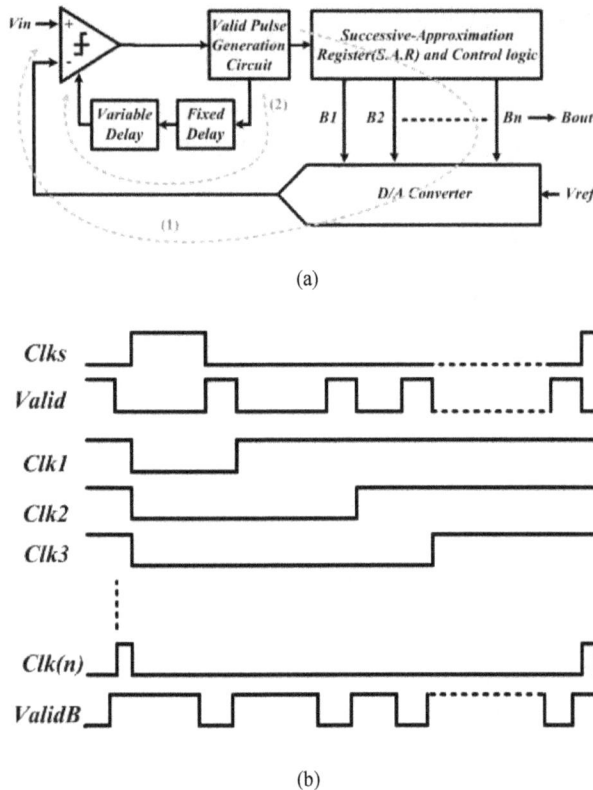

(a)

(b)

Fig. 4. *(a)Proposed asynchronous clock generator circuit (b) proposed asynchronous clock generator timing curves.*

Each buffer includes two cascaded inverters with the same sizes and power supply voltages that cause the same delay. In order to eliminate buffer delay and reduce the power of the delay line, a reset terminal is utilized for each buffer. Therefore, the total power consumption of the ADC reduces. Since resetting the buffer causes its output be zero, the input of the OR gate associated with this buffer output will be zero and the delay signal generated will be reduced. As depicted in Fig. 4, the conventional asynchronous clock generator make the reset signal required for the buffer line that additional circuit are not required.

As shown in Fig. 4, the operation of the asynchronous clock generator circuit is as follows: each buffer is active at the comparison phase and its output becomes 'high' until the reset signal is applied. Since each buffer in the chain has a specific delay and the outputs of the buffers are connected to an n-input OR gate, resetting each buffer causes reduction of the delay signal, which is called the *Valid B* here, generated by the OR gate.

The 2^n buffers in the chain are required to produce the bits

from MSB to LSB in the case of an *n-bit* asynchronous SAR ADC with unary delays. Therefore, the size and power consumption of the clock generator circuit increases drastically. Moreover, the switches in Fig. 2(a) provide an overshoot when the D/A voltage changes due to their gate-drain capacitance. The D/A array capacitors transfer the switches overshoot to the D/A array capacitors output which increases the capacitive settling time. The effect of overshoot is more on less significant bits since the overshoot is the same for different capacitances [9]. Therefore, it is necessary not to dedicate very small settling time to the less significant bits capacitors of the D/A array compared with MSBs capacitors. According to the number of the SAR ADC output bits, the number of buffers is set. Also, the total delay of the buffer chain can be designed based on the settling time of the MSB capacitor. Consequently, with eliminating each buffer from the chain, the total delay of the chain is reduced for different D/A capacitor. However, R. Sekimoto and et al. have recently reported a binary delay line which is proportional to D/A array settling time that the delay line has controlled with a separate circuit and consume extra power [10]. Also, no time has dedicated results some errors in the comparator decision for the overshoot effect and making the new decision.

4. Simulation Results

The proposed asynchronous clock generator and logic control circuit have been optimized for power consumption and area. The simulation results are presented as follows:

4.1. Time Domain Performance

Fig. 5 shows the comparison of the power dissipation and conversion time of proposed 10-bit 4-MS/s SAR ADC with conventional 10-bit 4-MS/s SAR ADC with a monotonic capacitor switching procedure has been implemented. As seen, the difference in the dedicated pulse width to the D/A capacitive array between the conventional and the proposed asynchronous process has been investigated. All of the SAR ADC conditions are the same for two structures but in the conventional one, none of the chain buffers are reset unlike the proposed one. Therefore, both the power dissipation and the conversion time are more compared to the proposed one.

4.2. Dynamic Performance

Fig.6 demonstrates the FFT spectrum with an input frequency of close to nyquist rate and a 4-MS/s sampling rate for power supplies voltage of 0.6 V and 1.2 V. Its SNDR and SFDR are 58.2 dB and 67.4 dB, respectively. The simulated ENOB, FOM and power dissipation values of the proposed SAR versus the input frequency at 4 MS/s are presented in the Fig. 7. The simulated ENOB and power dissipation are about 9.3 bits and 13.7 μw at input frequency close to 0.5MHz, respectively. Since the level shifters were unable to charge the D/A array capacitors, the maximum FOM is 6.8 fJ/step-conv at 0.1 sampling rate.

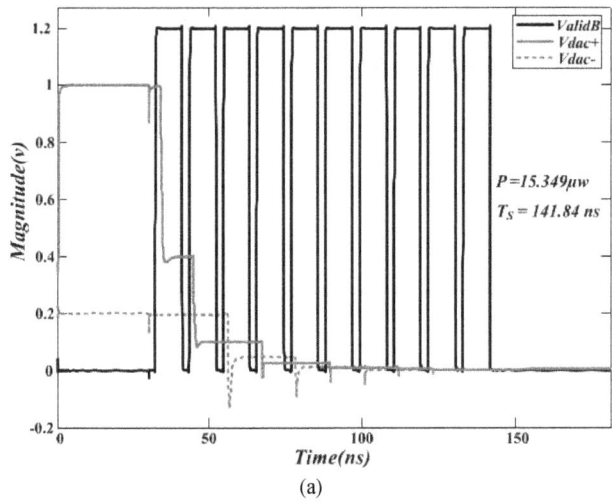

(a)

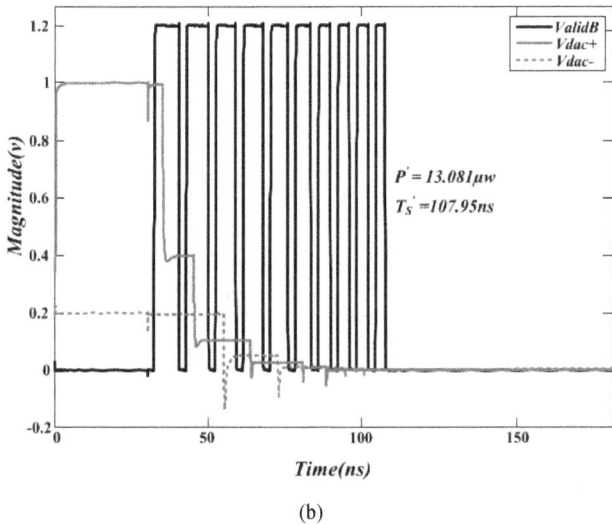

(b)

Fig. 5. *Comparison of the conversion time and power dissipation between (a) Conventional (b) Proposed 10-bit 4-MS/s SAR ADC with a monotonic capacitor switching procedure.*

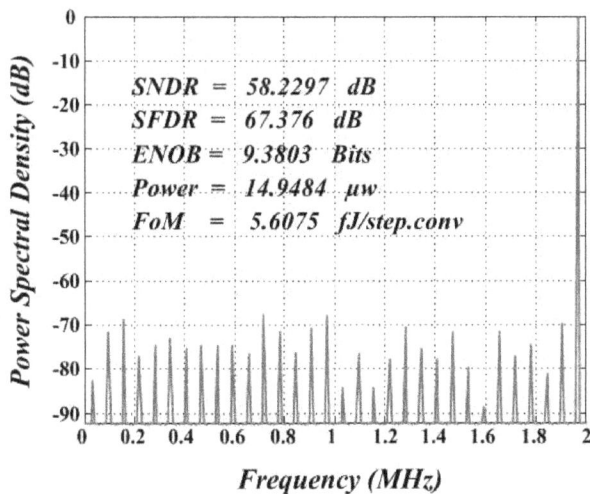

(a)

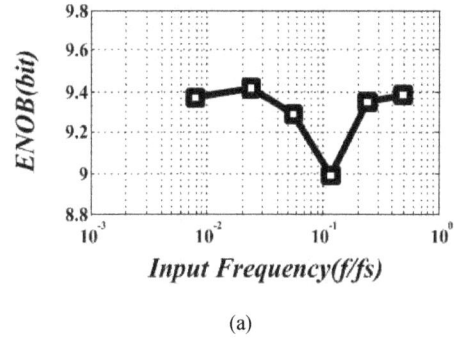

(b)

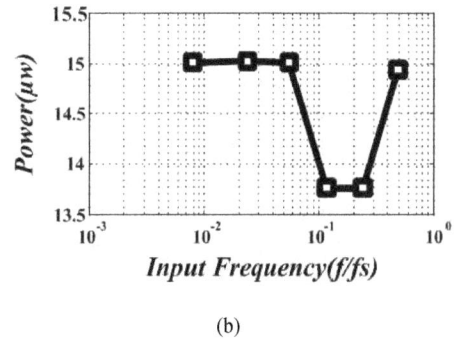

(c)

Fig. 7. *The parameters (a) ENOB (b) FoM (c) Power dissipation of the SAR ADC versus input frequency at 4-MS/s.*

Table 1. *Simulated dynamic performance at different corners.*

Corners	Dynamic Performance			
	SFDR(dBc)	ENOB(bit)	Power(μw)	FoM(fJ/c-s)
T.T	69.365	9.348	13.758	5.277
S.S	66.528	9.357	13.900	5.301
F.F	65.277	9.192	15.622	6.676

Table 2. *Specification summary versus sampling frequency.*

Specification Summary	Sampling Frequency			
	4	8	16	32
ENOB(Bits)	9.35	9.38	9.41	9.28
Power(μw)	13.76	31.17	59.74	185.80
FoM(fJ/c-s)	5.28	5.86	5.50	9.35
V_{DDL} (v)	0.6	0.6	0.75	1.2
V_{DDH} (v)	1.2	1.2	1.2	1.2

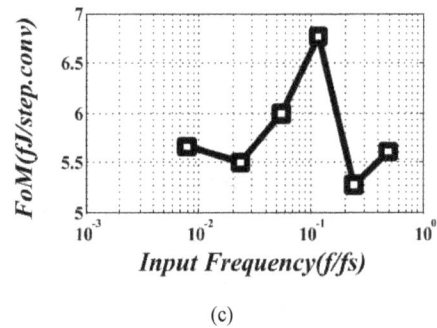

Fig. 6. *Simulated 2,048-point FFT spectrum at 4-MS/s.*

The simulated ENOB and Power Dissipation were about 9.4 bits and 15 μw, respectively, when the input frequency was decreased or increased to 0.5 MHz. Therefore, resultant FOM is about 5.5 fJ/step-conv.

Table 3. Comparison Summary to State-of-the-Art Works.

Specification	Papers					
	JSSC'10 [1]	ESSCIRC'11 [10]	ISSCC'12 [12]	MWSCAS'13 [13]	JSSC'12 [14]	*This Work*
Technology(nm)	130	40	90	130	40	130
Supply Voltage(V)	1.2	0.6&0.7	1.1	1.2	0.5	0.6&1.2
Sampling Rate(MS/s)	50	8.2	4	64	1.1	4
Resolution(Bits)	10	10	10	10	8	10
DAC unit Capacitor(fF)	4.8	0.5	0.6	5.5	0.5	1
ENOB(Bits)	9.2	7.7	9	8	7.5	9.35
Power(µw)	826	21.5	17.4	1325	1.2	13.76
FOM(fJ/c-s)	29	12.5	8.5	78.7	6.3	5.3

For the different process and temperature corners, all the simulations results of the 10-bit 4-MS/s SAR ADC have been summarized in Table I. Also, the simulated performances versus the sampling frequency with a 1-MHz sinusoidal stimulus are shown in Table II. The ENOB was still higher than 9 bits at sampling rate of 32 MS/s. Since the conversion time was insufficient, further increasing the sampling rate rapidly degraded the performance. The maximum sampling rate was limited to 32MS/s in our proposed SAR ADC due to the low supply voltage has been used for digital circuitry and asynchronous clock generator resulting in more delay times for digital circuits. As seen, reducing the sampling frequency (in this case up to 4MS/s) does not noticeably affect the performance.

4.3. Comparison and Discussion

In order to compare the proposed ADC to previous experimental results with different sampling rates and resolutions. At 4 MS/s and 0.6-V, 1.2-V supplies, the well-known figure-of-merit (FOM) equation [1] is used. The FOM of the proposed ADC is 5.3 fJ/conversion-step at 4 MS/s and 0.6-1.2 V supplies. Also, the FOM is 9.28fJ/conversion-step for 32 MS/s sampling rate. Table III compares the proposed ADC with other state-of-the-art previous experimental results [1], [10], [12], [13]-[14]. As can be seen from Table III, the proposed ADC has the lowest FOM compared to those of previous results with similar sampling rates and resolutions however it was simulated using 130-nm CMOS technology.

5. Conclusion

A new approach to asynchronous processing was proposed for extremely low power SAR ADC. Compared to the conventional method, it can further increase the speed of the converter with high power efficiency. The proposed circuit achieves a 4-32 MS/s operation speed with power consumption of between 14-186µW, resulting in a FOM of 5.3 fJ/conversion-step. The results indicate that proposed SAR ADC can has higher speed and power efficiency.

Acknowledgements

I'd like to thank Dr. Samaneh Babayan Mashhadi for their valuable help regarding advice on simulations.

References

[1] C.C. Liu, S.J. Chang, G.Y. Huang and Y.Z. Lin, "A 10-bit 50-MS/s SAR ADC With a Monotonic Capacitor Switching Procedure ", in *IEEE J. Solid-State Circuits*, vol.45, pp.731 - 740, April 2010.

[2] H. Wei, *et al.*, "A 0.024 mm² 8b 400 MS/s SAR ADC with 2 b/cycle and resistive DAC in 65 nm CMOS ", in *IEEE ISSCC Dig. Tech. Papers*, pp. 188-190, Feb. 2011.

[3] Y.M. Greshishchev, *et al.*, "A 40 GS/s 6 b ADC in 65 nm CMOS", in *IEEE ISSCC Dig. Tech. Papers*, pp.390-391, Feb. 2010.

[4] A. Arian, M. Saberi, R. Lotfi, S. Hosseini-Khayat and Y. Leblebici, "A 10-bit 50-MS/s SAR ADC with split capacitive-array DAC", *Analog Integrated Circuits and Signal Processing*, vol. 71, 583-589, Dec. 2011.

[5] Z. Cao, S. Yan, and Y. Li, "A 32 mW 1.25 GS/s 6 b 2 b/step SAR ADC in 0.13 m CMOS ", in *IEEE ISSCC Dig. Tech. Papers*, pp.542–543, Feb. 2008.

[6] P. Schvan, *et al.*, "A 24 GS/s 6 b ADC in 90 nm CMOS", in *IEEE ISSCC Dig. Tech. Papers*, pp.544–634, Feb. 2008.

[7] S.W.M. Chen and R.W. Brodersen, "A 6-bit 600-MS/s 5.3-mW Asynchronous ADC in 0.13-µm CMOS", in *IEEE J. Solid-State Circuits*, vol. 41, pp.2669-2680, Dec. 2006.

[8] J. Yang, T.L. Naing and R.W. Brodersen, "A 1 GS/s 6 Bit 6.7 mW Successive Approximation ADC Using Asynchronous Processing", in *IEEE J. Solid-State Circuits*, vol. 45, pp.1469-1478, Aug. 2010.

[9] Z. Huang, *et al.*, "Modeling the Overshooting Effect for CMOS Inverter Delay Analysis in Nanometer Technologies", in *IEEE Trans. Comput.-Aided Des. Integr. Circuits Syst*, vol.29, pp. 250-260, Feb. 2010.

[10] R. Sekimoto, A. Shikata, T. Kuroda and H. Ishikuro, "A40nm50S/s–8MS/sultralowvoltageSARADCwithtimingopti mizedasynchronousclockgenerator", in *Proc. ESSCIRC*, pp.471-474, Sept. 2011.

[11] J.M. Rabaey, A. Chandrakasan and B. Nikolic, Digital Integrated Circuits, 2nd ed., Prentice Hall: New Jersey, 2003, pp. 177-190.

[12] I.S. Jung, M. Onabajo and Y.B. Kim, "A 10-bit 64 MS/s SAR ADC using variable clock period method", in *IEEE International Midwest Symp. Circuits Syst*, pp. 1144-1147, Aug. 2013.

[13] A. Shikata, R. Sekimoto, T. Kuroda and H. Ishikuro, "A 0.5 V 1.1MS/sec 6.3fJ/Conversion-Step SAR-ADC with Tri-Level Comparator in 40 nm CMOS,", in *IEEE J. Solid-State Circuits*, vol. 47, pp. 1022-1030, April 2012.

[14] P. Harpe, Y. Zhang, G. Dolmans, K. Philips and H. de Groot, "A 7-to-10b 0-to-4MS/s Flexible SAR ADC with 6.5-to -16fJ/conversion-step," *IEEE ISSCC Dig. Tech. Papers,* pp. 472-474, Feb. 2012.

33

Effects of Distributed Generation on System Power Losses and Voltage Profiles (Belin Distribution System)

Chaw Su Hlaing, Pyone Lai Swe

Department of Electrical Power Engineering, Mandalay Technological University, Mandalay, Myanmar

Email address:

chawchaw.mht@gmail.com (C. S. Hlaing), pyonelai@gmail.com (P. L. Swe)

Abstract: In present times, the use of DG systems in large amounts in different power distribution systems has become very popular and is growing on with fast speed. Although it is considered that DG reduces losses and improves system voltage profile, this paper shows that this is usually true. The paper presents voltage stability index based approach which utilizes combine sensitivity factor analogy to optimally locate and size a multi-type DG in 48-bus Belin distribution test system with the aim of reducing power losses and improving the voltage profile. The multi-type DG can operate as; type 1 DG (DG generating real power only), and type 2 DG (DG generating both real and reactive power). It further shows that the system losses are reduced and the voltage profile improved with the location of type 2 DG than with the location of type 1 DG. It reaches a point where any further increase in number of DGs in the network results for minimizing power losses and voltage profiles improvement.

Keywords: Distributed generation (DG), Voltage stability index (SI), System Loss Reduction, Voltage Profiles Improvement, Optimal Locating and Sizing

1. Introduction

Distributed generation (DG) is small-scale power generation that is usually connected to distribution system. The Electric Power Research Institute (EPRI) defines DG as generation from a few kilowatts up to 50MW [1]. Ackermann et al. have given the most recent definition of DG as: "DG is an electric power generation source connected directly to the distribution network or on the customer side of the meter." [2].

In most power systems, a large portion of electricity demand is supplied by large-scale generators. This is because of economic advantages of these units over small ones. The distributed real power sources can be classified into two categories and referred in the following sections of this paper as type 1 DG and type 2 DG:

Type 1 DG: Distributed generations that supply real power, depending on the availability or demand, to the network without demanding any reactive power. Few examples of type 1 DG are photovoltaic cell, fuel cell, battery storage.

Type 2 DG: Distributed generations that supply both active and reactive power to the network. Type 2 is used for DG sources such as wind generation, combustion engines, and like synchronous generators [3].

Normally, the real power loss reduction draws more attention for the utilities, as it reduces the efficiency of transmitting energy to customers. Nevertheless, reactive power loss is obviously not less important. This is due to the fact that reactive power flow in the system needs to be maintained at a certain amount for sufficient voltage level. Consequently, reactive power makes it possible to transfer real power through transmission and distribution lines to customers [4]. System loss reduction by strategically placed DG along the network feeder can be very useful if the decision maker is committed to reduce losses and to improve network performance maintaining investments to a reasonable low level [5].

Studies indicate that poor selection of location and size of a DG in a distribution system would lead to higher losses than the losses without DG. In a power system, the system operator is obligated to maintain voltage level of each customer bus within the required limit [6]. Actually in practice, many electricity companies try to control voltage variations within the range of ±5% [7]. The DG units improve voltage profiles by changing power flow patterns. The locations and size of DGs would have a significant impact on the effect of voltage profile enhancement.

2. Voltage Stability Index

A system experiences a state of voltage instability when there is a progressive or uncontrollable drop in voltage magnitude following a disturbance, increase in load demand or change in operating condition. It is usually identified by an index called steady state voltage stability index, evaluated using sensitivity analysis. Sensitivity analysis is the computation of voltage stability index of all the nodes in RDS. Voltage stability index, SI can be computed as follows:

SI Index, proposed by [8], is utilized to find the weakest voltage bus in power system. This index will find the most optimum weakest link in the system which could lead to voltage stability in future, when the load will increase. The value of index is given by Eq. (1) and termed as Stability index (SI).

$$SI = |V_s|^4 - 4 \times [P_r x_{ij} - Q_r r_{ij}]^2 - 4 \times [P_r r_{ij} + Q_r x_{ij}] \times |V_s|^2 \quad (1)$$

where, SI is the stability index, V_s is the sending bus voltage, P_r is active load at receiving end, Q_r is the reactive load at receiving end, r_{ij} is the resistance of the line i-j and x_{ij} is reactance of the line i-j.

Under stable operation, the value of SI should be greater than zero for all buses. When the value of SI becomes closer to one, all buses become more stable. In the proposed algorithm, SI value is calculated for each bus in the network and sort from highest to lowest value. For the bus having the lowest value of SI, will be considered in fitness function.

3. Objective Function

As the main objective of this work is to determine the optimal location and sizing of the distributed generation in the distribution network to minimize the losses (active power loss), the following objective function is selected as [9]:

$$F_l = \min P_{loss} = \Sigma_{k=1}^{ntl} |I_j|^2 \cdot r_j \quad (2)$$

where, F_l is the objective function to minimize power losses. P_{loss} is the active power loss. ntl is the number of lines in the distribution system.

Subjected to constraints:

$$V_i^{min} \leq V_i \leq V_i^{max} \quad (3)$$

$$I_i \leq I_i^{max} \quad (4)$$

$$V_{DG}^{min} \leq V_{DG} \leq V_{DG}^{max} \quad (5)$$

$$P_{DG}^{min} \leq P_{DG} \leq P_{DG}^{max} \quad (6)$$

where,

P_{DG} =real power generations of DG
V_i =voltage magnitudes at bus i

V_{DG} =voltage magnitudes at bus i
I_i =ith feeder current loading

4. Problem Formulation

The problem formulation for the optimal location and sizing of the distributed generation in the distribution network to minimize the active power loss includes the power flow with and without distributed generation in the distribution system. The distributed generation is considered as active power sources at a particular voltage, which is at unity power factor. The well-known basis load flow equations are [10]:

$$S_i = P_i + jQ_i = V_i I_i^* \quad (7)$$

$$S_i = V_i \sum_{j=1}^{n} Y_{ij}^* V_j^* = \sum_{j=1}^{n} |V_i||V_j||Y_{ij}| \angle (\delta_i - \delta_k + \theta_{ij}) \quad (8)$$

Resolving into the real and imaginary parts, then the power flow equations without DG are given as:

$$P_i = \sum_{j=1}^{n} |V_i||V_j||Y_{ij}| \cos(\delta_i - \delta_k + \theta_{ij}) = P_{Gi} - P_{Di} \quad (9)$$

$$Q_i = \sum_{j=1}^{n} |V_i||V_j||Y_{ij}| \sin(\delta_i - \delta_k + \theta_{ij}) = Q_{Gi} - Q_{Di} \quad (10)$$

The basic power balance equations:

$$P_{Gi} = P_{Di} + P_L \quad (11)$$

$$Q_{Gi} = Q_{Di} + Q_L \quad (12)$$

The power flow equations considering losses with DG for the practical distribution system and the DG is an active power source at unity power factor (PV generator) then flow are given as:

$$P_i + P_{DGi} = P_{Di} + P_L \quad (13)$$

$$Q_i + Q_{DGi} = Q_{Di} + Q_L \quad (14)$$

The DG is active power source only at unity power factor, so $Q_{DG} = 0$.

$$P_i + P_{DGi} = P_{Di} + P_L \quad (15)$$

$$Q_i = Q_{Di} + Q_L \quad (16)$$

The final power flow equations for distribution system are:

$$\sum_{j=1}^{n} |V_i||V_j||Y_{ij}| \cos(\delta_i - \delta_k + \theta_{ij}) + P_{DGi} = P_{Di} + P_L \quad (17)$$

$$\sum_{j=1}^{n} |V_i||V_j||Y_{ij}| \sin(\delta_i - \delta_k + \theta_{ij}) = Q_{Di} + Q_L \quad (18)$$

$$\sum_{j=1}^{n} |V_i||V_j||Y_{ij}| \cos(\delta_i - \delta_k + \theta_{ij}) + P_{DGi} - P_{Di} - P_L = 0 \quad (19)$$

$$\sum_{j=1}^{n} |V_i||V_j||Y_{ij}| \sin(\delta_i - \delta_k + \theta_{ij}) - Q_{Di} - Q_L = 0 \quad (20)$$

$$P_i^{min} \le P_i \le P_i^{max} \quad (21)$$

$$Q_i^{min} \le Q_i \le Q_i^{max} \quad (22)$$

$$V_i^{min} \le V_i \le V_i^{max} \quad (23)$$

$$P_{DG}^{min} \le P_{DG} \le P_{DG}^{max} \quad (24)$$

where,

P_i, Q_i =real and reactive power flow at bus i

P_{Di}, Q_{Di} = real and reactive loads at bus i

V_i, V_k = voltage magnitudes at bus i and k

P_{DGi} =real power of DG at bus i

N=total number of buses

δ_i, δ_k=voltage angles of bus i and k

Y_{ik}= magnitude of the ik^{th} element in bus admittance matrix

θ_{ik} =angle of the ik^{th} element in bus admittance matrix

5. Solution Methodology

Following steps are involved in optimal siting and sizing of distributed generations:

Step: 1Determination of proposed locations for placing distributed generations

a) Perform load flow analysis to calculate the bus voltage magnitudes and total network power loss in the RDS.

b) Compute the voltage stability index (SI) using Eq. (1).

c) Arrange the buses in ascending order of the voltage stability index and select one or two buses with low value of voltage stability index from different laterals as the proposed locations for placing distributed sources.

Step: 2 Run the Base Case without DG using NR load flow using MATLAB software and calculate the bus voltage magnitude, angle, and real and reactive power loss respectively.

a) After Load flow identify the optimum sizing for each bus is calculated.

b) Find out the approximate losses for each bus by placing DG at the corresponding location with the optimum sizing obtains from the above step.

c) Check for constraint violation after DG placement.

d) Locate the bus at which the loss is minimum after DG placement and this is the optimum location for DG.

Repeat the above procedure till the termination condition is satisfied.

6. Results and Discussions

The solution methodology presented in this paper for

optimal siting and sizing of distributed power sources are analyzed using 132 kV, 33 kV and 11 kV, 48-bus Belin distribution system. The data for 48-bus Belin distribution system is given in Table 4 in the appendix.

This system is supplied from Yeywa Generation Station of 487 MW and 400 MVAR, 230 kV with a total peak load of 79.53 MW and 41.87 MVAR. The total system power loss at the peak demand without DG connection (base case scenario) is 1.351MW and 17.56 MVAR. The single line diagram of 48-bus Belin distribution system is given in Fig.1.

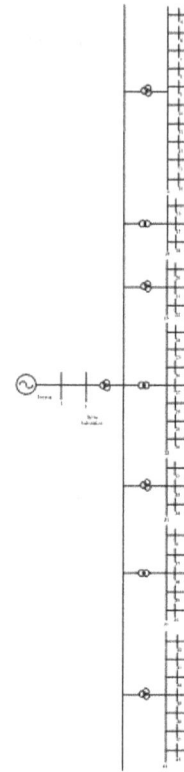

Figure 1. The 48-bus Belin distribution system.

6.1. Optimum Size Allocation

The number of DGs to be included in a power network can be limited by several factors. The two main factors are the undesirable effects on power system parameters and the economic factors. This research was mainly concerned with the system power losses and the voltage profile of the network and thus the effects of DG on these system parameters have been investigated. The DG limits were taken to be as follows; 0 MW – 53 MW for real power limit (Type 1, and 2 DGs), 0 MVAR – 31 MVAR for reactive power limit (Type 2 DG).

After calculating the combined sensitivity factors, the buses were arranged in order of sensitivity and those with a factor of less than 0.8 were selected as the proposed buses. Table 1 shows the results of the optimal DG sizes for each respective proposed location and the associated best fitness achieved for all the two types of DGs. Both real and reactive power losses are considered in while investigating the effect of DG on system power losses. The number of DGs was assumed to increase from one, two, three and then four. This was done sequentially ensuring that the proposed bus with the most optimal size was chosen first followed with the others in the same order. Thus the most optimal DG location and size was included in the four cases.

Table 1. Results for SI and optimal DG sizes for multi-type DGs located on chosen proposed buses.

Proposed Bus	SI index	Type 1 DG	Type 2 DG
		Optimal DG Size (MW)	Optimal DG size (MW+jMVAR)
10	0.7827	6.4635	10.5250+j5.5639
24	0.6577	6.0899	12.9831+j5.9928
25	0.7082	6.1199	10.0045+j5.9844
26	0.7750	9.7073	52.3274+j30.8305
27	0.7349	6.7413	10.3567+j5.4865
28	0.7688	3.4966	7.4777+j3.9912
29	0.7663	6.5099	9.0574+j4.8620
30	0.7607	5.2389	3.4092+j1.8000
36	0.7781	4.6756	17.6717+j9.3471
37	0.7813	1.8338	6.4406+j3.4067
38	0.7118	4.8256	4.9556+j2.6225
39	0.7613	2.5351	1.9437+j1.0251
40	0.7884	2.5446	1.0499+j0.5530
43	0.7953	2.8892	6.2551+j3.3038
46	0.7794	3.4975	2.3135+j1.2223

Table 2. Effects of type 1 DG on system power losses.

Number of DGs	Bus No.	DG Size	Power Losses	% Power Loss Reduction
		MW	MW+jMVAR	%(MW+jMVAR)
One	24	6.0899	1.125+j15.16	16.73+j13.67
Two	24	6.0899	0.971+j13.19	28.13+j24.89
	25	6.1199		
Three	24	6.0899	0.887+j11.91	34.34+j32.18
	25	6.1199		
	38	4.8256		
Four	24	6.0899	0.774+j10.18	42.71+j42.03
	25	6.1199		
	38	4.8256		
	27	6.7413		

As it can be seen from Table 2, the introduction of only one type 1 DG on bus 24 reduced the real power losses from the base case scenario of 1.351 MW to 1.125 MW and the reactive losses from 17.56 MVAR to 15.16 MVAR. The inclusion of the second DG in the system further reduced both real and reactive power losses to 0.971 MW and 13.19 MVAR. The introduction of the third DG reduces both real and reactive power losses to 0.887 MW and 11.91 MVAR. The inclusion of the fourth DG in the system results to 0.774 MW and 10.18 MVAR less than the case without DG in both real and reactive power losses.

From the results in Table 3, the introduction of the first optimally placed and sized type 2 DG in the network reduced the real power losses from the base case value of 1.351 MW to 1.135 MW and the reactive power losses from 17.56 MVAR to 12.45 MVAR. The inclusion of the second and third DG in the network further reduces the real power losses to 1.019 MW and 0.915 MW and the reactive power losses to 9.13 MVAR and 7.71 MVAR respectively. It is evident that the introduction

of the fourth DG in the network decreases both real and reactive power losses in the system from the previous case.

Table 3. Effects of type 2 DG on system power losses.

Number of DGs	Bus No.	DG Size	Power Losses	% Power Loss Reduction
		MW+jMVAR	MW+jMVAR	%(MW+jMVAR)
One	24	12.9831+j5.9928	1.135+j12.45	15.99+j29.10
Two	24	12.9831+j5.9928	1.019+j9.13	24.57+j48.01
	25	10.0045+j5.9844		
Three	24	12.9831+j5.9928	0.915+j7.71	32.27+j56.09
	25	10.0045+j5.9844		
	38	4.9556+j2.6225		
Four	24	12.9831+j5.9928	0.862+j5.49	36.20+j68.74
	25	10.0045+j5.9844		
	38	4.9556+j2.6225		
	27	10.3567+j5.4865		

6.2. Effect of DG on Bus Voltage Profile

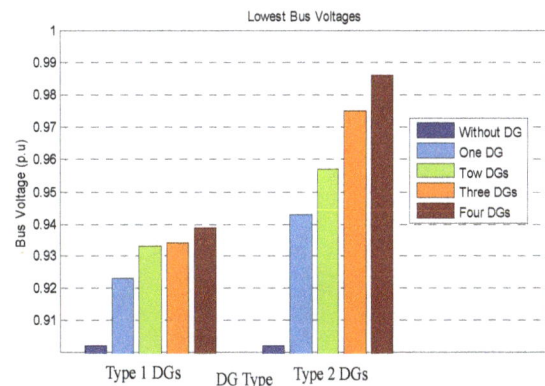

Figure 2. A graph of the lowest bus voltages for different DG types and DG numbers.

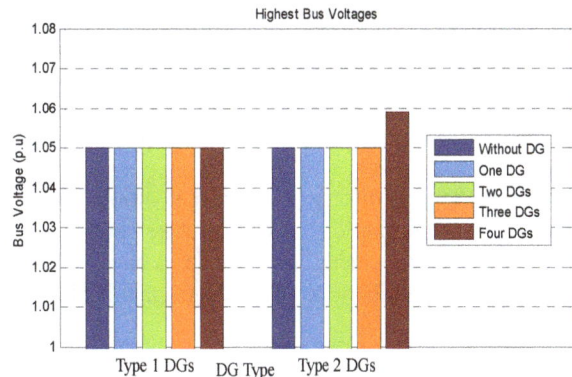

Figure 3. A graph of the highest bus voltages for different DG types and DG numbers.

From fig. 2 above it can be seen that all the two cases resulted to an increase in the lowest bus voltage level. It also important to note that there was an increase for each additional DG added in the network up to the fourth DG. Note that the minimum voltage level for the base case is about 0.902 p.u recorded at bus 24. Fig. 3 shows the highest bus voltages for different DG types and DG numbers.

After installation of type 1 DG, there are still buses which are lower than the pre-specified voltage limit of 0.95 p.u. After installation of type 2 DG, the voltage levels of these buses are improved with minimum 0.943 p.u of bus number 25 with one DG. Since the most ideal case was to have this voltage as close to 1 p.u as possible it can be concluded that type 2 DG performed better in this case compared to type 1 DG. This is because its bus voltage levels with three DGs in the system are within the range of ±5%.

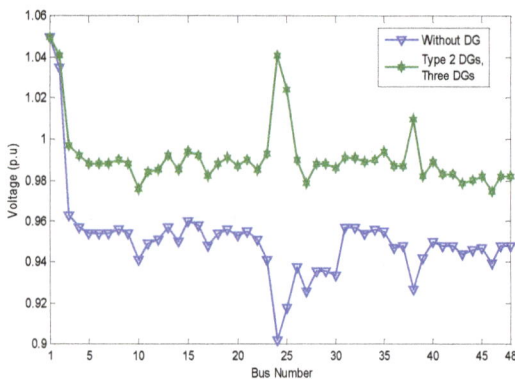

Figure 4. *Voltage profile before and after DG injection having optimum value.*

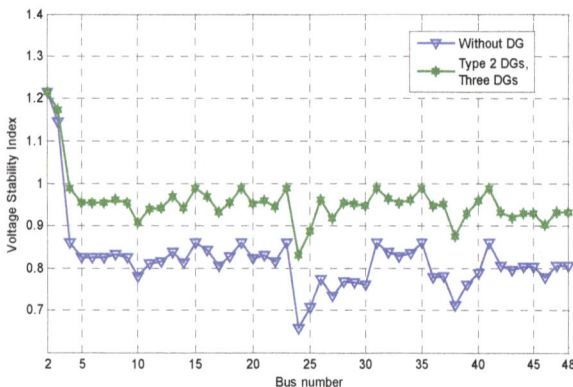

Figure 5. *Voltage Stability Index (SI) before and after DG installation at each bus of system.*

Fig.4. shows voltage profiles improvement at various nodes for 48-bus Belin distribution system before and after connecting DG. Improvement in voltage stability was observed from Fig.5. In this figure the voltage stability index at each buses of system is shown

7. Conclusion

The solution methodology for optimal siting and sizing of distributed generation in Belin distribution system, considering precise models for distributed generations is presented in this paper. The benefits and consequences of distributed sources for improvement in voltage profile, voltage stability index and reduction on total network power loss have been analyzed in detail. According to the objective function, the best location in 48-bus Belin distribution system is in the order 24, 25 and 38 corresponding type 2 optimal DG sizes of MW and MVAR are 12.9831 + j5.9928, 10.0045 + j5.9844 and 4.9556 + j2.6225. The results of the proposed system depict that the optimal size of type 2 DG with three DGs is the maximum possible penetration levels of distributed generation in Belin distribution system in terms of improved voltage profile and reduced total real and reactive power loss of 0.915 MW and 7.71 MVAR.

Acknowledgements

The author would like to express grateful thanks to her supervisor Dr. Pyone Lai Swe, Associate Professor, Department of Electrical Power Engineering at Mandalay Technological University for her valuable suggestions and supervision through this research work. She also wants to thank all her teachers from Department of Electrical Power Engineering at Mandalay Technological University. The author offers her appreciation to her parents and brothers for their supports, encouragement and valuable guidance.

Appendix

Table 4. *Data for 48-Bus Belin Distribution System.*

Sending Bus	Receiving Bus	R (p.u)	X (p.u)	Load at Receiving Bus	
				P (MW)	Q (MVAR)
1	2	0.00268	0.02232	0	0
2	3	0.00000	0.1418	0	0
3	4	0.01470	0.0356	0	0
3	15	0.02900	0.0818	0	0
3	19	0.02900	0.0818	0	0
3	23	0.02900	0.0818	0	0
3	31	0.03790	0.0920	0	0
3	35	0.03790	0.0920	0	0
3	41	0.08820	0.2486	0	0
4	5	0.36130	0.6974	0.45	0.24
4	6	0.30110	0.5812	0.61	0.32
4	7	0.37310	0.7199	0.44	0.23
4	8	0.07530	0.1453	1.14	0.60
4	9	0.15060	0.2906	1.1	0.58
4	10	0.16730	0.3228	4.5	2.37
4	11	0.18790	0.3626	2	1.05
4	12	0.17700	0.3416	1.8	0.948
4	13	0.00840	0.0161	0.15	0.079
4	14	0.08360	0.1614	3.94	2.075
15	16	0.09980	0.1294	0.925	0.487
15	17	0.30660	0.3975	2.204	1.158
15	18	0.23530	0.3051	1.429	0.753
19	20	0.03320	0.0719	4.52	2.38
19	21	0.13140	0.2844	0.375	0.197
19	22	0.06570	0.1153	4.146	2.185
23	24	0.30820	0.7736	4.846	2.554
23	25	0.25780	0.4522	4.279	2.254
23	26	0.02900	0.0559	4.4	2.317
23	27	0.13250	0.2868	4.85	2.554

Sending Bus	Receiving Bus	R (p.u)	X (p.u)	Load at Receiving Bus	
				P (MW)	Q (MVAR)
23	28	0.09840	0.2130	2.042	1.074
23	29	0.05910	0.1278	3.946	2.08
23	30	0.08860	0.1917	3.42	1.8
31	32	0.00840	0.0161	0.015	0.0079
31	33	0.04180	0.0807	3.129	1.648
31	34	0.01250	0.0242	3.713	1.954
35	36	0.18720	0.2426	2.308	1.216
35	37	0.43680	0.5662	0.868	0.457
35	38	0.49930	0.6472	3.049	1.606
35	39	0.49930	0.6472	1.422	0.748
35	40	0.24970	0.3236	1.051	0.553
41	42	0.13380	0.2583	0.065	0.034
41	43	0.11710	0.2260	1.46	0.769
41	44	0.05850	0.1130	1.925	1.016
41	45	0.07190	0.1388	0.58	0.305
41	46	0.17560	0.3390	2.21	1.164
41	47	0.10040	0.1937	0.102	0.054
41	48	0.07780	0.1501	0.12	0.063

References

[1]　Satish Kansal, B.B.R. Sai, Barjeev Tyagi, Vishal Kumar "Optimal placement of distributed generation in distribution networks", International Journal of Engineering, Science and Technology, Vol. 3, No. 3, 2011, pp. 47-55.

[2]　Ackermann, T., Anderson, G. and Soder, L., 2001. "Distributed Generation: A Definition", Electric Power System Research, 57(3): 195-204.

[3]　Soroudi. A and M.Ehsan. "Multi objective distributed generation planning in liberized electricity market", in IEEE Proc. 2008, PP.1-7.

[4]　Borges, C.L.T. and Falcao, D.M., 2006. "Optimal Distributed Generation Allocation for Reliability, Losses and Voltage Improvement." Electric Power and Energy System, 28: 413-420.

[5]　Kim, T.E., 2001a. "A method for determining the introduction limit of distributed generation system in distribution system." IEEE Trans. Power Delivery, 4(2): 100-117.

[6]　Kim, T.E., 2001b. "Voltage regulation coordination of distributed generation system in distribution system." IEEE Trans. Power Delivery, 6(3): 1100-1117.

[7]　Tautiva, C. and Cadena, A., 2008. "Optimal Placement of Distributed Generation on Distribution Network." Proceeding of Transmission and Distribution Conference and Exposition-IEEE/PES-Bogota.

[8]　Chakravorty M, Das D. Voltage stability analysis of radial distribution networks. Int J Electr Power & Energy Syst 2001;23:129–35.

[9]　C. L. Wadhwa, —Electrical Power system, New Age International Publication, 2010, Sixth Edition.

[10]　Hadi Sadat, —Power system analyses, TMH Publication, 2002 Edition.

34

Short-term Electrical Energy Consumption Forecasting Using GMDH-type Neural Network

Tsado Jacob[1], Usman Abraham Usman[1, *], Saka Bemdoo[2], Ajagun Abimbola Susan[1]

[1]Department of Electrical and Electronics Engineering, Federal University of Technology Minna, Niger State, Nigeria
[2]Transmission Company of Nigeria, TCN Abuja, Nigeria

Email address:
usman.abraham@futminna.edu.ng (U. A. Usman)

Abstract: Electric load forecasting plays an important role in the planning and operation of the power system for high productivity in any institution of learning. A short-term electrical energy forecast for Gidan Kwano campus, Federal University of Technology Minna, Nigeria was carried out using GMDH-type neural network and the result was compared to that of regression analysis. GMDH-type neural network was used to train and test weekly energy consumed in the campus from September 2010 to December 2014. The neural network was trained using quadratic neural function. Root mean square error (RMSE) and mean absolute percentage error (MAPE) were used as performance indices to test the accuracy of the forecast. The neural network model gave a root mean square error (RMSE) of 0.1189, a mean absolute percentage error (MAPE) of 0.0922 and a correlation (R) value of 0.8995 while the regression analysis method gave a standard error of 10968.1 and a correlation (R) value of 0.1137. Results obtained show the efficacy of the GMDH-type neural network model in forecasting over the regression analysis method.

Keywords: Group Method of Data Handling (GMDH), Polynomial Neural Network (PNN), Short Load Term Forecasting (STLF), Mean Absolute Percentage Error (MAPE), Root Mean Square Error (RMSE)

1. Introduction

Electricity is one of the vital utilities for the development and higher productivity in any academic institution. Federal University of Technology Minna like many other institutions of learning in a developing country like Nigeria is faced with inadequate electricity supply. In addition, the demand for electricity in *Gidan Kwano* campus of Federal University of Technology Minna has increased rapidly due to the growing population, increased infrastructure and social conditions in the university. To address the shortage of electricity in the university, large investment is required. Accurate load forecasting will ensure that these investments do not waste and at the same time help distribute electricity effectively in the university campus. It will also uncover inefficiencies in the system as well as determine where savings can be made. When load forecasting is not accurately done, it will lead to improper planning and sizing of the network.

Load forecasting can be classified into three categories

based on time namely: short-term load forecasting (STLF), medium term load forecasting (MTLF) and long term load forecasting (LTLF). Short-term load forecasting covers a period of one hour to one week. It is needed for efficient planning of the day-to-day operation and maintenance of the power system [1]. STLF is used to determine the electrical energy needed to meet the expected demand [2]. Medium term load forecasting covers a period ranging from one week to one year. Medium term load forecasting is important when carrying out power system maintenance, evaluation of economic dispatch and scheduling outage. Long term load forecasting covers a period of one year upwards and it is used for planning the network, future expansion plans and capital investments. This type of forecast is often complex in nature due to political and economic factors that are involved [1]. Our focus in this paper is restricted to short-term load forecasting.

Generally, load forecasting methods can be grouped into statistical and Artificial Intelligence (AI) methods. The statistical methods include regression analysis, time series modelling, Box-Jenkins ARIMA, exponential smoothing and similar day approach. Artificial Intelligence methods include Artificial Neural Networks (ANN), fuzzy logic, genetic algorithm (GA) and expert systems [3, 4].

The artificial intelligence methods hold an advantage over the statistical methods in their ability to predict accurately from non-linear data. Hence, neural network was used to forecast future energy consumption of Gidan Kwano campus.

2. Literature

A lot of work has been done on load forecasting using different techniques over the years. Generally, load forecasting methods can be grouped into conventional and artificial intelligent methods. The conventional methods include: regression analysis, time series modelling, Box-Jenkins ARIMA, exponential smoothing and similar day approach while the artificial intelligent methods include: Artificial Neural Networks (ANN), fuzzy logic, genetic algorithm (GA) and expert systems. In carrying out load forecast, some factors which affect it are also considered. Some of the factors considered for load forecasting include: historical load data, weather, time factor, economic growth and random disturbance [5, 16, 17]. In Gidan Kwano campus for example, more energy is consumed during the hot season than the cold season. This is so because when the weather is hot, more energy consuming devices like air conditioners, fridges and fans are put on while the reverse is the case during the cold season.

For this work however, artificial neural network (ANN) was used to forecast energy consumed due to itps superiority over conventional methods in forecasting accurately from nonlinear data. The energy data used is nonlinear hence the ANN was used over regression method.

The human nervous system inspired the development of Artificial Neural Networks [5]. Artificial Neural Networks (ANN) comprise of processing units (or neurons) that are linked via weighted interconnections. ANN consists of a processing unit which is known as the transfer function [6]. Figure 1 illustrates a simple neuron model of the human nervous system.

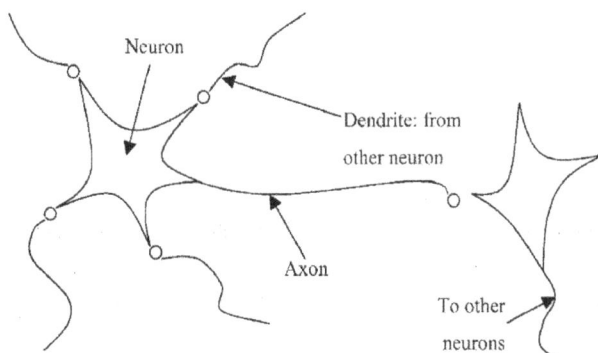

Figure 1. Neuron model of the human nervous system.

From the above model, each neuron is connected to

another neuron via the dendrite or the axon. A neuron takes input via the dendrites and presents an output to the next neuron via axon. The dendrites are the interconnections [6]. This is exactly the way ANN works. For this work, GMDH-type neural network is used to predict energy consumption.

GMDH was developed by Prof. Alexey G. Ivakhnenko in 1968 at the Institute of Cybernetics in Kiev (USSR). After the initial development, thee advanced models of GMDH was developed to express multi-variable and non-linear systems [12]. The idea of GMDH is to build a model that will predict a value that is close as possible to the actual value [13]. GMDH algorithm was developed to identify non-linear relationships between inputs and outputs. GMDH algorithm can be represented as a set of neurons in which different pairs of the neurons are connected through a quadratic polynomial and therefore produce new neurons in the next layer [15]. Polynomial neural network is a flexible architecture whose structure is developed through learning. Each node of the polynomial exhibits a high degree of flexibility and realizes a polynomial mapping (linear, quadratic or cubic) between the input and the output [13]. Neural networks developed using the GMDH algorithm are called GMDH-type neural networks and are classified within the group of polynomial neural networks (PNN) [14].

The GMDH is an inductive self-organizing data driven approach. Its basic equation is called Kolmogrov-Gabor polynomial and it is expressed as shown in equation (1) below [11].

$$y = a_0 + \sum_i^N a_i x_i + \sum_i^N \sum_j^N a_{ij} x_i x_j + \sum_i^N \sum_j^N \sum_k^N a_{ijk} x_i x_j x_k + \quad (1)$$

Where x_i (i=1, 2... N) and y represent input and output variables respectively. For example, neuron can be represented with a quadratic polynomial function as shown in equation (2),

$$y = f(x_i, x_j) = a_0 + a_1 x_i + a_2 x_i + a_3 x_i x_j + a_4 x_i^2 + a_5 x \quad (2)$$

In principle, constructed network is the composition of neurons with the mapping function f (x_i, x_j). The fixed number of neurons is selected at each layer and the output of these neurons is used on the next layer [11].

3. Materials and Methods

The daily energy consumption data was collected from Minna Transmission station while the monthly energy consumption data was collected from the Electrical works department of Gidan Kwano campus of Federal University of Technology, Minna. The data covers a period from 1st September, 2010 to 4th January, 2015. The gathered data was analysed, interpreted and assigned to the model in a simplified manner.

The daily energy consumption data was compiled using MS Excel spread sheet. The weekly energy consumption was gotten by adding the daily energy consumed for a week from

Monday to Sunday for the duration of time the data covered. The weekly energy consumption data of Gidan Kwano campus was typed in Microsoft excel. The excel file was imported to the modeling environment in CSV/XLS/XLSX format which was then imported into the neural network tool to train and validate a neural network model to forecast energy consumption of the campus. In this paper, the input to the neural network is the daily historical energy consumption data of the campus from 1st September 2010 to 30th December, 2014.

GMDH shell tool have various validation strategies such as k-fold cross-validation, split into training, testing and leave-one-out cross-validation. The k-fold cross-validation splits the whole dataset into k parts. The model is trained k times using k-1 parts. The independent model performance or the residuals obtained from the all the testing k parts are summarized in other to compare other competing models with it. In split training and testing cross-validation strategy the whole dataset is divided into two parts namely: training and testing part. The training part is used to obtain the model coefficients and the testing part is used to associate or compare all models generated. For this paper split training and testing cross-validation strategy were used. The datasets was divided in the ratio 65:35 respectively. This means 65 percent of the data was used for training and 35 percent for testing. Also, the core algorithm used for this work is GMDH neural network and the data was trained using quadratic neural function.

For the regression analysis method, the data was typed into MS excel and the regression toolbox from data analysis was used to predict energy consumption. The result from the analysis is discussed in the next section.

4. Results and Discussion

Result obtained from the trained model using neural network showed a correlation (R) of 0.8995. Since the R value is close to 1, it shows a close relationship between the actual and the predicted value, hence a good forecast. A Root Mean Square Error (RMSE) of 0.1189 and a Mean absolute percentage error (MAPE) of 0.0922 were obtained. The low error values indicate a high degree of forecasting accuracy.

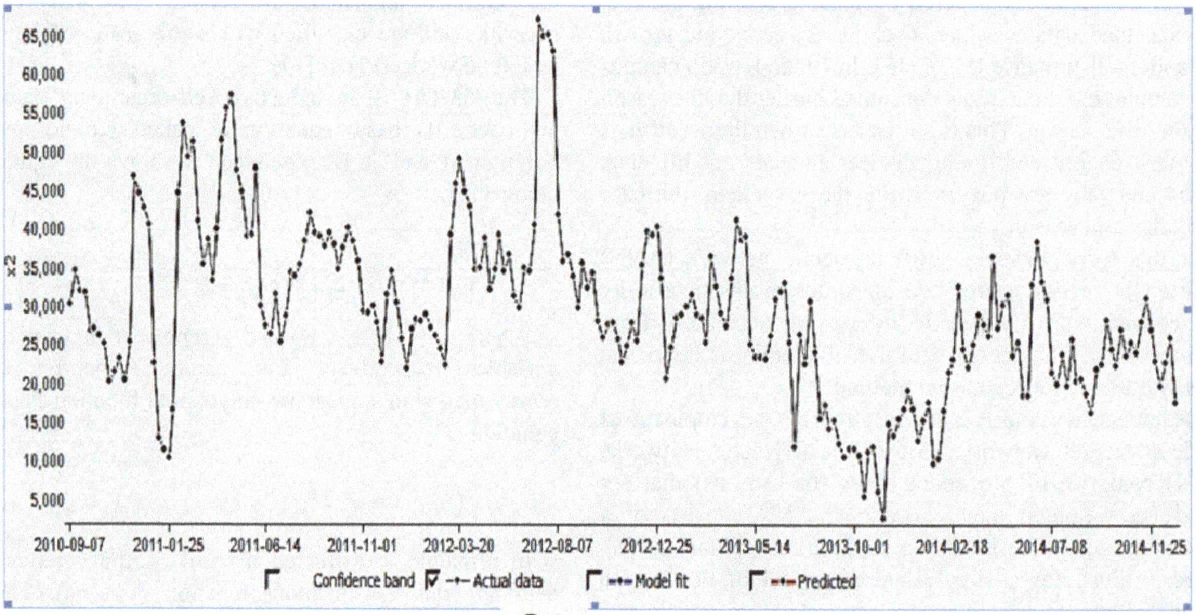

Figure 2. *Actual energy consumed by Gidan Kwano campus.*

The plot in figure 2 shows actual data (the historical energy consumed by Gidan Kwano campus) that was used by the neural network for training and testing.

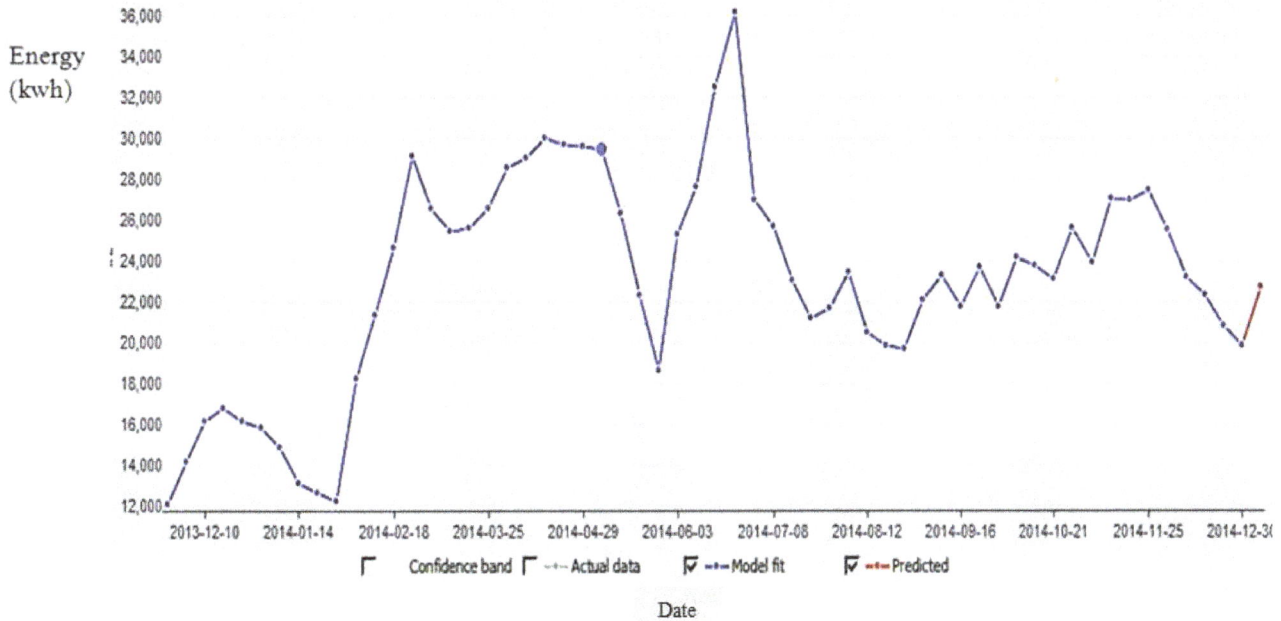

Figure 3. *Model fit and predicted energy consumed.*

Figure 3 show how the model values fit to the historical data. It also shows the next day prediction. The next day energy consumed from the plot is 22868kwh

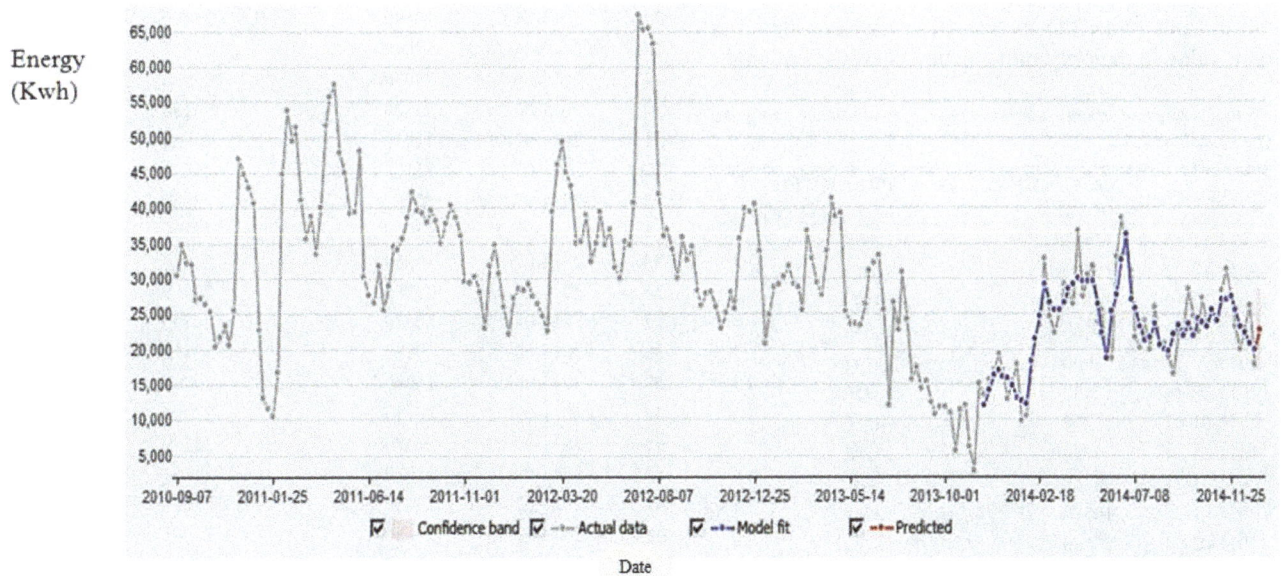

Figure 4. *The actual energy consumption, the model fit and the predicted data.*

Figure 4 shows the actual data, the model fit and the predicted energy consumed respectively. The actual data is the historical data that was used for training and testing and it is grey in colour. In this paper, 65 percent of the actual data was used for training while 35 percent of the actual data was used for testing. The model fit is a model value fitted to the data and is blue in colour. The predicted value is the value forecasted by the GMDH-type neural network model and is red in colour. In this paper, the next week (or step-ahead prediction) forecast is 22686kwh.

The prediction of energy consumed at Gidan Kwano was also done using the regression analysis method. From the results, the correlation between the output and the target was found to be 0.1137 which shows a poor relationship between the actual and the predicted energy consumption. To check the forecasting accuracy, the significance F and the P-value was considered. A small significance of F confirms the validity of the output and if this value is above 0.05, the forecasting accuracy is poor. In this case, the value of significance F was found to be 0.0899 which indicates an inaccurate prediction. Also, the P-value tells whether a variable has statistically significant predictive capability in the presence of other variables. For a good forecasting, the P-value should be less than 0.05. However, one of the values from our regression analysis was gotten to be 0.0899 which is above the 0.05 mark. This too indicates poor forecasting

results. Another drawback of using regression analysis in non-linear data of Gidan Kwano is the prediction error. The standard error gotten was 10968.1 which is very high thus affecting the forecasting accuracy.

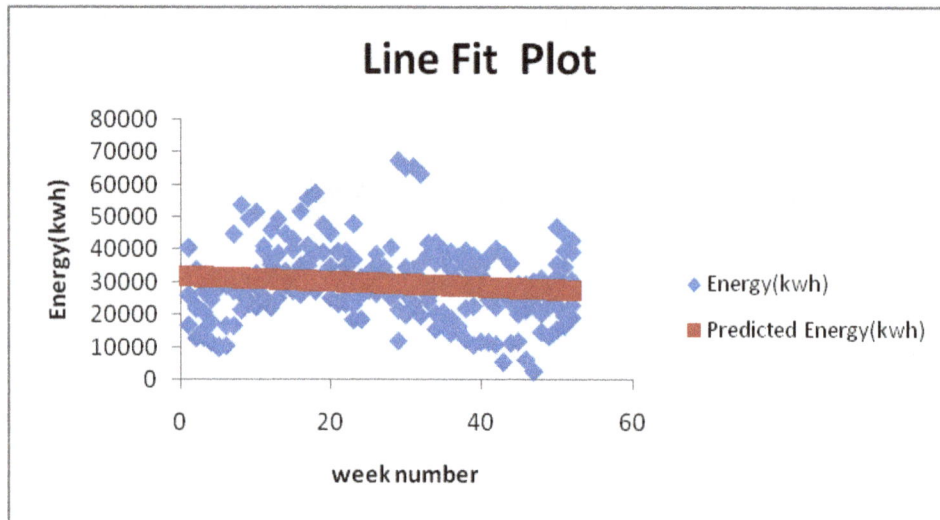

Figure 5. Line fit plot using regression method.

Figure 5 shows the plot of the actual weekly energy consumed by Gidan Kwano to the predicted energy consumed by the campus. From the plot, it can be seen that the model did not fit i.e the forecasted energy consumed is not close in value to majority of the actual energy consumed.

Table 1. Actual values of energy consumed and the predicted using the GMDH-type neural network.

S/N	DATE	ACTUAL ENERGY (Kwh)	PREDICTED ENERGY(Kwh)
1	2013-11-26	13480	12079
2	2013-12-03	15590	14141
3	2013-12-10	16870	16193
4	2013-12-17	19360	16833
5	2013-12-24	16900	16147
6	2013-12-31	12900	15856
7	2014-01-07	15390	14850
7	2014-01-14	17780	13083
8	2014-01-21	9870	12602
9	2014-01-28	10570	12268
10	2014-02-04	16780	18262
11	2014-02-11	21640	21386
12	2014-02-18	23340	24669
13	2014-02-25	32700	29126
14	2014-03-04	24650	26570
15	2014-03-11	22330	25409
16	2014-03-18	25350	25568
17	2014-03-25	29150	26535
18	2014-04-01	27900	28513
19	2014-04-08	26410	29036
20	2014-04-15	36640	30028
21	2014-04-22	27500	29643
22	2014-04-29	30490	29627
23	2014-05-06	31720	29466
24	2014-05-13	23780	26344
25	2014-05-20	25490	22330

S/N	DATE	ACTUAL ENERGY (Kwh)	PREDICTED ENERGY(Kwh)
26	2014-05-27	18640	18665
27	2014-06-03	18690	25299
28	2014-06-10	33040	27631
29	2014-06-17	38630	32500
30	2014-06-24	33320	36150
31	2014-07-01	31030	26964
32	2014-07-08	21570	25645
33	2014-07-15	20160	23061
34	2014-07-22	24020	21191
35	2014-07-29	19850	21640
36	2014-08-05	25880	23459
37	2014-08-12	20470	20490
38	2014-08-19	20820	19870
39	2014-08-26	18940	19704
40	2014-09-02	16530	22065
41	2014-09-09	22020	23259
42	2014-09-16	22670	21764
43	2014-09-23	28460	23657
44	2014-09-30	24680	21791
45	2014-10-07	22450	24161
46	2014-10-14	27130	23766
47	2014-10-21	23780	23131
48	2014-10-28	25490	25592
49	2014-11-04	24180	23891
50	2014-11-11	28290	27025
51	2014-11-18	31210	26938
52	2014-11-25	27120	27406
53	2014-12-02	23070	25506
54	2014-12-09	19850	23199
55	2014-12-16	23020	22291
56	2014-12-23	26230	20829
57	2014-12-30	17760	19856
58	2015-01-06		22686

Table 1 shows the actual and the predicted values. It also shows the next step prediction as 22686 kwh (total energy consumption for the next week).

5. Conclusion

The energy consumed at Gidan Kwano campus of Federal University of Technology, Minna was forecasted on a short-term basis using GMDH-type neural network and regression analysis method. The GMDH-type neural network gave better results when compared to the regression analysis method. The results are computed and the performance is measured using RMSE and MAPE. The GMDH-type neural network results show low RMSE and MAPE values while the regression analysis gave high RMSE and MAPE. These low values show a high degree of forecasting accuracy while the high values indicate a low degree of forecasting accuracy. The neural network model also gives a high correlation (R) value while the R value for the regression was low. This implies a close relationship between the output and the target.

From the results, the ANN performed better than the regression analysis hence, the great superiority of artificial intelligence in forecasting well from nonlinear data. The proposed STLF model can help in establishing operational plans for the University, load shedding, feeder reconfiguration and voltage control.

References

[1]   A.Indira , M. Prakash, S. Pradhan, S.S. Thakur and D.V. Rajan (2014). "Short-term load forecasting of an Interconnected Grid using Neural Network". American Journal of Engineering Research (AJER), e-ISSN: 2320-0847, p-ISSN: 2320-0936, volume-03, Issue-04, pp-271-280.

[2]   Seyed-Masoud Barakati, Ali Akbar Gharaveisi and Seyed Mohammad Reza Rafiei (2015). "Short-term load forecasting using mixed lazy learning method". Turkish Journal of Electrical Engineering & Computer Sciences, Turk J Elect Eng & Comp Sci (2015) 23: 201-211, doi: 10.3906/elk-1301-134.

[3]   Feinberg, E.A. and Genethliou D., (2005). Load Forecasting in: Applied Mathematics for Power Systems. State University of New York, Stony Brook, Chapter 12.

[4]   Isaac, A.S., Felly- Njoku, C.F., Adewale, A.A. and Ayokunle A.A., (2014). Medium-term load forecasting of Covenant University using the Regression analysis method. Journal of Energy Technologies and Policy, ISSN 2224-3232 (paper), ISSN 2225-0573 (online), vol. 4, No. 4, 2014.

[5]   Simaneka, A., (2008). Development of models for short-term load forecasting using Artificial Neural Network. Master's Thesis, Faculty of Engineering, Cape Peninsula University of Technology, November 2008.

[6]   Bougaardt, G., (2002). An Investigation into the application of Artificial Neural Networks and Cluster Analysis in Long-term load Forecasting. Master's Thesis, Department of Electrical and Electronic Engineering, University of Cape Town, 1st January, 2002.

[7]   Sanjoy Das (1995). "The Polynomial Neural Network", University of Carlifonia, California 94720 1995. Information Sciences 87, 231-246 (1995), SSDI 0020-0255 (95) 00133-A.

[8]   Ivan Galkin, U. Mass Lowell. "Crash Introduction to Artificial Neural Networks". Materials for UML 91.550 Data Mining Course.

[9]   E. Gomez-Ramirez, K. Najim and E. Ikonen (2007). "Forecasting time series with a new architecture for polynomial artificial neural network". Applied Soft Computing 7 (2007) 1209-1216.

[10] O.A Koshulko and G.A Koshulko (2011). "Validation Strategy in combinatorial and multilayered iterative GMHD Algorithm". The 4th International Workshop on Inductive Modelling IWIM 2011.

[11] Bon-Gil Koo, Sang-Wook Lee, Wook Kim and June Ho Park (2014). "Comparative Study of Short-term Electric Load Forecasting". 2014 Fifth International Conference on Intelligent Systems, Modeling and Simulation.

[12] Bon-Gil Koo, Heung-Seok Lee and June Ho Park (2015). "Short-term electric load forecasting based on wavelet transform and GMDH". J Electr Eng Technol. 2015; 10(?): 30-40, ISSN (Print) 1975-0102, ISSN (online) 2093-7423, http://dx.doi.org/10.5370/JEET.2015.10.2.030

[13] Huseynov A.F, Yusifbeyli N.A and Hashimov A.M (2010). "Electrical System Load forecasting with Polynomial Neural Networks (based on Combinatorial Algorithm". Modern Electric Power Systems 2010, Wroclaw, Poland, MEPS'10-paper 04.3

[14] Francisco Herrefa Fernández and Fidel Hernández Lozano (2010). "GMDH Algorithm Implemented in Intelligent Identification of a Bioprocess". ABCM Symposium series in Mechatronics, vol. 4-pp 278-287.

[15] Saeed Fallahi, Meysam Shaverdi and Vahab Bashiri (2011). "Applying GMDH-type Neural Network and Genetic Algorithm for stock price prediction of Iranian cement sector". Applications and Applied Mathematics: An International Journal (AAM), vol. 6, Issue 2 (December 2011), pp 572-591, ISSN: 1932-9466

[16] Samsher, K.S. and Unde, M.G., (2012). Short-term forecasting using ANN technique. International Journal of Engineering Sciences and Engineering Technologies, Feb. 2012, ISSN: 2231-6604, volume 1, issue 2, pp: 97-107 © IJSEST

[17] Sanjib, M., (2008). Short-term load forecasting using computational intelligence method. Master's Thesis, Electronics and Communication Engineering (Specialization in Telematics and signal processing), National Institute of Technology, Rourkela, 2008.

Permissions

All chapters in this book were first published in JEEE, by Science Publishing Group; hereby published with permission under the Creative Commons Attribution License or equivalent. Every chapter published in this book has been scrutinized by our experts. Their significance has been extensively debated. The topics covered herein carry significant findings which will fuel the growth of the discipline. They may even be implemented as practical applications or may be referred to as a beginning point for another development.

The contributors of this book come from diverse backgrounds, making this book a truly international effort. This book will bring forth new frontiers with its revolutionizing research information and detailed analysis of the nascent developments around the world.

We would like to thank all the contributing authors for lending their expertise to make the book truly unique. They have played a crucial role in the development of this book. Without their invaluable contributions this book wouldn't have been possible. They have made vital efforts to compile up to date information on the varied aspects of this subject to make this book a valuable addition to the collection of many professionals and students.

This book was conceptualized with the vision of imparting up-to-date information and advanced data in this field. To ensure the same, a matchless editorial board was set up. Every individual on the board went through rigorous rounds of assessment to prove their worth. After which they invested a large part of their time researching and compiling the most relevant data for our readers.

The editorial board has been involved in producing this book since its inception. They have spent rigorous hours researching and exploring the diverse topics which have resulted in the successful publishing of this book. They have passed on their knowledge of decades through this book. To expedite this challenging task, the publisher supported the team at every step. A small team of assistant editors was also appointed to further simplify the editing procedure and attain best results for the readers.

Apart from the editorial board, the designing team has also invested a significant amount of their time in understanding the subject and creating the most relevant covers. They scrutinized every image to scout for the most suitable representation of the subject and create an appropriate cover for the book.

The publishing team has been an ardent support to the editorial, designing and production team. Their endless efforts to recruit the best for this project, has resulted in the accomplishment of this book. They are a veteran in the field of academics and their pool of knowledge is as vast as their experience in printing. Their expertise and guidance has proved useful at every step. Their uncompromising quality standards have made this book an exceptional effort. Their encouragement from time to time has been an inspiration for everyone.

The publisher and the editorial board hope that this book will prove to be a valuable piece of knowledge for researchers, students, practitioners and scholars across the globe.

List of Contributors

Bindu S. J. and C. A. Babu
Department of Electrical and Electronics Engineering, School of Engineering, CUSAT, Kochi, Kerala, India

Monica-Adela Enache, Sorin Enache, Ion Vlad and Gheorghe-Eugen Subtirelu
University of Craiova, Faculty of Electrical Engineering, Craiova, Romania

Dhanesh Kumar Sambariya
Department of Electrical Engineering, Rajasthan Technical University, Kota, India

Xuegang Hu
College of Computer Science and Technology, Chongqing University of Posts and Telecommunications, Chongqing, China
Research Center of System Science, Chongqing University of Posts and Telecommunications, Chongqing, China

Lei Li
College of Computer Science and Technology, Chongqing University of Posts and Telecommunications, Chongqing, China

Asif Ahmed
Department of EEE, American International University Bangladesh, Dhaka, Bangladesh

Abu Jahid, Sanwar Hossain and Raziqul Islam
Dept. of Electrical and Electronic Engineering, Bangladesh University of Business and Technology (BUBT), Dhaka, Bangladesh

Yao Hong Guang
School of Air Transportation / Flying, Shanghai University of Engineering and Science, Shanghai, China

Ankamma Rao Jonnalagadda and Gebreegziabher Hagos
Department of Electrical & Computer Engineering, School of Engineering & Technology, Samara University, Semera, Afar Region, Ethiopia

Mousaab M. Nahas
Electrical and Computer Engineering Department, Faculty of Engineering, University of Jeddah, Jeddah, Saudi Arabia

Mohammad Mahdi Share Pasand
Department of Electrical and Electronics Engineering, Standard Research Institute - SRI, Alborz, Iran

Mehdi Hamidkhani and Behdad Arandian
Department of Electrical Engineering, Dolatabad Branch, Islamic Azad University, Isfahan, Iran

Mousaab M. Nahas
Electrical and Computer Engineering Department, Faculty of Engineering, University of Jeddah, Jeddah, Saudi Arabia

Behrouz Alfi
Department of Electrical Engineering College of Engineering Ardabil Branch, Islamic Azad University, Ardabil, Iran

Tohid Banki
Department of Electrical Engineering College of Engineering Bilasouvar Branch, Islamic Azad University, Bilasouvar, Iran

Faramarz Faghihi
Department of Electrical Engineering College of Engineering Science and Research Branch, Islamic Azad University, Tehran, Iran

Xu Shi-hong
Department of Aeronautic Electronic Engineering, the First Aeronautical College of Air Force, Xinyang, China

Zhao Wei-bin
School of Information Engineering, Zhengzhou University, Zhengzhou, China

Huang Guo-qing
Department of Aeronautic Electronic Engineering, the First Aeronautical College of Air Force, Xinyang, China
School of Information Engineering, Zhengzhou University, Zhengzhou, China

Lei Cui , Tele Tan and Khac Duc Do
Department of Mechanical Engineering, Curtin University, Perth, Australia

Peter Teunissen
Department of Spatial Sciences, Curtin University, Perth, Australia

Alireza Hassanzadeh and Ahmad Shabani
ECE Dept., Shahid Beheshti University, Tehran, Iran

M. Dashtbayazi and S. Marjani
Department of Electrical Engineering, Ferdowsi University of Mashhad, Mashhad, Iran

M. Sabaghi
Laser and Optics Research School, Nuclear Science and Technology Research Institute (NSTRI), Tehran, Iran

M. Rezaei
Department of Electrical Engineering and Computer Engineering, Laval University, Quebec City, QC, Canada

Ouarda Barkat
Department of Electronics, University of Constantine 1, Constantine, Algeria

Thet Mon Aye and Soe Win Naing
Dept. of Electrical Power Engineering, Mandalay Technological University, Mandalay, Myanmar

Pengcheng Xu and Zhigang Han
Department of Electronic Science and Technology, Tongji University, Shanghai, China

Okcana Kharchenko
National Technical University "Kharkiv Polytechnic Institute», Kharkiv, Ukraine

Vladislav Tyutyunnik
Scientific Center, Kharkiv Air Force University named I. Kozhedub, Kharkiv, Ukraine

Zaiming Fan
Faculty of Health and Science, University of Cumbria, Lancaster, United Kingdom

Xiongwei Liu
Entrust, The Innovation Centre, Science Technology Daresbury, Cheshire, United Kingdom

Fei Peng and Lin Cheng
School of Electrical and Electronic Engineering, Huazhong University of Science and Technology, Wuhan, China
Wuhan Nari Group Corporation of State Grid Electric Power Research Institute, Wuhan, Hubei, China

Kaikai Gu and Zhenbo Du
Wuhan Nari Group Corporation of State Grid Electric Power Research Institute, Wuhan, Hubei, China

Jiang Guo
School of Power and Mechanical Engineering, Wuhan University, Wuhan, Hubei, China

Nan Win Aung and Aung Ze Ya
Department of Electrical Power Engineering, Mandalay Technological University, Mandalay, Myanmar

M. Dashtbayazi and S. Marjani
Department of Electrical Engineering, Ferdowsi University of Mashhad, Mashhad, Iran

M. Sabaghi
Laser and Optics Research School, Nuclear Science and Technology Research Institute (NSTRI), Tehran, Iran

Parung J., Santoso A., Prayogo D. N. and Angelina M.
Industrial Engineering Department, University of Surabaya, Surabaya, Indonesia

Tayibnapis A. Z. and Djoemadi F. R.
Economic Department, University of Surabaya, Surabaya, Indonesia

Wang Xue-Jun
School of Electronic Information Engineering, Beihang University, Beijing, China

Taku Saiki
Department of Electrical and Electronic Engineering, Faculty of Engineering Science, Kansai University, Suita, Japan

Shi Mingming
Jiangsu Electric Power Research Institute, Nanjing, China

Lu Wenwei and Ge Le
Nanjing Institute of Technology, Nanjing, China

Huang Jing, He Huiying, Zhou Shiwan and Zhang Tao
Electrical Engineering Department, School of Electrical Engineering, Naval University of Engineering, Wuhan, China

Hossein Karimianfard
Department of Electrical Engineering, Jahrom Branch, Islamic Azad University, Jahrom, Iran

Chi-Yen Shen, Yu-Min Lin and Rey-Chue Hwang
Department of Electrical Engineering, I-Shou University, Kaohsiung, Taiwan

M. Dashtbayazi and S. Marjani
Department of Electrical Engineering, Ferdowsi University of Mashhad, Mashhad, Iran

M. Sabaghi
Laser and Optics Research School, Nuclear Science and Technology Research Institute (NSTRI), Tehran, Iran

Chaw Su Hlaing and Pyone Lai Swe
Department of Electrical Power Engineering, Mandalay Technological University, Mandalay, Myanmar

Tsado Jacob, Usman Abraham Usman and Ajagun Abimbola Susan
Department of Electrical and Electronics Engineering, Federal University of Technology Minna, Niger State, Nigeria

Saka Bemdoo
Transmission Company of Nigeria, TCN Abuja, Nigeria

Index

www.ingramcontent.com/pod-product-compliance
Lightning Source LLC
Chambersburg PA
CBHW080536200326
41458CB00012B/4455